Fisheries Acoustics

Fisheries Acoustics

David N. MacLennan

Deputy Director
Marine Laboratory
Scottish Office Agriculture and Fisheries Department
Aberdeen, Scotland

E. John Simmonds

Head of Sonar Section
Marine Laboratory
Scottish Office Agriculture and Fisheries Department
Aberdeen, Scotland

CHAPMAN & HALL
London · Glasgow · Weinheim · New York · Tokyo · Melbourne · Madras

Published by Chapman & Hall, 2-6 Boundary Row, London SE1 8HN, UK

Chapman & Hall, 2-6 Boundary Row, London SE1 8HN, UK

Blackie Academic & Professional, Wester Cleddens Road, Bishopbriggs, Glasgow G64 2NZ, UK

Chapman & Hall GmbH, Pappelallee 3, 69469 Weinheim, Germany

Chapman & Hall USA., 115 Fifth Avenue, New York, NY 10003, USA

Chapman & Hall Japan, ITP-Japan, Kyowa Building, 3F, 2-2-1 Hirakawacho, Chiyoda-ku, Tokyo 102, Japan

Chapman & Hall Australia, 102 Dodds Street, South Melbourne, Victoria 3205, Australia

Chapman & Hall India, R. Seshadri, 32 Second Main Road, CIT East, Madras 600 035, India

First edition 1992
Reprinted 1995

Typeset in 10/12 Photina by Interprint Ltd, Malta
Printed in Great Britain by St Edmundsbury Press, Bury St Edmunds, Suffolk

ISBN 0 412 33060 1

A Catalogue record for this book is available from the British Library

Library of Congress Cataloging-in-Publication Data

MacLennan, D.N.
Fisheries accoustics / David N. MacLennan, E. John Simmonds
p. cm.
Includes bibliographical references and index.
ISBN 0-442-31472-8
1. Fish detection. I. Simmonds, E. John, 1952- . II. Title
SH344.2.M33 1991
639.2—dc20 91-24132
CIP

Printed on acid-free text paper, manufactured in accordance with ANSI/NISO Z39.48-1992 and ANSI/NISO Z39.48-1984 (Permanence of Paper).

To Sheila and Chris

Contents

Colour plate section appears between pages 176 and 177

Acknowledgements

We wish to thank all the authors, editors and colleagues who willingly gave us permission to redraw or reproduce published material, Tony Pitcher for his advice and support at all stages from conception to completion of this book, Sinclair Forbes for many hours of helpful discussion and advice on a wide range of topics, Chuck Hollingworth for his painstaking and constructive editing, and Anne-Marie Meconi who word-processed the manuscript through many drafts with skill and patience. We are especially grateful to the following colleagues who read parts of the text and provided helpful criticisms: Rob Fryer (Chapters 7 and 8); Dave Reid (Chapters 2, 3, 5 and 6); Tim Stanton (Chapter 5); Jim Traynor (Chapter 6); Clem Wardle (Chapter 4). We have taken account of most (but not all) of the comments received, and the responsibility for any errors or omissions that remain is ours alone.

Acknowledgements

Series foreword

Among the fishes, a remarkably wide range of biological adaptations to diverse habitats has evolved. As well as living in the conventional habitats of lakes, ponds, rivers, rock pools and the open sea, fish have solved the problems of life in deserts, in the deep sea, in the cold Antarctic, and in warm waters of high alkalinity or of low oxygen. Along with these adaptations, we find the most impressive specializations of morphology, physiology and behaviour. For example we can marvel at the high-speed swimming of the marlins, sailfish and warm-blooded tunas, air-breathing in catfish and lungfish, parental care in the mouth-brooding cichlids and viviparity in many sharks and toothcarps.

Moreover, fish are of considerable importance to the survival of the human species in the form of nutritious and delicious food of numerous kinds. Rational exploitation and management of our global stocks of fishes must rely upon a detailed and precise insight of their biology.

The Chapman and Hall *Fish and Fisheries Series* aims to present timely volumes reviewing important aspects of fish biology. Most volumes will be of interest to research workers in biology, zoology, ecology and physiology, but an additional aim is for the books to be accessible to a wide spectrum of non-specialist readers ranging from undergraduates and postgraduates to those with an interest in industrial and commercial aspects of fish and fisheries.

This volume, the fifth in the *Fish and Fisheries Series*, reviews fisheries acoustics, an area which has been overdue for a thorough coverage. At last the mysteries of this arcane art are laid bare by two workers with international reputations from the famous Marine Laboratory at Aberdeen, Scotland. The book reviews the fundamental principles and the practical techniques by which acoustic echoes are transformed into measurements of fish stock biomass, introduces various types of acoustic instruments used by commercial fishermen and research scientists, and surveys the biological basis of sound production and reception underwater, including the effects and possible mitigation of underwater explosions.

The emphasis is on graphical explanations, and consequently readers do not require advanced mathematics to understand this book. It should therefore be accessible to a wide range of non-specialist workers interested in the management and conservation of commercial fish stocks. Nevertheless, the treatment is comprehensive so that it is suitable for reference by those involved in fisheries assessment for both marine and freshwater environments. Acoustic estimates of fish stock size are widely applied: this book will enable all to appraise their scope and validity.

Dr Tony J. Pitcher
Editor, Chapman and Hall Fish and Fisheries Series
Special Research Fellow, Imperial College, London

Preface

Fisheries research is an interesting field in which to work, among other reasons because it calls for many different kinds of expertise. The team approach is central to success, involving cooperation and mutual understanding between experts in various fields – biology, engineering, physics and statistics to mention but a few. The whole is greater than the sum of the parts, but only if the parts fit together. The problems of multi-disciplinary science are well illustrated by the theme of this book, underwater acoustics and how it has been applied to the remote detection, observation and study of aquatic life. Our approach to the subject is founded on the principles and theory of acoustics, interpreted with the numerate life scientist in mind, not those whose interests are narrowly confined to the physics of sound. We have tried to maintain a broad scope of interest to fishery scientists in the widest sense, giving statements of the necessary theory rather than rigorous derivations, and discussion of relevant non-acoustical topics such as fish behaviour, fishing gear, design of experiments and statistical analysis. The many references we cite may be consulted by readers who wish more detail than we have been able to provide owing to constraints of space and time.

This book has developed from a series of lectures and practical training exercises which we have given over the past ten years or so, in one form or another, to postgraduate students and professional fishery scientists. In this context, we would particularly mention the 'acoustics modules' which have been part of fishery-related MSc courses at the University College of North Wales and the University of Aberdeen. We have also participated from time to time in more specialized courses which have concentrated on the application of acoustics to surveys of fish populations, most recently the 'Curso Hydroacustico' held at Guayaquil, Ecuador, in November 1990. We have found the interaction between instructor and student to be most helpful in teaching us, we hope, how to express a technically difficult subject in clear, meaningful terms.

David N. MacLennan
John Simmonds
Marine Laboratory, Aberdeen

Chapter one

Introduction

The world production of fish has grown steadily in recent times, and now it stands at more than 90 million tonnes per annum. However, the rising global production conceals the fact that individual populations rise and fall. On occasions, the fall is a dramatic collapse which has severe consequences for those who are supported by the local fishery (Beverton, 1990). We shall not discuss the causes of such events, except to say that long-term success in fishing depends upon knowledge of the exploited population, whose size and distribution may change unpredictably from year to year. Much of this knowledge comes from the investigation of fish in their natural environment. This book is primarily concerned with the use of sound for the observation and monitoring of aquatic life in the sea and fresh waters.

At the wavelengths which we can see, light does not penetrate more than a few metres below the water surface, and much less when the medium is loaded with suspended solids or biota such as plankton. However, sound waves travel much longer distances through water. Thus acoustic instruments which transmit and receive sound waves are capable of detecting fish or other objects far beyond the range of vision. Because of this, acoustic technology has had a major impact on fishing. The information provided by sonars and echosounders is an important factor in the efficiency of modern fishing operations. Purse seining and pelagic trawling are two methods which depend upon the skilful interpretation of acoustic data to ensure success.

In fisheries research too, acoustic techniques have become increasingly important over the years. With sonar it is possible to search a large volume of water in a short time; other sampling methods such as trawl fishing are very slow by comparison. Acoustic echoes may be observed from fish anywhere in the water column, except the near-surface region and just above the sea-bed. Sonar has contributed greatly to our understanding of life in the sea and fresh waters, especially the distribution of animals in space and time. Many experimental investigations have been reported in the literature. As a few examples, we note the sizing of fish by exciting resonance of the swimbladder (Holliday, 1977a), the use of automatic transmitters or pingers (Hawkins *et al.*, 1974) and transponders (Arnold and Metcalfe, 1989) to monitor the movements of fish, and the study of fish behaviour (Olsen *et al.*, 1983a,b).

The measurement of fish abundance is probably the most important application of acoustics in fisheries research (MacLennan, 1990). The management of most fisheries depends upon controlling the quantity caught in relation to the size of the exploited population (Gulland, 1983). To do this, it is necessary to estimate the current size (abundance) of the population, or that in the recent past. The acoustic method records the fish echoes and estimates the abundance as the quantity of fish which would be expected to produce such echoes. This is the technique of acoustic survey. It is useful only when the fish of interest are conveniently located: they must not be too close to the surface or the bottom, where the fish echoes are obscured by much stronger reflections from these boundaries. Thus acoustic methods of observation are unsuited to the flatfish and other species which live in close association with the sea-bed. Fortunately, many important species are found in mid-water, notably the herring family (Clupeidae), the mackerels (Scombridae), the anchovies (Engraulidae) and the salmonids (Salmonidae). In appropriate circumstances, an acoustic survey provides a synoptic estimate of the abundance which is independent of the fishery.

Sound also has a natural importance for fish and the mammals that live in water. They use it as a means of communication and detection of prey. This aspect of underwater acoustics has also been the subject of much research (Hawkins, 1986).

1.1 A BRIEF HISTORY

Leonardo da Vinci is credited with the earliest known reference to underwater sound. In a notebook dated 1490, he observed that by listening to one end of a long tube, with the other end in the sea, 'you will hear ships at a great distance' (Urick, 1983). The speed of underwater sound, about $1450\,m\,s^{-1}$, was first measured by Colladon and Sturm in 1827. They deduced the sound speed by reference to the time delay when they transmitted both a flash of light and the sound of an immersed bell across Lake Geneva in Switzerland. However, practical applications had to await the development of electrical transducers and receivers. In 1918, as a result of research motivated by the First World War, it was discovered that submarines could be detected by listening for the echo of a sound transmission.

Kimura (1929) described the first successful experiment on the acoustic detection of fish. He installed a transmitter and a separate receiver in a pond used to cultivate fish (Fig. 1.1). The sound was transmitted in a 20° beam and detected after reflection from the opposite side of the pond. The transmission was continuous at 200 kHz frequency, but modulated at 1 kHz so that the rectified signal would be audible. The pond contained a school of *Pagrosomus major*, about 25 fish of length 40–50 cm. Kimura found that the received sound was noticeably disturbed when fish were in the beam, although reliable detection depended on the surface of the pond being flat calm. In this experiment, the fish were detected by the fluctuation of the forward transmission caused by their movement. The alternative method of echo detection was first reported by Sund (1935) who published echograms of the cod, *Gadus morhua*, in the Lofoten region of Norway. Sund used a 16 kHz echosounder with a magneto-strictive transducer installed on the research vessel *Johan Hjort*. This equipment revealed many unsuspected features of the fish distribution, notably that the cod were confined to a narrow layer 10 m thick at a constant distance below the surface.

Another period of rapid development occurred during the Second World War, after which fishermen soon discovered the civilian potential of the instruments developed by the military. The pioneering work of the British skipper Ronnie Balls is especially notable in this context. The power and resolution of sonars continued to improve as new instruments were designed specifically for fish detection. Many different kinds of sonar are now employed in fishing, from the simple echosounder to scanning sonars which

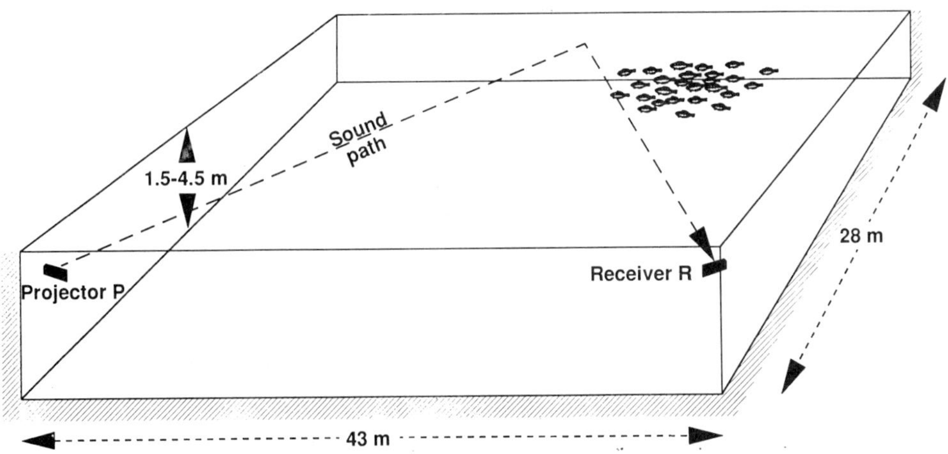

Fig. 1.1 Apparatus used by Kimura (1929) in the first acoustic detection of fish. Sound from the projector P is reflected by the far side of a pond 1.5–4.5 m deep and detected by the receiver R. Fish passing through the beam cause the amplitude of the received signal to fluctuate.

provide a radar-like display in two dimensions. The introduction of echograms displayed in colour has greatly improved the perception of signal strength. Fishermen have exploited these developments with great skill. To some extent they can deduce the species composition of targets, usually by interpreting the echograms in conjunction with other information about the local population. However, the problem of target identification has not been completely solved.

Acoustic methods of abundance estimation were first applied around 1970, notably the technique of echo integration proposed by Dragesund and Olsen (1965). In the early days, the results that could be obtained were subject to large errors. The calibration methods of the time were imprecise, and the target strength of important species was uncertain. Intensive theoretical and experimental investigations in the 1970s and 1980s led to a better understanding of what acoustic techniques could and could not do. High-performance scientific echosounders were introduced with larger dynamic range, more stable gain characteristics and better compensation of the propagation losses. Calibration is no longer a limiting factor, provided it is done in the recommended way (Foote *et al.*, 1987). New types of acoustic instruments have been developed which allow the target strength of wild fish to be measured directly, notably the dual-beam and split-beam echosounders. These instruments have also been used for abundance estimation by echo counting, a technique which is applicable to dispersed populations of fish. Echo-counting has been particularly successful in the study of fish in lakes and rivers.

1.2 SYNOPSIS

This book has been written primarily for postgraduate students, professional scientists and administrators concerned with fishery management. The principles of underwater acoustics are covered to the extent necessary to show the advantages and disadvantages of acoustic methods in fishery investigations. Bearing in mind that many readers will have a background in the biological sciences, and may be less familiar with mathematics, we have tried to explain the subject in words and illustrative graphics, to reduce the reliance on mathematical shorthand which is rather too common in acoustic publications. However, some theory is essential for the proper understanding of sound, fish as targets and the statistical problems which are common to acoustic and other kinds of survey. A list of symbols is provided at the beginning of each chapter. Throughout the book, the emphasis is on acoustics as an applied science. Practical advice is given on the use of acoustic instruments in the field and the solution of problems which are often ignored by those who prefer the purely theoretical approach.

Chapter 2 is an introduction to the concept of energy transmission by sound waves and the remote detection of targets. We discuss the generation of sound by transducers, the propagation of waves, the scattering properties of targets and the formation of echoes. The competing requirements of resolving close targets and detection at long range are considered in relation to the frequency of the sound waves.

The operating principles of modern acoustic instruments are described in Chapter 3. The simple echosounder transmits sound in a single beam, which permits the range, but not the direction, of targets to be determined. More sophisticated devices are capable of angular resolution. They include dual-beam, split-beam, multi-beam and sector-scanning sonars. Wide-band sonar is a recent development which offers the possibility of identifying targets from the spectrum of echoes. The calibration of acoustic instruments is discussed. Practical advice is given on how to perform the calibration in accordance with internationally accepted standards.

In Chapter 4, we review a number of diverse topics to do with physiology and the behavioural importance of sound for aquatic life. We discuss the sensitivity of hearing, the production of sound by animals and the remarkable sonar capability of the Cetacea. The biological consequences of underwater explosions have been studied experimentally, and we show how the risk of fish mortality can be assessed. Another topic is the phenomenon of swimbladder resonance which has been proposed as a means of determining fish size. The same theory is relevant to the acoustics of plankton, which may be studied by the simultaneous use of sonars operating at widely separated frequencies.

Chapter 5 is concerned with the observation and measurement of fish, i.e. how to interpret the information provided by acoustic instruments. We begin with the simple echogram and the technique of school counting. The measurement of fish density relies upon more advanced techniques, notably echo counting and echo integration. The density is supposed to be proportional to the integral of the echo energy returned from the depth channel of interest. This assumption depends upon the linearity principle, which is central to the theory of fisheries acoustics. The evidence in support of linearity is discussed.

The target strength of fish is reviewed in Chapter 6. Experimental techniques for measuring the target strength are described. The swimbladder is the dominant sound reflector in those species having one. Accordingly, fish targets may be classified as physoclists (bladder closed), physostomes (bladder open) or those with no bladder. Within each category, fish of the same size have similar target strengths on average, but there is much residual variation which emphasizes the stochastic nature of target strength. Published experimental results are presented in a form suitable for easy reference.

The last two chapters deal with the practice of acoustic surveying to measure the abundance and distribution of fish. The emphasis is on the practical problems that arise in applying the theoretical principles discussed earlier. In Chapter 7, we consider the equipment required for acoustic surveys, the design of the cruise track and the sampling strategy to make the best use of the available time. Live-fish calibration and inter-ship comparison are two methods for testing the overall performance of the survey equipment in the field, although they are not a substitute for the recommended calibration procedure. In Chapter 8, we discuss the analysis of the data collected during the survey. The aim is to calculate abundance estimates within defined confidence limits. First, the echo-integrator data must be partitioned between species, and the surveyed area may have to be stratified depending on the stationarity of the fish density. An echo-integrator conversion factor is calculated for each species and stratum from which the density samples are obtained. The total abundance is estimated from the observed densities. There are a number of approaches to this problem, which is complicated by the statistics of spatial correlation. We describe contour mapping, kriging and numerical methods based on rectangular grid strata. The various factors which contribute error are discussed and we show how the overall accuracy of the abundance estimate can be assessed. The results obtained from acoustic surveys are compared with those of alternative methods, again to indicate the accuracy that can be expected in typical circumstances.

In compiling the references, we have concentrated on the many publications which have appeared in the past 15 years or so, avoiding obsolete material but including key references to earlier work that laid the foundations of modern fisheries acoustics. See Verema (1985) for a more complete bibliography of earlier work.

Chapter two

Underwater sound

2.1 INTRODUCTION

The theory of sound is similar in many respects to that of light. Both phenomena consist of waves which propagate through a medium. Both are also subject to scattering, reflection and absorption, which complicate the simple idea of waves transmitting energy through a homogeneous, lossless medium as it is explained in elementary physics textbooks. In water, sound can be transmitted over much longer distances than light, but water is nevertheless an imperfect acoustic medium. Energy is removed (scattered) from the sound wave by suspended solids, biota or entrained gas, or simply converted to heat by physical absorption. All these effects need to be considered in fishery applications, where the usual intention is to deduce features of remote targets from the acoustic signals detected by a sonar or other instrument.

This chapter provides a largely qualitative view of the generation, transmission and detection of sound in water. Mathematical descriptions are kept to the minimum necessary as background for the discussion of practical applications later on. We begin with the concept of sound waves and their physical properties – pressure, particle velocity and intensity. Next we consider the transducers which generate sound from electrical energy, and the manner in which the sound is projected in particular directions as a beam. The transmission of sound is discussed with particular reference to energy losses through beam spreading and absorption. Empirical equations are given for calculating the sound speed and absorption coefficient in water. We go on to deal with acoustic scattering and the nature of echoes reflected by various kinds of target. Finally, we consider the detection of acoustic signals and the importance of noise as the limiting factor in the performance of acoustic instruments.

The reader who wishes to know more about the physics of sound may consult one of the following references. Rayleigh (1945) is the classic and still much-quoted work on the subject; see also Morse (1948), or Officer (1958), Rschevkin (1963), Mason (1964). Urick (1983), Clay and Medwin (1977), and Tucker and Gazey (1966) are textbooks on underwater acoustics, while Forbes and Nakken (1972), Burczynski (1979), Johannesson and Mitson (1983) and Mitson (1984a) deal specifically with fishery applications.

List of symbols

a	Radius of a sphere, or the side of a square
$\mathbf{b}$	The beam pattern; function of direction which describes the amplitude sensitivity
B	Bandwidth of the sonar receiver
c	Sound speed in water
c_1	Speed of longitudinal sound waves in a solid
c_2	Speed of transverse sound waves in a solid
d	Characteristic linear size of a target
D	Depth below the water surface
E	Energy
EBA	Equivalent beam angle in dB relative to 1 steradian
f	Frequency (cycles per second)
f_0	Frequency of the sine waves within a pulse
f_1	Frequency of the lowest resonance in a solid target
F	Form function (ratio of acoustic and geometric cross-sections)
I	Intensity of a sound wave (power per unit area)
I_0	Intensity normalized to 1 m range
I_b	Intensity of the backscattered wave
I_i	Intensity of the incident wave
J	Flux of energy transmitted per unit area
k	The wavenumber ($2\pi/\lambda$)
L	Pulse length in water
n	Number of targets, cycles, etc.
N_{dB}	Ratio of two intensities expressed in decibels
p	Instantaneous pressure of a sound wave
$\|p\|$	Amplitude of the sound pressure variation
$\|p_0\|$	Pressure amplitude normalized to 1 m range
P	Ambient pressure in the water
$P(\omega)$	Power response of the sonar at frequency ω
r	Reflection coefficient; proportion of incident energy reflected at a boundary
R	Range or distance
R_b	Range at the boundary between the near and far fields
$s(\omega)$	Acoustic cross-section at frequency ω

S	Salinity of water
SL	Source level of a transducer (sound pressure at 1 m, in dB)
t	Time
T	(1) Time between the transmitter pulse and the echo being received (2) Temperature
TS	Target strength, a logarithmic measure of σ
$\mathbf{v}$	Particle velocity associated with a sound wave
$\lvert\mathbf{v}\rvert$	Amplitude of the particle velocity variation
V_s	Sampling volume
x	Distance along the propagation path, or a general variable
Z	Acoustic impedance, equal to the density times the sound speed
Z_r	Acoustic impedance of a reflector
Z_w	Acoustic impedance of water
α	Absorption coefficient in dB per unit distance (1 Np = 8.686 dB)
β	Absorption coefficient in nepers (Np) per unit distance
θ	Angle from the acoustic axis
ϕ	Azimuthal angle in the transducer plane
λ	Wavelength, distance between successive peaks of a sound wave
ρ	Density of water
ρ_1	Density of a solid target
σ	The total acoustic cross-section, effective value averaged over the sonar bandwidth
σ_b	The backscattering cross-section
τ	Pulse duration, time from start to finish
ψ	Equivalent beam angle in steradians
ω	Angular frequency (radians per second)

2.2 SOUND WAVES

Sound is transmitted by the periodic compression and expansion which is permitted by the elasticity of water. This results in a travelling pressure wave as illustrated in Fig. 2.1 for the case of sinusoidal variations at one frequency. At any point in space, the pressure p relative to the ambient level varies as $\sin(\omega t)$ where t is the time and ω is the angular frequency of the oscillation. $f = \omega/(2\pi)$ is the frequency in hertz (Hz), the number of cycles per second. When the wave is plane, this means that changes occur in one direction only, the direction in which the wave propagates. p is the same everywhere in any plane perpendicular to this direction.

The wave-fronts are those planes in which p is maximal. More generally, each wave-front is a surface joining continuous loci of the peak pressures. If x is the distance along the direction of propagation of a plane wave, then at any instant p changes as $\sin(kx)$. k is called the wavenumber. The pattern of pressure changes in space is also cyclical and it repeats at intervals of the

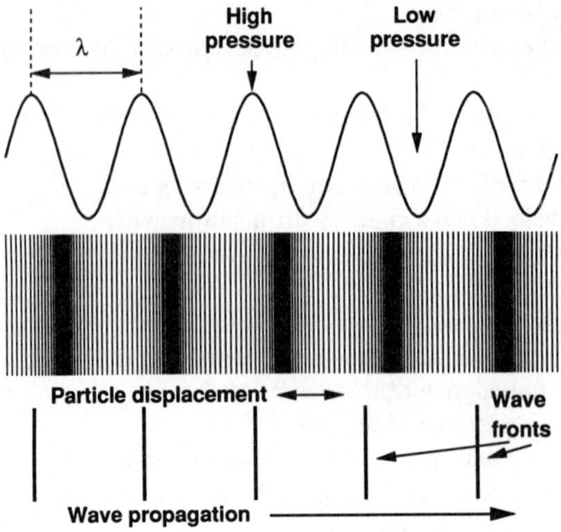

Fig. 2.1 Sound waves. The pressure (top) varies cyclically as a sine wave; λ is the wavelength. The particle displacement (middle) is out of phase with the pressure. The wave-fronts (lower) are lines which follow the maximum pressure.

wavelength $\lambda = (2\pi/k)$. Thus λ is the distance between successive wave-fronts (Fig. 2.1). Combining the changes in space and time, we can say that p varies as $\sin(kx - \omega t)$. This implies that the wave-fronts move forward at the sound speed $c = \omega/k$. Another important relationship which immediately follows from these definitions is that the sound speed is equal to the product of the wavelength and the frequency:

$$c = \lambda f \tag{2.1}$$

The sound speed describes the movement of the pressure peaks (wave-fronts) through the medium and should not be confused with the particle velocity, which applies to the local movement of molecules. For water, c is generally in the range 1450 to 1550 m s^{-1}, depending on the temperature, ambient pressure and the salinity. The wavelength is important as the fundamental limit on the spatial resolution of targets. The smaller the wavelength, or the higher the frequency, the easier it is to discriminate targets that are close together. If c is 1500 m s^{-1}, a typical value in the sea, then sound of 10 kHz frequency has a wavelength of 15 cm. If the frequency is 500 kHz, the wavelength is reduced to 3 mm.

The continuous sine wave is a convenient and simple description of sound, but in practice we need to consider more complicated waveforms such as pulses and echoes whose amplitude changes with time. However, any sound wave may be conceived as the sum of continuous sine waves over a spectrum of frequencies. The pulses transmitted by sonars comprise a few cycles of a sine wave which lasts for a finite time, the pulse duration. The frequency of the sonar as quoted by the manufacturer is that of the sine wave, f_0, but some of the energy is transmitted at other frequencies within a band centred on f_0. The width of the band depends upon the pulse duration. The shorter the pulse, the wider is the spectrum of frequencies transmitted by the sonar. For example, if the pulse duration is $\tau = 1$ ms, the bandwidth is about $B = 1$ kHz; if $\tau = 2$ ms, then $B = 500$ Hz.

Pressure and displacement

In addition to the pressure changes, the wave causes the water molecules to vibrate. The amplitude of the molecular movement is called the particle displacement, and the rate of change is the particle velocity. In a plane wave, the particle velocity and the sound pressure are in phase, meaning that the maximum values coincide in time and space. They both vary as $\sin(kx - \omega t)$, and the amplitudes are proportional. If ρ is the density of the water, then

$$p = \rho c \mathbf{v} \tag{2.2}$$

Now consider a small source which generates sound at one frequency. If the source is remote from any reflecting objects or surfaces, the waves are said to propagate in free-field conditions. The wave-fronts are now spherical and they travel away from the source in any direction. The curved wave-fronts result in a more complicated relationship between the pressure and the particle velocity, compared with the plane wave (Harris, 1964; Siler, 1969). At distance R from the source,

$$|p| = \rho c |\mathbf{v}| / \surd[1 + (\lambda / 2\pi R)^2] \tag{2.3}$$

The modulus symbols are used to indicate the amplitude of a sinusoidal variable. Thus the amplitude ratio $|p|/|\mathbf{v}|$ changes with range when R is the order of a wavelength or less. This phenomenon, known as the near-field effect, has implications for the hearing ability of fish (section 4.2).

Since water is a fluid, it cannot sustain shear forces, and the particle displacement is always in the direction of sound propagation. This type of sound wave is described as longitudinal. More complicated sound fields occur in solids where transverse (shear) waves can propagate in the direction normal to the particle displacement. The speeds of transverse and longitudinal waves in solids are not generally the same.

Energy and intensity

An important feature of the travelling wave is that it transports energy from one place to another. The flux J is the energy of the waves passing through a unit area perpendicular to the wave-front. The intensity I is the energy flux per unit time. It is convenient to use the intensity to describe the power of continuous waves, or long pulses within which the amplitude is constant for many cycles. In the case of short pulses, the intensity may change from one cycle to the next and it is better to describe the pulse in terms of the total energy transmitted over the finite pulse duration (Craig, 1983). J is simply the integral of I with respect to time. The total energy carried by the wave, E, is the integral of J with respect to area over the surface of the wave-front. In the case of the scattered waves spreading outwards from a small target, the wave-fronts are spherical or nearly so, and E is finite. The definition of the echo strength in terms of the total energy, rather than the intensity, is better suited to the modern theory of echo integration as discussed in Chapter 5.

The instantaneous intensity is the product of the pressure and the particle velocity. In a plane wave, combining the definition with Equation 2.2, we find that the intensity is proportional to the pressure squared:

$$I = p^2/\rho c \tag{2.4}$$

More usually, it is the average intensity over one or more cycles that is required, in which case Equation 2.4 still applies if the mean squared sound pressure is substituted for p^2. The quantity $Z = \rho c$ is called the acoustic impedance, which is constant within a few percent over the sound path in typical oceanic or freshwater conditions. If the variation of Z is ignored, the echo energy obtained by integrating Equation 2.4 is proportional to p^2. This is the reason why the echo integrator operates by first squaring the voltage at the echosounder output (which is proportional to the sound pressure at the transducer) before the integration is performed. It is the echo energy, not the pressure amplitude, which is believed to be proportional to the observed quantity of fish.

The decibel

Acoustic measurements are often quoted in decibel (dB) units rather than the more formally correct SI units for pressure, intensity, etc. This is done because the numbers involved can be very large or very small, covering many orders of magnitude. The decibel is a logarithmic measure of the ratio of two intensities, e.g. I_1 and I_2. The ratio is expressed in dB by the formula

$$N_{dB} = 10 \log(I_2/I_1) \tag{2.5}$$

'Log' means the logarithm to the base 10, as elsewhere in this book. A change in the sound level covering many orders of magnitude may be expressed within a modest range of decibels. Because the dB unit is so commonly used in acoustics, it is important to understand how it is applied.

In principle, I_1 should be a reference level of intensity against which I_2 is measured. For example, suppose a transducer generates a source level of 200 000 Pa (1 Pa = 1 Nm^{-2}, the SI unit of pressure). The source level is defined as the sound pressure at 1 m from the transducer. The reference pressure used for transducer measurements is commonly 1 μPa. Because the intensity is proportional to the pressure squared, the source level in dB is

$$\begin{aligned} SL &= 10 \log(200\,000/0.000001)^2 \\ &= 226 \text{ dB relative to } 1\ \mu\text{Pa} \end{aligned} \tag{2.6}$$

It is important to include the reference pressure or intensity when describing the measurement of an absolute quantity such as the source level. Otherwise the dB number is meaningless. The above source level might equally be referenced to 1 Pa, in which case it would be written as '106 dB relative to 1 Pa', commonly shortened to '106 dB//1 Pa'.

The decibel unit is also used to describe the acoustic reflectivity of targets, namely the target strength (*TS*). The reflectivity is defined by the ratio of the reflected intensity at 1 m from the target (I_2) and the incident intensity (I_1). Since I_2 is proportional to I_1, the target strength is a true ratio and there is no need to quote a reference level in this case. For example, if I_2 is 0.0002 I_1, then

$$\begin{aligned} TS &= 10 \log(0.0002) \\ &= -37 \text{ dB} \end{aligned} \tag{2.7}$$

We discuss the important subject of target strength in more detail in section 2.5 and Chapter 6.

2.3 TRANSDUCERS AND BEAMS

The transducer supplied with fishery sonars has two functions. Firstly, it converts electrical energy into the transmitted acoustic pulse, sometimes called the ping. Secondly, when targets reflect the ping, the transducer converts the acoustic echoes to electrical signals which are applied to the receiving amplifier. The most common type of transducer is made from ceramic materials which are piezo-electric. When pressure is applied to this material, a voltage is generated, and the effect is reversible: when an oscillating external voltage is applied, the ceramic expands and contracts, radiating sound as it does so. The magneto-strictive transducer is another type and is found particularly in low-frequency applications. However, the ceramic transducers have a higher efficiency, and they are preferred for fishery applications. More complete descriptions of the physics of transducers are given by Mason (1964) and Tucker and Gazey (1966).

Ceramic and magneto-strictive transducers project modest amounts of energy into the water. Various other techniques have been used to produce the high-energy pulses required for geophysical surveys, such as the airgun and explosive charges. These devices are described in Chapter 4.

Transducers are often constructed as an array of individual elements. Figure 2.2 illustrates the design of a typical ceramic element and the arrangement of 34 elements which form the transducer array. Each element consists of four ceramic tubes with steel head and tail masses which are designed to ensure efficient transfer of energy into the water. The ceramic tubes are held together by a prestressing bolt. The low-density backing material ensures that most of the acoustic energy is transmitted in the forward direction into the water. This type of transducer is reversible: it may be used either to transmit or to receive sound waves.

A transducer that is used to listen to underwater sound, like a microphone in air, is called a hydrophone. This device is often a single piece of piezo-electric ceramic. If the ceramic is shaped as a spherical shell or a small cylinder, then in theory the hydrophone is equally sensitive to sound waves coming from any direction. However, the sensitivity of most transducers is strongly dependent on the wave direction because of the phase differences between the signals produced by the elements or parts of the transducer. The effect is illustrated in Fig. 2.3 for the simple case of a transducer consisting of two spherical hydrophones. When the wave-fronts are parallel to the line joining the hydrophones, the two signals are in phase and their sum (the transducer output) is maximal (Fig. 2.3(a)). The direction of propagation in this case is called the acoustic axis. As the sound

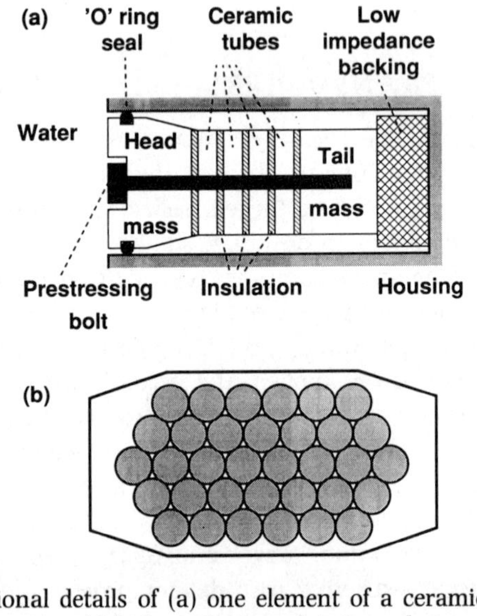

Fig. 2.2 Constructional details of (a) one element of a ceramic transducer 35 mm dia. and (b) 34 elements which form the complete transducer.

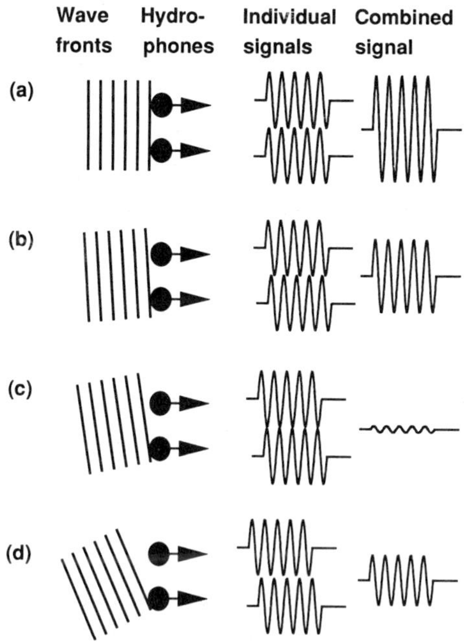

Fig. 2.3 Combination of signals from two hydrophones. (a) The signals are in phase and the summed amplitude is maximal; (b) the amplitude is reduced as the source moves off the acoustic axis; (c) the signals nearly cancel at the null; (d) after the null, the amplitude increases again.

source moves away from the axis (Fig. 2.3(b)), the phase difference increases and the amplitude of the sum is less than the maximum. In Fig. 2.3(c), the path lengths between the sound source and the hydrophones differ by nearly half a wavelength. The two signals cancel each other and the amplitude of the sum is very small. As the sound source moves yet further from the axis, the summed amplitude increases once again (Fig. 2.3(d)). This effect may be used to determine the direction of a sound source, by comparing the phases of the signals produced by the individual elements of a transducer.

The function **b** which describes the change of sensitivity with direction is called the beam pattern. The same beam pattern normally applies both to transmission and to reception of sound. By convention, **b** is unity in the direction of greatest sensitivity, the acoustic axis. For a point source or receiver, one whose size is much smaller than the wavelength, **b**$=1$ everywhere and the sensitivity is omni-directional. In the case of a square transducer of side length a, the acoustic axis is perpendicular to the radiating surface. To describe the sound field produced by such a transducer, consider the plane through the acoustic axis and parallel to one side of the transducer. If θ is any angle from the acoustic axis in this plane, the beam pattern in that direction is

$$\mathbf{b}(\theta) = \sin[(\pi a/\lambda) \sin \theta] / [(\pi a/\lambda) \sin \theta] \qquad (2.8)$$

When transmitting, the sound pressure at a fixed range $R \gg a$ is proportional to $\mathbf{b}(\theta)$, and the intensity varies as $\mathbf{b}^2(\theta)$. We see from this equation that **b** is zero in directions where $\sin \theta = n\lambda/a$, n being a non-zero integer. These directions are said to be 'nulls' in the beam pattern. Between adjacent nulls the region of higher sensitivity is called a lobe. The larger the transducer the more lobes there are, and the more directional is the sensitivity. The beam pattern is normally represented as a polar diagram of sensitivity against direction (Fig. 2.4).

If the transducer is a circular disc of diameter a, the beam pattern is symmetrical. θ is now the angle from the axis in any plane normal to the disc, and

$$\mathbf{b}(\theta) = 2J_1[(\pi a/\lambda) \sin \theta] / [(\pi\ a/\lambda) \sin \theta) \qquad (2.9)$$

$J_1(x)$ is the first-order Bessel function of the argument x (Abramowitz and Stegun, 1964). Again, **b** goes through a pattern of maxima and nulls as θ increases from zero. The first null occurs when $\sin \theta = 1.2\ \lambda/a$.

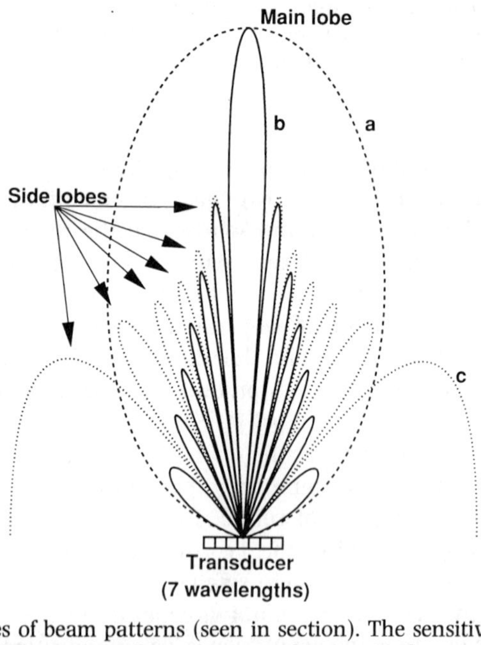

Fig. 2.4 Examples of beam patterns (seen in section). The sensitivity is proportional to the distance of the pattern from the central point. Broken oval, square element one wavelength long; solid curves, transducer made from seven such elements; dotted curves, transducer made from seven point sources one wavelength apart.

Any transducer may be considered as being made from a number of elements. If the beam pattern of each element is known, that for the complete transducer is obtained by adding the signals from each element, taking account of the phase differences between them (Tucker and Gazey, 1966). The signal phase is determined by the path length from the element to the sound source (or the target in the case of transmission). The elements may be individual units as in Fig. 2.2, but the same principle may be applied to a transducer whose radiating surface is large and continuous, however complicated its shape may be. The irregular surface is considered as a composite of simple elements which might be flat squares, in which case the elemental beam pattern is given by Equation 2.8.

Figure 2.4(a) shows a section through the beam pattern of a square element which is one wavelength long. In this case there is only one null, at $\theta = 90°$. Now consider a larger transducer which consists of seven of these elements side by side. The interference between the signals at the seven elements results in the beam pattern of the complete transducer having the more complicated shape shown in Fig. 2.4(b). It is most sensitive in a narrow cone centred on the acoustic axis, called the main lobe, and there are six side lobes at larger values of θ, separated by nulls. Figure 2.4(c) shows the beam pattern of a transducer whose elements are point sources in a line, again spaced one wavelength apart. In this case the side lobes are larger and there is no null at $\theta = 90°$.

Figure 2.5 is a pictorial representation of the wave-fronts generated by the point sources when the transducer is used as a transmitter. It illustrates how the seven omni-directional sound sources combine to produce the lobe structure of the beam pattern. The main lobe is evident as the near coincidence of the wave-fronts when they are far from the transducer. The picture is more confused in the region close to the transducer face, due to the near-field effect which we discuss later (p. 21). The coincidence of the wave-fronts is less complete in the side lobes, except for the one at 90°.

The beam width of a transducer is commonly described by the angle between the directions on opposite sides of the main lobe where $\mathbf{b} = 1/\sqrt{2}$, and the intensity is 3 dB less than that on the acoustic axis. Numerical methods for calculating the beam width are given by Urick (1983). Beam width depends upon the frequency and the size of the transducer (examples, Table 2.1).

The equivalent beam angle

Another measure of the beam pattern is the equivalent beam angle ψ, sometimes called the reverberation angle of the transducer. ψ is the solid angle at the apex of the ideal conical beam which would produce the same echo integral as the real transducer when the targets are randomly

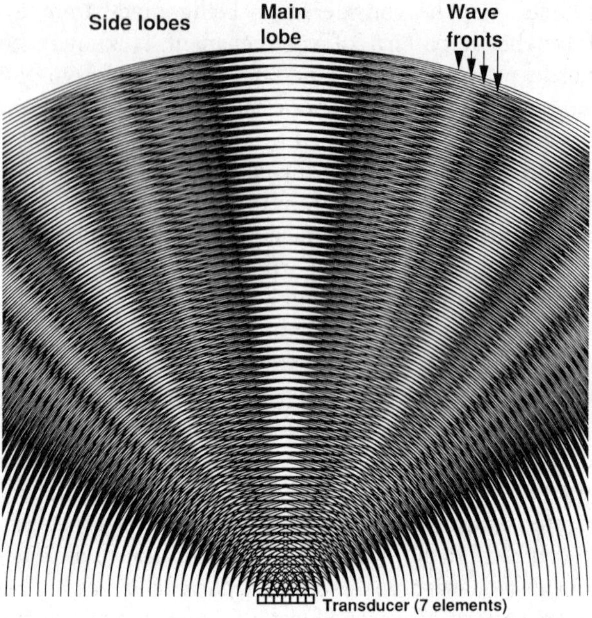

Fig. 2.5 Illustration of the beam pattern for a transducer consisting of seven point sources one wavelength apart. The lines are the wave-fronts from each source.

Table 2.1 Directional properties of circular disc transducers. The beam width is the angle between the directions on opposite sides of the main lobe where the intensity is half that on the acoustic axis. The equivalent beam angle is the solid angle of the equivalent ideal conical beam. Calculations assume that the sound speed is $1500\,\mathrm{m\,s^{-1}}$

Frequency (kHz)	*Diameter (cm)*	*Beam width (degrees)*	*Equivalent beam angle*	
			(sr)	*(db* re *1 sr)*
10	30	29.9	0.149	−8.3
40	30	7.4	0.009	−20.4
120	30	2.5	0.001	−29.9
120	6	14.8	0.025	−16.0
200	4	11.1	0.021	−16.9
10	50	17.8	0.053	−12.8
1	500	17.8	0.053	−12.8

distributed in space. The conical beam is like an ideal searchlight. $\mathbf{b}=1$ for any direction within the cone, and $\mathbf{b}=0$ elsewhere (Simmonds, 1984a). To describe the beam pattern, we use spherical polar coordinates, which require two angles (θ, ϕ) to determine the direction of any point in space relative to the transducer. θ is the angle of the point from the acoustic axis, and ϕ is the angle of the point projected onto the plane of the transducer face. ψ is defined mathematically as

$$\psi = \int_{\theta=0}^{\pi} \int_{\phi=0}^{2\pi} \mathbf{b}^4(\theta, \phi) \sin\theta \, d\theta \, d\phi \tag{2.10}$$

The integral is taken over the entire beam pattern, which is normally supposed to be the hemisphere in front of the transducer ($\theta=0$ to π; $\phi=0$ to 2π). The fourth power of **b** appears in the definition because ψ is defined in terms of the echo intensity, involving the combined transmitting and receiving sensitivities. The importance of ψ in the theory of echo integration is discussed in Chapter 5. The equivalent beam angle describes the volume effectively insonified (p. 17) by the transducer and it takes account of signals from targets in the side lobes. However, for most transducers less than 1% of the transmitted energy is projected outside the main lobe. Thus the side lobes have very little effect on the value of ψ.

ψ is a solid angle which is measured in steradians. Some writers use logarithmic units for the equivalent beam angle, defined as $EBA = 10 \log \psi$, which is in dB relative to 1 steradian.

Beam shape

We have already described how a number of similar elements may be combined to determine the beam pattern of the complete transducer. In the examples shown in Fig. 2.4, the elements are connected in parallel. When transmitting, the same signal is applied to all the elements, and when receiving the element signals are summed to produce one output. It is possible to modify the resultant beam pattern by controlling the electrical gain applied to the signal at each element. This technique is called shading. For a given size of transducer, shading may be used to reduce the strength of the side lobes, in which case the width of the main lobe increases. Conversely, the shaded transducer might be designed to reduce the beam width if larger side lobes can be accepted. In the case of the transducer array, the strength of the side lobes also depends upon the proportion of the transducer surface that is filled by the radiating parts of the elements. In a good transducer design, this proportion should be at least 70%, and in addition the spacing of adjacent elements should be less than a wavelength.

A simple form of shading is to separate the signals from two or more parts of the transducer. The transducer is used as though it were several independent devices having different beam patterns. This technique is the basis of the dual-beam and split-beam echosounders which are described in Chapter 3.

The scanning sonar is another useful application of shading. In this device the signals from adjacent elements are delayed before they are summed, so that the beam is tilted from the acoustic axis. By varying this delay rapidly, the beam is made to rotate within the pulse duration and targets may be viewed in azimuth as well as in range, i.e. in two dimensions as on a radar display.

2.4 ACOUSTIC PROPAGATION

In this section we consider the sound field in the water between the transducer and any target which generates an echo. Much of the literature on acoustic propagation is concerned with horizontal transmission over long distances, when reflections from the sea surface and the bottom are important. There is also the phenomenon of refraction, which causes the direction of propagation to change if the sound speed varies across the wave-front. In fisheries acoustics, however, we are more often concerned with short-range or vertical transmission in which refraction is unimportant.

Beam spreading

As the wave-fronts travel outwards from the transducer, they spread over a larger area. The total energy of the transmission is fixed, so the intensity (power transmitted through unit area) decreases as the beam spreads (Fig. 2.6(a)). At ranges that are large compared to the transducer size, said to be in the 'far field', the range dependence of the intensity obeys the inverse square law:

$$I = I_0/R^2 \qquad (2.11)$$

It follows that the pressure amplitude is inversely proportional to the range.

$$|p| = |p_0|/R \qquad (2.12)$$

The constants I_0 and $|p_0|$ are respectively the intensity and the pressure amplitude normalized to unit range. However, the actual intensity at unit range may not be the same as I_0. There is a region immediately in front of the transducer where the range dependence of the intensity is more complicated, called the near field or the Fresnel zone. This occurs at ranges where the wave-fronts produced by the transducer elements are not parallel, a state which alters the phase relationships compared with the far field (Fig. 2.5). In the near field, the intensity varies rapidly with the range in an oscillatory manner (Fig. 2.6(b)). It is only in the far field (also known as the Fraunhofer zone), when the element wave-fronts are nearly parallel, that

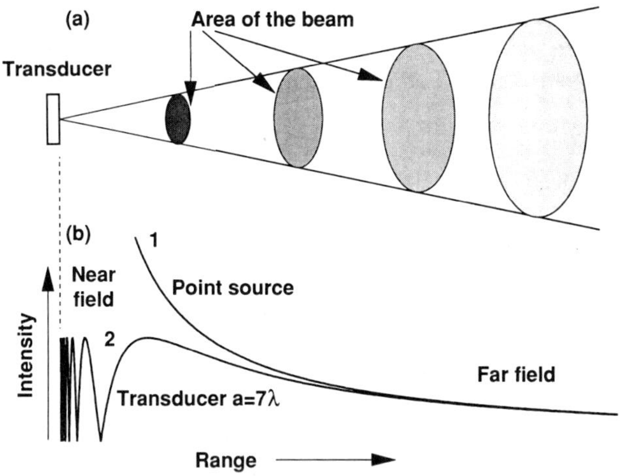

Fig. 2.6 Propagation of acoustic energy generated by a transducer. (a) Spherical spreading reduces the intensity at large ranges. (b) The intensity of a point source follows the inverse square law (curve 1). The near field limits the intensity of a finite transducer, $a=7\lambda$ in this case (curve 2).

the beam is properly formed and the inverse square law applies. If a is the linear distance across the transducer face, the boundary between the near and far fields is approximately at the range

$$R_b = a^2/\lambda \tag{2.13}$$

For example, the wavelength in seawater is about 3.95 cm at 38 kHz. If $a=7\lambda$ as in Fig. 2.5, the near field extends to $R=1.94$ m. A higher-frequency transducer designed to have the same beam width would have a smaller near field. The transition from near to far-field conditions occurs gradually around R_b. For this reason, acoustic measurements that depend upon the assumption of far-field conditions should be done at ranges of at least 2 R_b.

It is possible to calculate the spatial variation of the intensity in the near field (Rschevkin, 1963). However, such calculations are likely to be inaccurate when applied to real transducers, because small and unpredictable differences between the elements can result in large changes of the intensity. Acoustic measurements to calibrate the transducer must be performed in the far field where the intensity is more predictable.

Absorption

As sound propagates through water, acoustic energy is lost through the process of absorption. This causes the intensity of a plane wave to decrease exponentially with the distance x along the propagation path. β is the absorption coefficient in nepers per unit distance, which means that the amplitude decreases by the exponential constant e = 2.718 over the distance $(1/\beta)$. Since the intensity is proportional to the amplitude squared, we have

$$I(x) = I_0 \exp(-2\beta x) \tag{2.14}$$

It is more usual to express the absorption coefficient as the energy loss in dB per unit distance, for which we use the symbol α. Thus

$$10 \log[I(x)/I_0] = -\alpha x \tag{2.15}$$

Comparing Equations 2.14 and 2.15, the two measures are simply related as $\alpha = 8.69\ \beta$.

Several mechanisms contribute to the absorption. Viscous (frictional) losses occur in both fresh and saline water. This part of the absorption coefficient is proportional to the square of the frequency. In the sea, there are additional losses owing to the molecular relaxation of certain compounds. For each compound, there is a critical 'relaxation' frequency below which the losses occur. Relaxation is a pressure-induced reduction of molecules to ions which takes a certain time to occur. At high frequencies, the sound pressure cycles too quickly for the reduction to occur and thus no energy is absorbed by this process. The magnesium sulphate relaxation dominates the absorption in the frequency range 2–500 kHz, and at lower frequencies there is a further loss associated with boric acid. The frequency is the main determinant of the absorption, but α also depends upon the water temperature and salinity (Fig. 2.7).

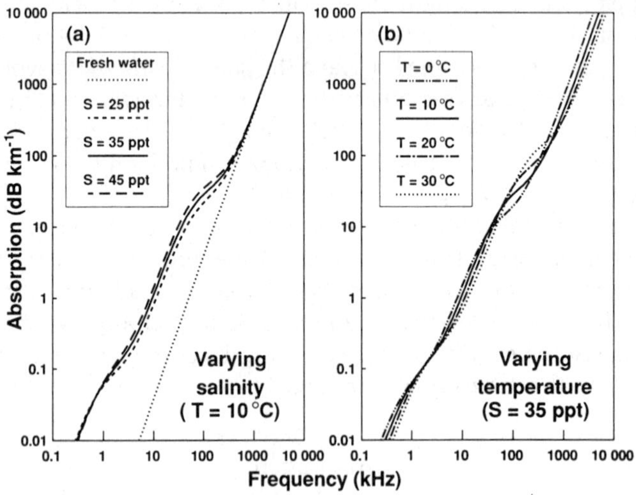

Fig. 2.7 Absorption of acoustic energy in water v. frequency, calculated from Francois and Garrison (1982). S is the salinity and T is the temperature. (a) Various salinities at $T = 10\,°C$; (b) various temperatures at $S = 35$ ppt.

Table 2.2 Absorption losses in water at various frequencies. Calculations (based on Francois and Garrison, 1982) assume temperature 10°C, water pH 8 and surface pressure. *S*, salinity; α, absorption coefficient; $\Delta E/E$, proportion of energy lost by a plane wave over 100 m of the propagation path; <signifies value less than 0.001

	Seawater (S=35 ppt)		*Fresh water (S=0 ppt)*	
Frequency (kHz)	α *(dB km^{-1})*	$\Delta E/E$ *per 100 m*	α *(dB km^{-1})*	$\Delta E/E$ *per 100 m*
1	0.06	<	0.0002	<
3	0.18	<	0.0019	<
10	0.95	0.022	0.021	<
30	6.76	0.144	0.19	0.004
38	10.0	0.206	0.31	0.007
50	15.1	0.294	0.53	0.012
100	32.6	0.528	2.14	0.048
120	37.3	0.576	3.08	0.068
200	50.5	0.687	8.55	0.179
300	64.3	0.772	19.2	0.358
500	100.3	0.901	53.4	0.708
1000	261.4	0.998	213.8	0.993

Various equations have been developed for predicting α at a given frequency, temperature, salinity and depth, notably by Shulkin and Marsh (1963) and Fisher and Simmons (1977). The more recent equations of Francois and Garrison (1982) are the most accurate available (Appendix A). Francois and Garrison maintain that the predicted α is within $\pm 5\%$ of the true value, assuming that the temperature and other parameters are known exactly.

The rapid increase of absorption with frequency is the limiting factor which determines the highest useful frequency for the detection of targets at a given range. Above this limit, the absorption loss is so great that the received signal is below the noise level and cannot be detected. Some examples of the absorption loss are shown in Table 2.2. For targets at 100 m range in the sea, absorption is unimportant below 10 kHz, but at 1 MHz the useful range is only a few metres. The absorption in fresh water is relatively small at low frequencies, up to 200 kHz or so, since the molecular relaxation loss is absent. Thus a higher frequency may be used in freshwater applications to achieve the same maximum range of target detection.

Sound speed

As with the absorption, the sound speed c depends upon the water temperature, salinity and depth. It may be predicted by empirical equations,

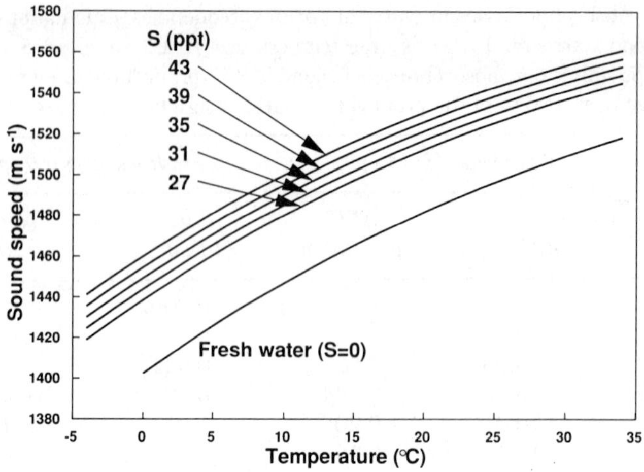

Fig. 2.8 Sound speed in water as a function of the temperature and the salinity *S*.

and that of Mackenzie (1981) is perhaps the most accurate of those published so far. Simpler equations are good enough for most fishery applications, notably Leroy's (1969). The relevant formulas are given in Appendix B. The variation of c with temperature and salinity is illustrated in Fig. 2.8. Mackenzie's formula is accurate to $\pm 1\%$ in the predicted value of c.

In the case of sonars which transmit horizontally, refraction is important because the sound speed changes with depth. The temperature is normally highest at the surface, so the sound speed decreases with depth and the rate of change is most rapid at the thermocline. As a result, waves propagating near the surface bend downwards because the lower part of the wave-front is in cooler water where it travels more slowly (Fig. 2.9). When considering

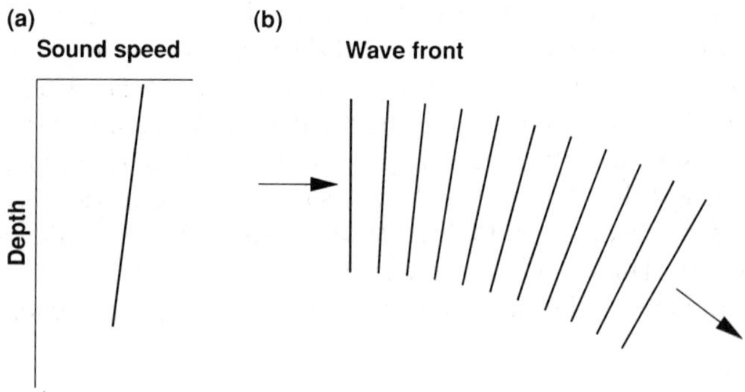

Fig. 2.9 Refraction of sound waves. (a) The sound speed decreases with depth; (b) this causes the wave-fronts to incline progressively along the propagation path (arrowed), which is deflected downwards.

this effect, it is easier to understand what happens to the sound if we think of the acoustic signals as travelling along rays. A ray is a line which is everywhere perpendicular to the wave-front. The downward bending caused by refraction limits the horizontal range at which a target can be detected. This is seen by drawing rays in various directions from the sonar transducer (Fig. 2.10). At each point on the diagram, the ray is shown as a circular arc whose radius is proportional to the gradient of c across the wave-front. More precisely, the centre of the arc is at a point where c would be zero if the gradient were constant along the radius. When a ray meets the surface or the bottom, it continues as a reflection. The angles from the surface to the incident and reflected rays are the same. Some energy will be reflected (scattered) in other directions, especially at the sea-bed if it is rough, but the simple reflection law is good enough for the ray diagram to show any 'hidden zone' where targets will not be detected. It is possible for bottom-reflected rays to penetrate a zone which is inaccessible to the direct unreflected rays (Fig. 2.10(c)).

Figure 2.11 shows two examples of measured sound speed profiles and the calculated surface-grazing ray which defines the maximum horizontal range of sonar detection. The transducer is 5 m below the surface and the range depends upon the depth of the target. When the gradient of the sound speed is uniform (Fig. 2.11(b)), the ray-bending is particularly severe. In Fig. 2.11(a), most of the change in c occurs at the strong thermocline above 25 m and the maximum range is greater. The position and strength of the thermocline change with the time of year, and so does the performance of

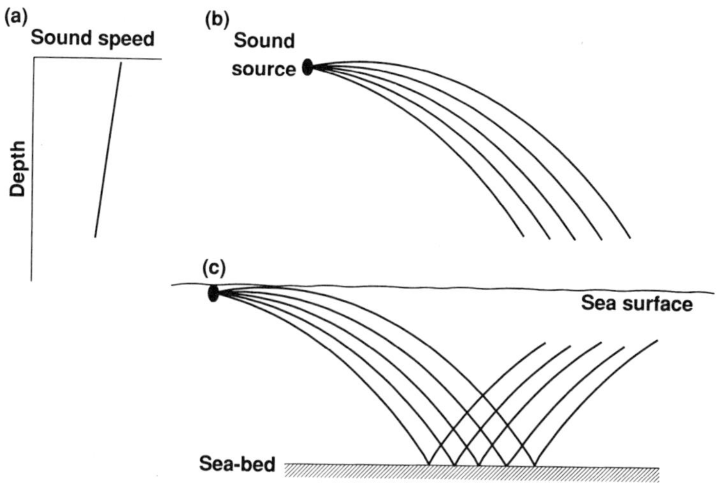

Fig. 2.10 Ray-tracing. (a) The sound speed decreases with depth; (b) the rays are projected in a horizontal beam which is deflected downwards by refraction, limiting the range; (c) reflection from the sea-bed increases the effective range.

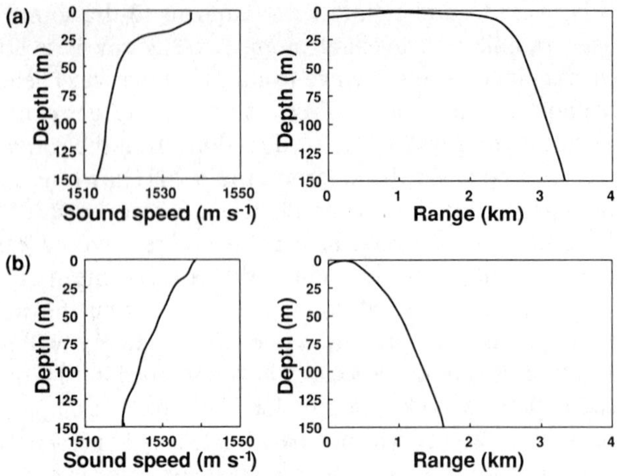

Fig. 2.11 Variation of sound speed with depth and the maximum range for sonar detection. (a) Persian Gulf; (b) Gulf of Oman.

sonar. Figure 2.12 shows the maximum range that can be achieved in different months. In the Persian Gulf (Fig. 2.12(a)), the surface water becomes very hot in summer, up to 35°C, while the temperature below 50 m is less variable. Thus the thermocline is strongest in the summer, when the conditions for sonar observations are much worse than in other seasons.

Deep oceanic water is nearly isothermal, and below 500 m it is the depth dependence that controls the variation of *c*. The sound speed increases with depth, so the negative gradient near the surface (controlled by the tempera-

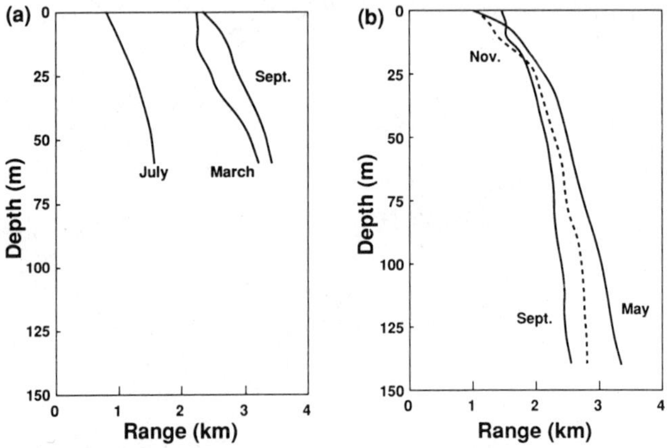

Fig. 2.12 Maximum range for sonar detection at different times of year v. depth of targets. (a) Persian Gulf; (b) Gulf of Oman.

ture) changes to a positive gradient further down. The refraction by these gradients has the strange effect of bending shallow waves downwards and deep waves upwards, confining the acoustic energy to an interval of depth called the SOFAR channel. Low-frequency sound propagates in the SOFAR channel without much loss over hundreds of kilometres, since the propagation is two-dimensional and no energy is lost in reflections from boundaries. The great whales are believed to use the SOFAR channel to communicate over long distances (Schevill *et al.*, 1954).

Pulses and ranging

The pulsed sonar transmits a short burst of sound, called the pulse or the ping, consisting of several cycles at the sonar operating frequency. Figure 2.13 shows a pulse of 19 cycles generated by a 38 kHz sonar. The pulse duration (time from start to finish) is $\tau=(19/38)=0.5$ ms. If the sound speed is $1500\ \mathrm{m\,s^{-1}}$, the pulse length in the water at any instant is $L=c\tau=75$ cm. The envelope of the pulse is the curve which shows the amplitude of the oscillations. Thus the envelope in Fig. 2.13 is a rectangle because the amplitude is constant during the transmission.

The transmitted pulse travels away from the transducer. When it encounters a target, some energy is reflected as the echo which travels back to the transducer (Fig. 2.14). The echo is received at time T after the transmission. The distance R between the target and the transducer is estimated by measuring T. The two-way path length is $2R$, so

$$T=2R/c \tag{2.16}$$

$$R=cT/2 \tag{2.17}$$

Suppose there are two targets at ranges R_1 and R_2. In order to resolve the targets and measure them individually, the range difference (R_2-R_1) must be large enough for the two echoes not to overlap. The echo from the nearer target is detected first, at time $T_1=2R_1/c$, and it continues for the pulse duration, that is until time $(T_1+\tau)$. The second echo produces a signal which begins at $T_2=2R_2/c$. To resolve the targets, we must have $T_2>T_1+\tau)$ or

$$R_2-R_1>c\tau/2 \tag{2.18}$$

Thus targets must differ in range by at least $c\tau/2$, half the pulse length in water, to produce separate echoes. Another way of looking at this problem is to say that if there are many targets at all ranges, then at any time, echoes are being received from those targets within a shell whose thickness is half the pulse length. Furthermore, the width of the shell is limited by the transducer beam because targets outside the beam produce no echo. This leads to the concept of the sampling volume V_s, which is the shell thickness

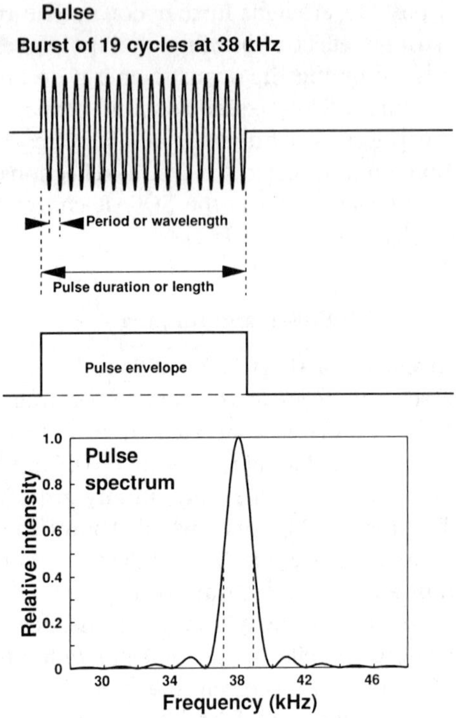

Fig. 2.13 Illustration of the transmitted pulse. As observed at the transmitter output, the pulse is 19 cycles at 38 kHz and the duration is 0.5 ms. In water the pulse length is 0.75 m when $c = 1500\ \mathrm{m\,s^{-1}}$. The envelope is the curve of maximum amplitudes. The spectrum shows the frequency composition as the intensity or power relative to the peak value. The broken vertical lines indicate the bandwidth between half-power points, 1.8 kHz in this case.

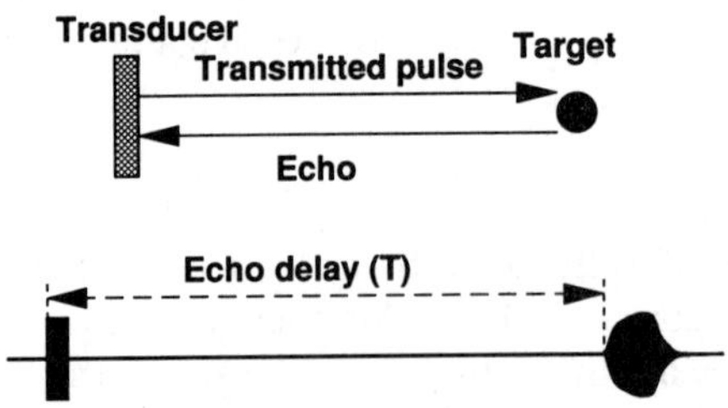

Fig. 2.14 Estimation of the target range from the time delay of the echo. T is the time delay and the range is $cT/2$. The lower figure represents the pulse waveforms that would be seen on an oscillogram.

multiplied by the effective cross-sectional area of the beam. The latter is equal to the equivalent beam angle times the square of the range, so

$$V_s = c\tau\psi R^2/2 \tag{2.19}$$

The sampling volume indicates the space that is contributing echoes at any instant. It is a useful measure for comparing the performance of different sonars.

2.5 ACOUSTIC SCATTERING

When acoustic waves encounter a target, some of the incident energy is scattered, generating a secondary wave which propagates in all directions away from the target. The reflection which occurs at a large smooth surface is a special kind of scattering, in which the secondary (reflected) wave is confined to one direction such that the incident and reflected wave-fronts are at the same angle to the surface. More generally, scattering occurs wherever there is a spatial change of the acoustic impedance Z. The reflected proportion of the incident energy is determined by *r*, the coefficient of reflection (Tucker and Gazey, 1966). If Z_w and Z_r are respectively the acoustic impedances of the water and of the reflector, then

$$r = (Z_r - Z_w)/(Z_r + Z_w) \tag{2.20}$$

The greater the change of Z across a boundary, the stronger is the scattered wave. The remaining energy passes through the boundary as the transmitted wave. The strength of the scattered wave is determined by boundary conditions such as the continuity of the sound pressure. The theory of scattering by simply shaped reflectors is well established and need not be repeated here (Morse, 1948; Officer, 1958). The important problem of scattering by fish is discussed at length in Chapter 6.

The energy reflected back towards the sound source is said to be backscattered. It is this component of the total scattering that provides the sonar echo. Small targets reflect energy in all directions. The component that propagates in the same direction as the incident wave is called the forward scatter. When the transmission is a short pulse, the incident and backscattered pulses occur at different times in the region between the source and the target, and there is no interference. On the far side of the target, the sound field is the sum of the incident and forward-scattered waves which coincide in time. After passing the target, the incident wave continues to propagate but the amplitude (pulse shape) is distorted by the addition of the forward scatter.

Targets large and small

A target is any object with a boundary across which there is a discontinuity in the acoustic impedance. If the target is very small compared with the wavelength, then as the incident wave arrives, the whole target is subjected to the same sound pressure. The target contracts and expands in response to the pressure oscillation and it acts as a point source of the scattered waves, which spread spherically in all directions (Fig. 2.15). However, the scattered intensity is not the same in all directions. It is the volume rather than the shape of the target that determines the scattering. If d is the characteristic size, the cube root of the volume, then the scattered energy is proportional to $(d/\lambda)^4$ when $d \ll \lambda$. This is the Rayleigh scattering law, which applies to light as well as to sound.

Now consider the other extreme, where the target is much larger than the wavelength. In this case the incident sound is scattered by the surface rather than the volume of the target. When the surface is smooth, the scattering is simply a reflected wave whose direction is determined by the usual rule that the angles of incidence and reflection are equal (Fig. 2.16). If the target is spherical, the scattered energy increases approximately as the square of the sphere radius. When this relationship applies, we say that the scattering is geometric. The shape of the target is important as well as the size. Thus a flat disc and a sphere of the same radius have different scattering properties.

Reflections at the sea surface and the bottom are of particular interest. The surface is a good reflector because the impedance mismatch between water and air is very great, and $Z_r \ll Z_w$. For this condition, Equation 2.20 gives the result $r \cong -1$. This means that the intensities of the incident and reflected waves are nearly the same, but the negative coefficient of reflection

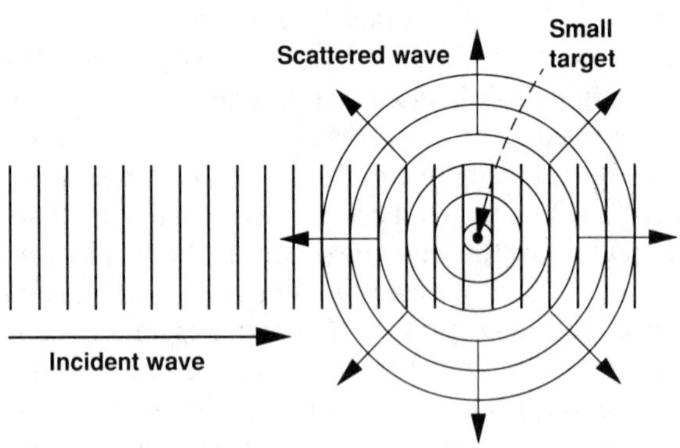

Fig. 2.15 Scattering of sound by a small target. The scattered waves propagate outward from the target in all directions.

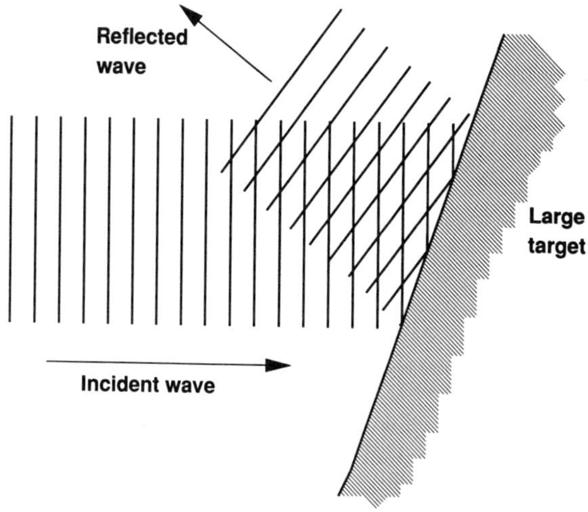

Fig. 2.16 Scattering of sound by a large target. The scattered waves form a specular reflection in one direction such that the angles of incidence and reflection are the same.

indicates that the two waves are in antiphase; the reflected pressure is minimal when the incident pressure is maximal, and vice versa. In the case of the bottom reflection, the acoustic impedance is higher in the ground than the water ($Z_r > Z_w$). Referring again to Equation 2.20, r is now positive and the reflected and incident waves are in phase. If the bottom is rough, meaning that it is uneven on the scale of the wavelength, some energy is scattered in directions other than that of the specular reflection.

At intermediate sizes where the target dimensions and the wavelength are similar, the scattering depends upon the geometric structure and the material properties of the target in a rather complicated way. Resonances occur which cause the strength of scattering to change rapidly with frequency. In the case of gas bubbles in shallow water, there is one resonance near the high-frequency limit of the Rayleigh scattering region (Fig. 2.17). At higher frequencies, in the geometric region, the scattering strength of the gas bubble is almost constant. The resonance behaviour of solid targets is more complicated. Many sharp resonances occur at intervals above a certain frequency which is normally within the geometric scattering region. This phenomenon is discussed further in section 2.5.

In summary, scattering by a small target increases rapidly with frequency; for large targets the frequency has little effect; at intermediate sizes of the order of a wavelength, resonances occur which make it difficult to predict the scattering accurately.

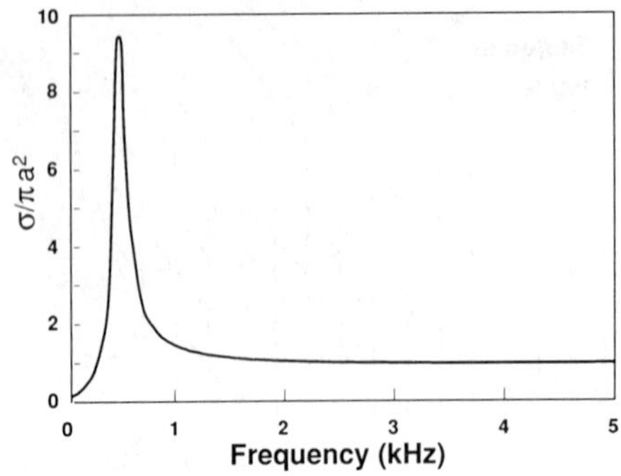

Fig. 2.17 Scattering by a gas bubble v. frequency. σ is the acoustic cross-section and a is the bubble radius. The scattered intensity increases rapidly to a peak at the resonance frequency of the bubble.

Target strength

Most fisheries applications depend upon one transducer, which is used both to transmit acoustic pulses into the water and to detect echoes from targets within the transducer beam. In that case, only the backscattered waves are important, namely those travelling in exactly the opposite direction to the incident waves generated by the transmitter.

The target strength is a logarithmic measure of the proportion of the incident energy which is backscattered by the target. To understand the nature of target strength, it is better to begin with the related quantity σ, the acoustic cross-section, which is a more meaningful parameter in physical terms. σ is measured in units of area, square metres in the SI system. It is defined in terms of the intensities of the incident and the backscattered waves.

Suppose I_i is the intensity of the incident waves at the target. The fact that the transmission is a pulse complicates matters, but suppose for the present that the pulse is long and I_i refers to the intensity at the midpoint of the pulse. I_b is the intensity at the midpoint of the backscattered pulse. I_b will depend upon the distance R from the target at which the intensity is measured. If R is large enough to be outside the near field of the target, which means that R has to be much greater than the linear size of the target, but not so large that absorption losses are important, then the cross-section σ is defined by the relationship

$$\sigma = 4\pi R^2 I_b / I_i \tag{2.21}$$

This equation can be interpreted in the following manner. The intensity is the power transmitted per unit area, so σI_i is the power in the incident wave passing through an area of size σ. The surface area of a sphere of radius R is $4\pi R^2$, and if the scattered intensity were the same in all directions, the total scattered power would be $4\pi R^2 I_b$. Multiplying both sides of Equation 2.21 by I_i, we see that σI_i is equal to $4\pi R^2 I_b$. In other words, σ may be conceived as the area intercepting part of the power transmitted by the incident wave, which power is retransmitted as the scattered wave. This interpretation is strictly true only if the scattering is omni-directional, but the same definition of the acoustic cross-section is used for any target, since the scattering in other directions is not relevant to the backscattered echo strength. Furthermore, in the case of pulsed transmissions, it is more appropriate to describe the scattering in terms of the energy transmitted by the pulse. The energy is the integral of the power over a period of time that covers the pulse duration.

σ is more correctly described as the total cross-section, to distinguish it from another quantity, the so-called backscattering cross-section or σ_b, which is also an area (Clay and Medwin, 1977). The defining equation for σ_b is

$$\sigma_b = R^2 I_b / I_i \tag{2.22}$$

Thus σ is 4π times σ_b. In this book, all further references to the acoustic cross-section mean the value of σ defined by Equation 2.21. We do not use σ_b.

Equation 2.21 is rigorously correct only for the case of continuous transmission at constant amplitude and in the absence of absorption. It is also assumed that the target is isolated; no other targets or reflecting boundaries are close enough to perturb the scattered sound field at the position where I_b is measured. The problem with continuous transmission is that the incident and scattered waves coexist, so it is not possible to measure them separately. When the transmission is a pulse of finite duration, this problem is overcome because the incident and backscattered pulses are separated in time. If the pulse is long enough, intensity measurements in midpulse will be a good approximation to the continuous-wave case. If the pulse is short, however, the definition of σ is more complicated. It involves the frequency response of the sonar as well as scattering properties of the target, because a short pulse may be conceived as the sum of many continuous waves covering a spectrum of frequencies (MacLennan and Forbes, 1984). Suppose that $s(\omega)$ is the cross-section as defined by Equation 2.21 when the transmission is continuous at frequency ω, and $P(\omega)$ is the power response of the sonar. $P(\omega)$ is the signal power which would be observed at the output of the sonar receiver if the transmissions were applied directly to the receiving transducer. According to the definition of Foote (1982), σ is a weighted average of $s(\omega)$ over the bandwidth of the sonar. The formal definition is

$$\sigma = \int_0^\infty s(\omega)P(\omega)\,\mathrm{d}\omega \Bigg/ \int_0^\infty P(\omega)\,\mathrm{d}\omega \tag{2.23}$$

Note that the frequency response of the sonar is unimportant if $s(\omega)$ is constant over the bandwidth, in which case Equation 2.23 reduces to $\sigma = s$. One complication is the absorption loss, which is ignored in the defining equations. However, a simple correction can be made if necessary. The measured I_b is multiplied by the factor $\exp(\beta R)$ to give the intensity that would be observed in a lossless medium.

The target strength is the acoustic cross-section expressed in decibels, according to the formula

$$TS = 10 \log(\sigma/4\pi) \tag{2.24}$$

When this formula is used, the value substituted for σ must be in m^2. *TS* and σ are simply alternative ways of describing the same physical property of the target, namely the strength of backscattering. The logarithmic *TS* is convenient because of the great difference between aquatic organisms at opposite ends of the size scale, which covers many orders of magnitude, from microscopic plankton to the great whales. A modest span of decibels is sufficient to describe the target strengths of all these creatures. For almost all fish, *TS* is within the range −60 dB to −20 dB. The equivalent cross-sections span four orders of magnitude, from 0.000013 to 0.13 m^2. To calculate the cross-section equivalent to a given target strength, the formula is

$$\sigma = 4\pi\, 10^{(TS/10)} \tag{2.25}$$

The concept of logarithmic units may seem a little strange to those not familiar with fields such as electrical engineering where their use is common. Comprehension is not helped by the fact that in the case of fish, *TS* is always negative, and −60 is a smaller number than −20. In thinking about the nature of acoustic scattering and its theoretical description, it is easier to work with the cross-section σ, which can be conceived as a physical area which is related to the size of the target. Target strength is more often used in practical work which involves calculations with real data, to provide numerical values within a convenient range. The important point to remember is that the decibel is a measure of ratio. To say that one target is 3 dB more than another means that the stronger target scatters twice as much energy; if the difference were 6 dB, the scattered energy ratio would be four times, and so on. The −20 dB target, a large tuna perhaps, produces an echo 10 000 times as strong as one at −60 dB, such as a small sprat around 4 cm in length.

Standard targets

One method of calibrating the sonar is to measure the echo from a standard target whose acoustic scattering properties are known. Early proposals for standard targets included table tennis balls (Welsby and Hudson, 1972) and solid spheres made from various materials such as steel or brass (MacLennan, 1982). However, more recent work has shown that spheres made from tungsten carbide (MacLennan and Armstrong, 1984) or copper (Foote, 1982; Foote and MacLennan, 1984) give the best results.

The scattering properties of the homogeneous sphere do not change as the target rotates. Thus the orientation of the sphere is unimportant, and the echo is determined only by the position of the centre relative to the transducer. There is a well-established theory for calculating the acoustic cross-section from the material properties of the sphere expressed as proportions of those in water (Hickling, 1962; Dragonette *et al.*, 1974; Neubauer *et al.*, 1974; MacLennan, 1981b). The functional dependence is

$$\sigma(\omega) = \pi a^2 F(\omega a/c,\ \rho_1/\rho,\ c_1/c,\ c_2/c) \tag{2.26}$$

There are two sound speeds within the sphere, c_1 and c_2, which are the speeds of longitudinal and transverse waves respectively. ρ_1 is the sphere density and a is the radius. ρ and c are respectively the density of water and sound speed in water. Thus changes in the medium influence the scattering properties. F is the so-called form function, which is the ratio of the acoustic and geometric cross-sections.

The variation of σ with frequency is shown in Fig. 2.18 for tungsten carbide and copper spheres. At very low frequencies, Rayleigh scattering applies and σ varies as f^4. Once the wavelength is commensurate with the sphere size, σ varies cyclically and smoothly at first, but extreme variations occur at frequencies above a certain limit f_1. In the high-frequency region, there are sharp maxima and minima of the cross-section which correspond to resonances between the elastic vibrations of the sphere and the sound field in the water. The resonances occur at discrete frequencies which depend upon the sound speed ratios. The density ratio also has an effect but it is small. The resonances are controlled by the number of wavelengths across the sphere, since $\omega a/c = 2\pi a/\lambda$. This means that for the same material and sound speed, the resonance frequencies are inversely proportional to the sphere radius. The larger the sphere, the lower is f_1 which marks the beginning of the resonance region.

The standard target must be large enough to provide a strong echo well above the background noise level and of similar amplitude to the echoes that would be received by the sonar in the intended application. This dictates the approximate cross-section of the target. To meet this requirement, the sphere may have to be so large that the sonar frequency f_0

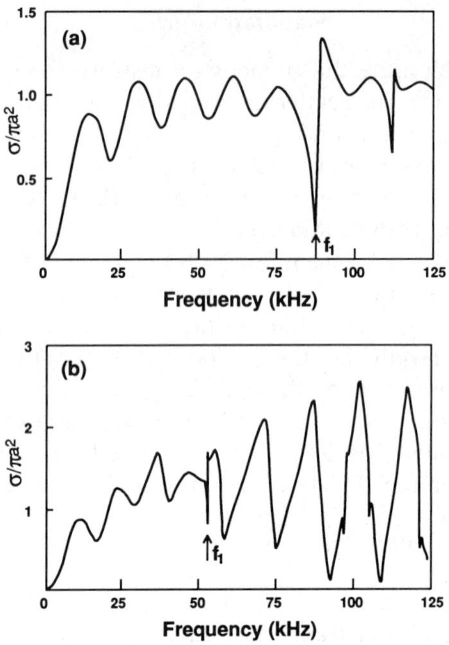

Fig. 2.18 Scattering strength of homogeneous solid spheres of radius *a* v. frequency. (a) Tungsten carbide, $a = 19.05$ mm; (b) copper, $a = 30.0$ mm. The vertical axis (note differing scales) is the ratio of the acoustic and geometric cross-sections. The arrows indicate f_1, the lowest resonance frequency.

is greater than f_1. It is important to ensure that no resonance is within the frequency band of the transmission. Otherwise the target strength will be sensitive to small changes in the environmental conditions, notably the water temperature, which determines c, and the sphere will be unsuitable as a standard for calibration purposes. More generally, the optimum sphere size for a particular application can be determined as follows (Foote, 1982; Foote *et al.*, 1983). A first approximation to *a* is the size which gives the required target strength at frequency f_0 on the assumption that the form function $F = 1$. The nominal value of the target strength is less important than the need to avoid resonances. Calculations are performed for a range of sizes around the approximate value, to determine how the target strength changes with the sound speed, the pulse duration and other parameters. The optimum sphere size is the one whose target strength is least sensitive to changes in the parameters. However, it should not be necessary for the users of acoustic instruments to perform these rather complicated calculations. The manufacturers of scientific echosounders should be able to recommend optimum spheres for the calibration of their own equipment.

The harder the material of the standard target, the higher f_1 is for the same size of sphere. For this reason tungsten carbide is particularly suitable for standard targets. It is durable and does not corrode in the sea. The spheres are manufactured by sintering the tungsten carbide with a small amount (about 6%) of cobalt as a binder. The density varies slightly with the proportion of cobalt, but ρ_1 is easily measured. MacLennan and Dunn (1984) have determined the sound speeds in tungsten carbide spheres from which the target strength for any size and frequency may be calculated precisely.

Multiple targets

We have seen that to resolve the echoes from two targets, they must be separated in range by at least $c\tau/2$, half the pulse length in water. In the case of a layer containing many scatterers in close proximity, $c\tau/2$ is the width in range of the sampling volume from which echoes are being received at any instant.

The echo waveform produced by an isolated small target is rather similar to the waveform of the incident pulse. The echo is modified by the frequency response of the target, but if the sonar has a narrow bandwidth (10% of f_0 or less), the frequency response of the target can generally be ignored, unless f_0 happens to be close to a resonance. However, when the received signal contains the superimposed echoes from many targets, the different phases of each echo cause the signal amplitude to fluctuate. The echoes are said to interfere with one another. Furthermore, the targets may move (this is certainly true of fish) and the amplitude fluctuations are not consistent from one ping to the next. Examples of this effect are shown in Fig. 2.19. Because of the interference, one sample of the signal amplitude provides little information about the density of targets in the sampling volume.

If the targets are positioned randomly on the scale of a wavelength, the phases are random and the instantaneous amplitudes are described by the Rayleigh distribution (Rayleigh, 1945). The probability that the instantaneous intensity (amplitude squared) exceeds I is $\exp(-I/\bar{I})$ where $\bar{I}$ is the mean intensity. Furthermore, the mean and standard deviation of the intensity are equal. This distribution has the important property that the shape of the PDF (probability density function) is independent of n, the number of scatterers contributing to the received signal. The mean intensity is proportional to n, but the proportional fluctuation of the signal is the same for $n=10$ or $n=100$.

When the density of targets is very high, the received signal is further modified by acoustic extinction and multiple scattering, which remove energy from the direct path of the incident and backscattered waves. These problems are discussed in Chapter 5.

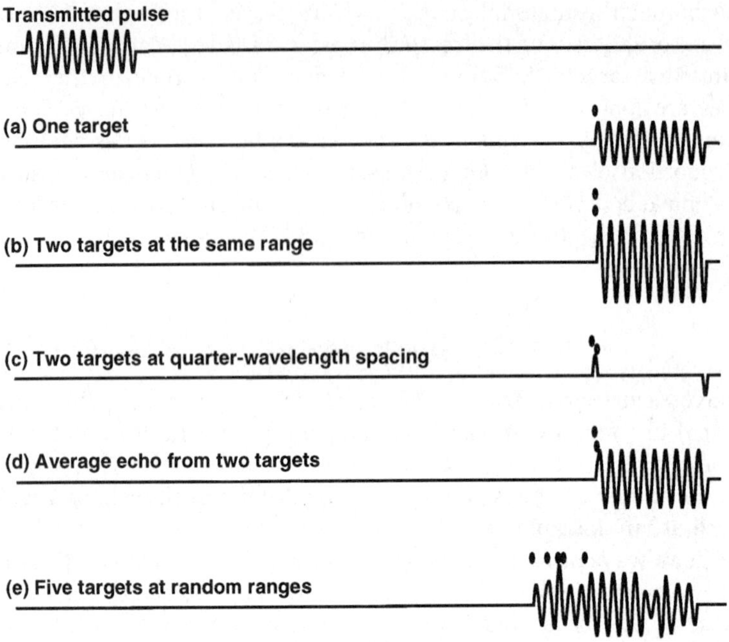

Fig. 2.19 Combination of echoes from multiple targets. (a) An isolated target reproduces the transmitted pulse; (b) when two targets are at the same range, the amplitudes sum; (c) with two targets at quarter-wavelength spacing, the amplitudes cancel; (d) with two targets at random ranges, the average energy sums; (e) five targets at random ranges give an irregular signal amplitude.

The inverse scattering problem

Much of sonar theory is formulated in terms of the scattering problem. The transmitted energy propagates through the water and is scattered by one or more targets, producing echoes which propagate back to the transducer where they combine to form the received signal. The standard theory describes the received signal as a unique result of the scattering properties of the targets, their location in space, the propagation losses and so forth. In practice, we are able to measure the received signal. The performance of the sonar is measured by calibration and the propagation losses are calculated from knowledge of the sound speed and absorption coefficient. The question then is, what do these measurements tell us about the unknown targets? This is the inverse scattering problem. In general, it does not have a unique solution because different sets of targets might produce the same signal.

Much of this book is concerned with establishing the rules and assumptions required to solve the inverse scattering problem in fisheries applications, in particular to identify and quantify the aquatic organisms which have generated the signals observed by sonar.

2.6 ECHO DETECTION

The receiver is an electronic amplifier between the transducer and the sonar output. The transducer converts the acoustic echoes to small electrical signals which must be amplified many times before they can be measured or displayed. The receiver has two other functions. Firstly, it may include time-varied gain (TVG, section 3.7) to compensate the signal for spreading and absorption losses, so that the sonar output becomes independent of the target range. Secondly, the receiver filters the transducer signal, rejecting components of the signal at frequencies outside the pass-band of the amplifier. The filtering is described by the bandwidth B, which normally means that signals at frequencies $f_0 \pm B/2$ are attenuated by 3 dB relative to the response at the sonar frequency f_0.

The transmitted pulse is a number of cycles at frequency f_0. The frequency spectrum of the pulse depends upon the pulse duration τ. The shorter the pulse, the wider is the spectrum; there is more energy at frequencies further from f_0. The same is true of the echo. Thus B must be large enough to pass most of the echo energy. On the other hand, a narrow bandwidth reduces the broad-band noise in the signal and makes it easier to detect small echoes.

The width of the transmitted spectrum is inversely proportional to τ. As a general rule, to process the signal without loss of echo energy, the bandwidth should be such that $B\tau = 3$. For example, if $\tau = 1$ ms, then B should be 3 kHz. If the echo detection is limited by broad-band noise, then the bandwidth might be reduced to $B\tau = 1$. Some of the echo energy would then be removed by the filter, but the signal-to-noise ratio (SNR) would nevertheless be increased to its optimum level.

Reverberation

Reverberation is a general term used to describe the echoes from unwanted targets. Thus it depends upon the purpose of the observation. If the intention is to detect fish, the echoes from plankton are part of the reverberation. If the intention is to study plankton, there are still likely to be unwanted signals from solid particles in suspension, and of course the large echoes from fish are no longer wanted. It is important to remember that one person's reverberation may come from targets that another might wish to observe.

The design of the sonar has little bearing on the reverberation problem. Increases in transmitted power or receiver gain will change the target echoes and reverberation to the same extent, with no improvement in discrimination. Changes of pulse duration or the transducer beam width may help, but only if the targets of interest are spatially distributed in a manner different from the sources of reverberation. However, the sonar frequency is important in this context because the relative scattering strength of two targets is unlikely to be the same at widely separated frequencies, especially if the targets are very different in size. As a general proposition, if the intention is to detect large targets in a background of small echoes, a lower frequency should reduce the reverberation. Conversely, if small targets are the primary interest, then a higher frequency might give better results.

Noise

Noise is any unwanted signal other than the reverberation. Thus it is caused by independent effects and is present in the receiver output even when the transmitter is switched off. There are many sources of noise in the sea and they may be classified as (a) physical – wind, breaking waves, turbulence; (b) biological – animal sounds and movement; or (c) artificial – shipboard machinery, the propeller, water flow around the hull. These sources are a few examples which are not intended to be a complete list. There is also the electrical noise in the receiver which adds to the acoustic noise, although the latter is likely to dominate in fishery sonars.

It is usual to describe noise in terms of its frequency composition. The spectrum level is the noise power per unit frequency interval as would be observed by passing the signal through an ideal filter of bandwidth 1 Hz. If the spectrum level is constant at all frequencies, the noise is said to be white, but more generally there is a systematic variation with frequency. Peaks in the noise spectrum may occur due to narrow-band sources such as rotating machinery (Wenz, 1972).

Figure 2.20 illustrates the ambient acoustic noise which is always present in the sea. The spectrum level decreases with frequency to a minimum at a few tens of kilohertz. At very low frequencies, below 20 Hz, the noise is mostly caused by large-scale oceanic turbulence, while at middle frequencies (0.5–25 kHz), wind-induced waves are the main source of noise whose level depends upon the sea state. At high frequencies, it increases again due to the thermal motion of the water molecules.

For an echo to be detected, it must appear in the output signal at a power level which is at least commensurate with the combined power of all noise sources. There are several ways of improving the SNR if the noise is too great. The transmitter power might be increased – a simple remedy, but there is a physical limit to the power which can be transmitted by a given transducer. The pulse duration might be increased, matched to a narrower bandwidth so that the echo strength is maintained but the noise is reduced. However, this is achieved at the cost of a larger sampling volume and it is

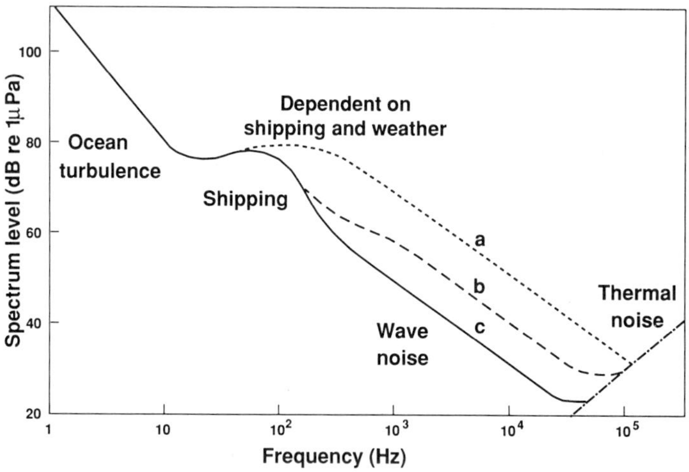

Fig. 2.20 General characteristics of ambient noise in the sea, redrawn from Wenz (1972). The spectrum level is the noise per 1 Hz bandwidth. The curves (a–c) represent the variation of ambient noise according to weather conditions and amount of shipping.

less easy to distinguish targets at nearly the same range. The background noise in the sea is omni-directional. Thus a transducer with a narrower beam will suffer less from extraneous sources, although noise generated within the sonar (the self-noise) is unaffected. If machinery noise is a problem, moving the transducer to another position on the hull might improve matters.

2.7 THE OPERATING FREQUENCY

Depending on the application, fishery sonars may operate at frequencies from 1 kHz to 1 MHz. Several factors need to be considered in deciding the most suitable frequency for a particular application.

There is an element of compromise in choosing the frequency. It is necessary to balance the competing requirements of long range and the resolution of close targets. As the frequency increases, so does the absorption which limits the maximum range of observations. On the other hand, high-frequency sonars can transmit short pulses because the normal requirement is for the pulse to consist of a certain number of cycles. The shorter the pulse, the better is the resolution.

The target strength of the species of interest is another consideration. If this has been measured in experiments at one frequency, it cannot be assumed that the same value will apply in a survey conducted with equipment operating at a different frequency. More generally, high fre-

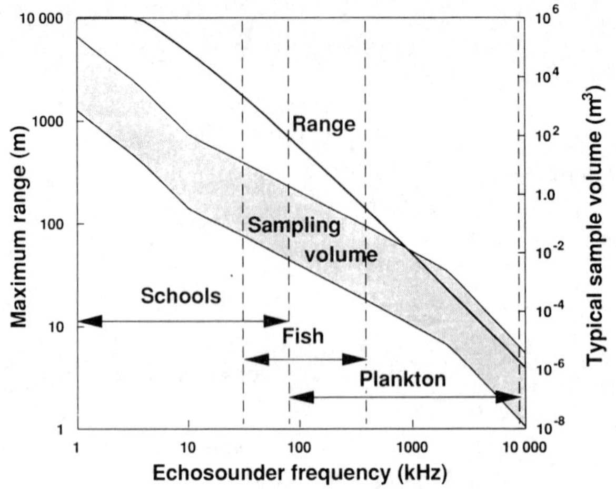

Fig. 2.21 Dependence of sonar parameters on frequency. Range is the maximum distance at which targets might be distinguished from noise. Sampling volume (shaded region) is $c\tau\psi/2$. Suitable frequencies for the observation of schools, individual fish and plankton are indicated.

quencies are necessary when the targets are small, to avoid the Rayleigh scattering region where TS increases rapidly with frequency. The scattering is more predictable when the wavelength is similar to or smaller than the target size, excepting the problem of resonance (p. 31). Thus sonars for detecting fish might operate in the range 30–200 kHz, while instruments designed for plankton studies generally work at high frequencies, perhaps 100–10 000 kHz. Very-low-frequency sonars (1–10 kHz) are unsuitable for resolving individual fish, but can be used to observe schools at a range of many kilometres.

One practical consideration is the size of the transducer. For the same beam width, the size decreases as the frequency goes up. Low-frequency transducers are large and heavy. Special handling machinery is required on the ship to deploy the largest transducers safely.

The various factors to be considered in choosing the operating frequency are illustrated in Fig. 2.21, which gives an indication of the practical performance to be expected from sonar equipment.

APPENDIX A Calculation of the absorption coefficient

The recommended method for calculating the absorption coefficient α from knowledge of the environmental conditions in the water is the formula developed by Francois and Garrison (1984). α is obtained as the sum of three components representing the absorption owing to boric acid, magnesium sulphate and viscosity in the order shown below.

$$\alpha = f^2\{[A_1 f_1/(f_1^2+f^2)] + [A_2 P_2 f_2/(f_2^2+f^2)] + A_3 P_3\} \qquad \text{(A2.1)}$$

In fresh water, the first two components are negligible and may be ignored; the absorption is entirely viscous and $\alpha = f^2 A_3 P_3$. In seawater, all the factors in Equation A2.1 need to be evaluated as functions of the temperature T(°C), salinity S (parts per thousand or ppt), depth D (m) and the acidity pH. The following equations give α in units of dB per kilometre. c is the sound speed in m s^{-1} and may be calculated separately (Appendix B).

$$A_1 = (8.86/c)\,10^{(0.78\,\mathrm{pH}-5)} \qquad \text{(A2.2)}$$

$$f_1 = 2.8\,(S/35)^{0.5}\,10^{[4-1245/(T+273)]} \qquad \text{(A2.3)}$$

$$A_2 = 21.44\,(S/c)\,(1+0.025\,T) \qquad \text{(A2.4)}$$

$$P_2 = 1 - 1.37\times10^{-4} D + 6.2\times10^{-9} D^2 \qquad \text{(A2.5)}$$

$$f_2 = 8.17\times10^{[8-1990/(T+273)]}/[1+0.0018(S-35)] \qquad \text{(A2.6)}$$

$$P_3 = 1 - 3.83\times10^{-5} D + 4.9\times10^{-10} D^2 \qquad \text{(A2.7)}$$

The value of A_3 depends upon the temperature. If $T \leqslant 20$ °C,

$$A_3 = 4.937\times10^{-4} - 2.59\times10^{-5} T + 9.11\times10^{-7} T^2 - 1.5\times10^{-8} T^3 \qquad \text{(A2.8)}$$

and if $T > 20$ °C,

$$A_3 = 3.964\times10^{-4} - 1.146\times10^{-5} T + 1.45\times10^{-7} T^2 - 6.5\times10^{-10} T^3 \qquad \text{(A2.9)}$$

The alternative coefficient β, in nepers per kilometre, is equal to α divided by 8.69.

The Francois–Garrison formula is claimed to predict α to an accuracy of 5% for temperatures from −1.8 to 30 °C, salinities from 30 to 35 ppt, and frequencies from 400 Hz to 1 MHz.

The pH of seawater is normally in the range 7.8 to 8.2. If the precise value is unknown, it is reasonable to assume pH = 8 for the purpose of calculating α.

APPENDIX B Calculation of the sound speed

Speed in seawater

The recommended method for accurate calculation of the sound speed c, from knowledge of the environmental conditions in seawater, is the following nine-term equation described by Mackenzie (1981):

$$c = 1448.96 + 4.591\,T - 0.05304\,T^2 + 2.374\times10^{-4} T^3$$

$$+1.34\,(S-35)+0.0163\,D+1.675\times10^{-7}\,D^2$$
$$-0.01025\,T\,(S-35)-7.139\times10^{-13}\,T D^3 \qquad \text{(A2.10)}$$

where T is the temperature (°C), S is the salinity (parts per thousand or ppt), and D is the depth (m), giving c in m s^{-1}. Mackenzie's equation is valid for temperatures from -2 to 30 °C, salinities from 25 to 40 ppt and depths from 0 to 8000 m. Within these ranges, the standard error of the predicted sound speed is ± 0.07 m s^{-1}. In depths less than 1000 m, the final term of Equation A2.10, the one in TD^3, is very small and may be ignored.

For many fishery applications, a less accurate but simpler formula will be good enough. The following formula devised by Leroy (1969) is suggested for rapid evaluation of the sound speed.

$$c=1492.9+3(T-10)-0.006(T-10)^2-0.04(T-18)^2$$
$$+1.2(S-35)-0.01(T-18)(S-35)+D/61 \qquad \text{(A2.11)}$$

Leroy's formula predicts c correctly within ± 0.1 m s^{-1} for temperatures from -2 to 23 °C, salinities from 30 to 40 ppt and depths not exceeding 500 m.

Speed in fresh water

The above equations are not valid for fresh water. Del Grosso and Mader (1972) have described the following equation which predicts the sound speed in pure water at atmospheric pressure to an accuracy of 0.015 m s^{-1}, over the temperature range 0–95 °C.

$$c=1402.388+5.03711\,T-0.0580852\,T^2$$
$$+0.3342\times10^{-3}\,T^3$$
$$-0.1478\times10^{-5}\,T^4+0.315\times10^{-8}\,T^5 \qquad \text{(A2.12)}$$

There is a slight depth dependence of c (Chen and Millero, 1976). At 10 °C and 10 m depth, c is 0.161 m s^{-1} greater than the near-surface value given by Equation A2.12.

Chapter three

Acoustic instruments

3.1 INTRODUCTION

Sonar is a general term for any device that uses sound for the remote detection or observation of objects in water. The echosounder is a particular kind of sonar, one whose acoustic beam is directed vertically downwards. Many types of sonar and echosounder are used for the detection and observation of fish. We shall consider the practical application of these devices later. Here we discuss the principles of operation and the facilities for underwater observation which modern acoustic instruments provide.

This chapter begins with the single-beam echosounder and the display of echoes on the echogram as a pictorial representation of the insonified water. We describe a number of other instruments which are in common use: the netsonde, which is attached to fishing gear; dual-beam and split-beam echosounders, which measure the target strength directly; the sector scanner and other sonars, which produce radar-like displays in two dimensions. Then there is Doppler sonar, which measures the speed as well as the position of targets. Another development with potential for the future is wide-band sonar, a technique which provides more information about targets from the frequency composition of echoes. Pingers and transponders are useful ancillary devices which are attached to fish or other objects to make them more visible on the sonar display. The calibration of acoustic instruments is an important topic which we consider in some detail. Practical advice is given on how to perform the calibration of echosounders and echo integrators in accordance with internationally accepted standards.

List of symbols

$a(t)$	Ideal TVG function for exact range compensation
$A(t)$	Actual TVG function of an echosounder
$\mathbf{b}_n$	One-way amplitude sensitivity of the narrow beam of a dual-beam echosounder

$\mathbf{b}_w$	One-way amplitude sensitivity of the wide beam
$\boldsymbol{B}$	Two-way energy sensitivity of the transducer
c	Sound speed in water
C	Calibration factor for the on-axis sensitivity
c_1	Longitudinal sound speed in a standard target
c_2	Transverse sound speed in a standard target
d	Linear size of the transducer face
E	Echo integral
E_t	Echo integral of a standard target
E_0	Echo integral of a target on the acoustic axis
f_0	Frequency of the echosounder or sonar transmission
f_r	Frequency of the received signal
f_1	First resonance frequency of the standard target
F	Fish density
$\bar{g}$	Time-varied gain correction factor
$\boldsymbol{I}_n$	Echo intensity on the narrow beam of a dual-beam echosounder
$\boldsymbol{I}_w$	Echo intensity on the wide beam
R_{opt}	Optimum distance between the transducer and the standard target
R_t	Actual range of the standard target
t	Time after the start of the transmitter pulse
t_{del}	Electronic delay in the receiver
t_h	Time at which the echo amplitude is half the peak value
t_0	Optimum start time of the ideal TVG function
T	Duration of the transmitted pulse
TS_t	Target strength of the calibration sphere
v	Radial speed of a target relative to the sonar transducer
V_B	Variance of **b** over the transducer beam
V_s	Sample variance of the observed acoustic cross-section
V_σ	True variance of σ in the target population
w_i	Weight given to the i'th sensitivity measurement
β	Acoustic absorption coefficient in nepers per unit distance
ϕ	Phase difference between two signals
σ	Acoustic cross-section of a target
ψ	Equivalent beam angle of the transducer
ρ_1	Density of a standard target
ω	Angular frequency of a sine wave
$\Delta\Omega$	Element of solid angle
θ, ϕ	Angles defining target direction in a split-beam echosounder
λ	Wavelength of sound

3.2 ECHOSOUNDERS

Figure 3.1 shows the basic components of the echosounder. The transmitter produces a burst of electrical energy at a particular frequency.

Typical frequencies of the echosounders used in fisheries applications might be 38 kHz (38 000 cycles per second), 120 kHz or 400 kHz. The transmitter output is applied to a transducer, which converts the electrical energy to acoustic energy in the water. The transducer projects sound in a directional beam, similar to the beam of light produced by a search-light, as described in section 2.3. The width of the beam is inversely proportional to the frequency of the sound. Thus for the same physical size of transducer, a 400 kHz echosounder would have a beam width one-tenth that of a 40 kHz echosounder. Conversely, for a particular beam width, a transducer one-tenth of the size would be required at 400 kHz compared to 40 kHz. The beam from a conventional fishing echosounder is typically between 5 and 15 degrees wide, but it is not necessarily symmetrical.

The transmitted pulse of sound propagates through the water away from the transducer. It may encounter various targets, e.g. fish or the sea-bed. These targets reflect or scatter the pulse, and some energy returns towards the transducer. This backscattered sound (the echo) is detected by the transducer and converted to electrical energy as the received signal. As described in section 2.4, the time at which the echo is received tells us the distance of the target from the transducer. The received signal is amplified by electrical circuits in the receiver and displayed on the echogram. On the monochrome display, high-amplitude signals appear as dark marks on a white background. Thus a 'picture' of the targets is generated which shows their depth or range as the distance from the transmission mark on the echogram (Fig. 3.1).

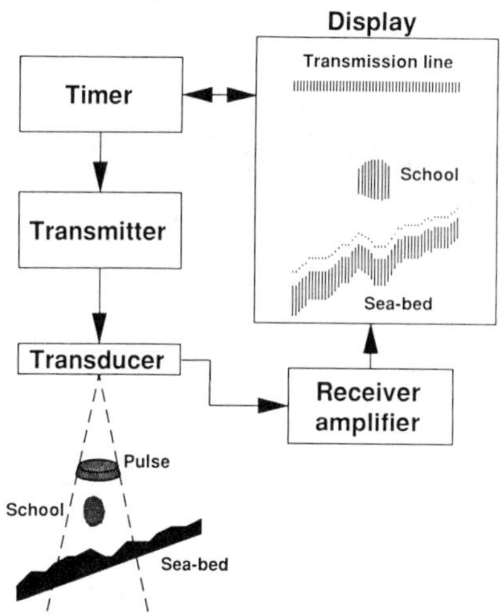

Fig. 3.1 Principles of echosounding. Echoes from the fish school and the sea-bed below the transducer are displayed on the echogram.

After some time, the transmitter produces another pulse and the whole process is repeated. This results in a two-dimensional picture of targets as connected echo traces. The vertical extent of the picture is used to define the depth of a target, while the horizontal position shows changes in time if the echosounder is stationary, or in space if the echosounder is mounted on a moving vessel. Figure 3.1 shows how targets in the water, a fish school and the rising sea-bed appear as the echosounder moves.

Early types of display used mechanically operated pens which marked paper to provide the recording. Two types of paper were used. The first was a dry paper on which dark recordings indicated the strength of targets as a grey scale on a white background (Fig. 3.2), but this paper had poor dynamic range which made it difficult to see the difference between dense and diffuse fish traces. The top of Fig. 3.2 shows the transmitter marks which correspond to the depth of the transducer, close to the surface. The diffuse marks around 25 m depth are caused by plankton. A few fish schools

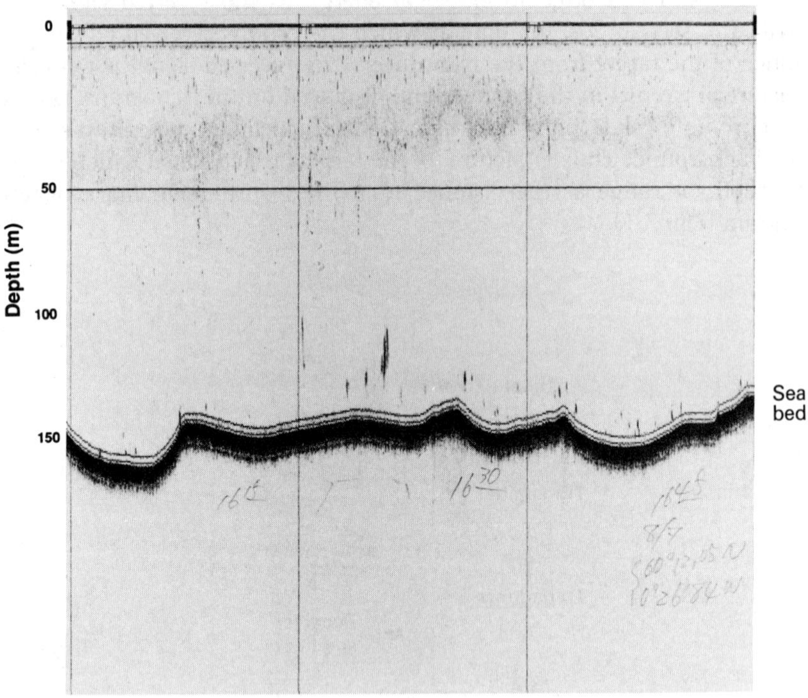

Fig. 3.2 Monochrome display on dry paper. Dark marks correspond to strong echoes. Simrad EK400 echosounder, 38 kHz. The horizontal line at 50 m is a depth marker. Above this are diffuse traces caused by plankton. A few fish schools (short vertical lines) can be seen just above the sea-bed. The two curves following the sea-bed at 1 and 3 m distance mark a channel for echo integration. This display extends for 45 minutes from left to right.

can be seen as black traces below the plankton layer, just clear of the broad dark line at about 125 m which indicates the sea-bed. Fishing on these schools established that they were herring of mean length 27.5 cm. In addition to the representation of real targets, various artificial marks are included to annotate the echogram. There is a horizontal depth marker at 50 m, vertical time markers at 16.15 and 16.30 hrs, and the two curves following the sea-bed, at 1 and 3 m distance, represent a channel used for echo integration.

Secondly, there was a wet paper which had better dynamic range. An example of this type of record is shown in Fig. 3.3; the original would be in brown and white. Fig. 3.3 shows a similar pattern of plankton and fish traces to that in Fig. 3.2. In addition there is a 'white-line' feature on the sea-bed trace. This feature is provided to highlight the sea-bed, and is particularly useful when fish schools are found close to the bottom. The strong sea-bed echo is suppressed on the display for a short interval after it initially occurs, giving a thin line which marks the sea-bed, followed by a narrow white band and then the remaining part of the sea-bed echo. The white-line feature is triggered when the amplitude of the received signal exceeds a preset limit, and it sometimes occurs erroneously in very dense fish schools.

Modern echosounders often use a colour monitor to show the echoes. At any one time, the last few hundred transmissions appear on the monitor screen, but earlier echoes are not permanently recorded. The various colours represent codes for different signal amplitudes; strong signals are red whereas weak signals are blue. Plate 1 shows an example of an echogram

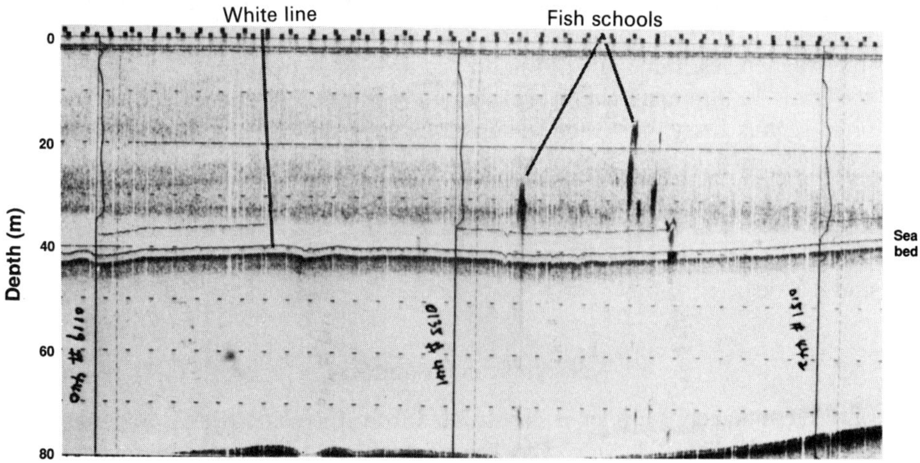

Fig. 3.3 Monochrome display on wet paper, showing fish schools and the 'white line' feature, which marks the sea-bed. The vertical lines are produced at intervals by an echo integrator to show the volume scattering strength v. depth. Furuno FQ50 echosounder, 50 kHz.

on a colour monitor. Inexpensive colour printers are now available which permit permanent records to be made. Plates 2 and 3 show coloured paper representations of two very different kinds of traces, both recorded in Norwegian fjords. In Plate 2, many individual targets are seen around 80 m depth, while Plate 3 shows a large school of overwintering herring which is so dense that the individual fish cannot be distinguished.

The simplest echosounders are used primarily to locate aggregations of fish and to determine the depth of the sea-bed. They provide little information about the quantity of targets detected as echoes, for which purpose more sophisticated instruments have been developed, the scientific echosounders and other devices described later in this chapter.

There are some general considerations in designing echosounders to obtain the best performance. The absorption of sound in water increases rapidly with frequency (section 2.4), so the maximum range of an echosounder is determined by the frequency of operation. High-frequency echosounders are limited to short ranges. We can improve the detection capability of the echosounder by reducing the width of the beam, but the physical size of the transducer presents a problem if too narrow a beam is required at low frequency. We can improve the range resolution by reducing the pulse duration. This is easier to do at higher frequencies because the duration is shorter for the same number of cycles within the pulse. If the pulse is very short, the receiver must have a wide bandwidth to receive the echoes and it is more vulnerable to noise. To overcome noise it is helpful to generate more power in the transmission, but there are limitations in both the electronics and the sound levels that can be transmitted in the water. It is easier to generate higher power levels at lower frequencies. From these general considerations we see that long range is obtained at low frequencies, but high definition requires high frequency, and for any specified requirement there will be an optimum frequency. This compromise has been illustrated in Fig. 2.21. We find in practice that echosounders used in continental shelf areas, at depths up to 400 m, operate in the 20 to 50 kHz range. Work in the deep ocean requires lower frequencies, while the echosounders used in shallow lakes or to observe plankton near the surface normally operate at frequencies between 100 and 400 kHz.

Scientific echosounders

The scientific echosounder is constructed from the same basic components as shown in Fig. 3.1. However, the electronics have been designed with amplitude stability in mind and with some additional features, notably an accurate time-varied gain (TVG) amplifier. This provides compensation for the beam-spreading and absorption losses described in section 2.4. An example of the scientific echosounder (Simrad EK400) is shown in Fig. 3.4

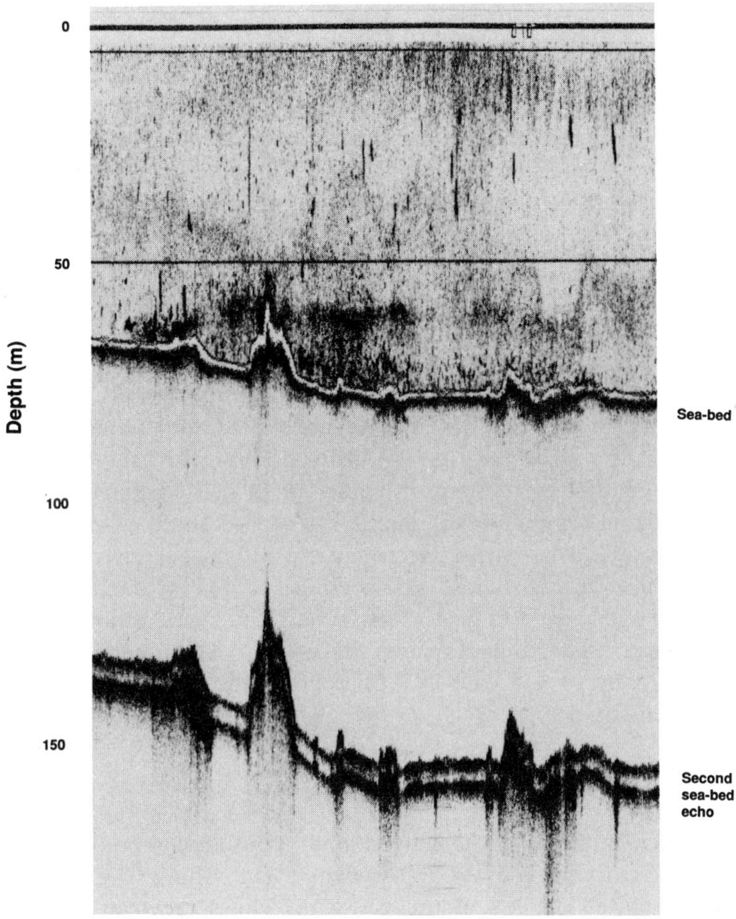

Fig. 3.4 Simrad EK400 echosounder, example of dry-paper recording. There is a depth marker at 50 m. Fish traces above this are sandeels, *Ammodytes* sp., those below mainly pout, *Gadus esmarkii*. Diffuse background traces from plankton appear throughout. The sea-bed is marked by the white-line feature. The second sea-bed echo results from the first echo being reflected back from the sea surface and again from the sea-bed.

(see also section 7.2). The output signal of such an echosounder is sufficiently accurate to count individual fish or to measure the density of aggregations. An echo integrator may be included within the echosounder, or it may be connected as a separate device, to provide an output proportional to the fish density. Some echosounders can determine the direction of targets, and they provide outputs proportional to the target strength when individual fish are detected. Two kinds of instrument have been employed for this purpose, the dual-beam and the split-beam echosounders, and they are described later (section 3.3).

The echo integrator

This device is connected to the output of an echosounder. It sums or integrates the energy in echoes returned from one or more depth channels which are pre-selected by the operator. In practice, this is done by first squaring the voltage produced by the echosounder, then integrating the squared voltage with respect to time, over intervals corresponding to the ranges of the depth channels. The theory of echo integration is discussed in section 5.4 and the principles involved are illustrated in Fig. 5.13.

The implementation of the time-varied gain (TVG) is an important part of the system design. Early instruments used analogue electronic circuits to control the gain and also to square the voltage. However, modern instruments perform these functions using microprocessors or digital circuits. The accuracy and stability of the digital echo integrators are much superior.

It is necessary to exclude the sea-bed echo from the echo integral when fish are being surveyed. Some fish may be in dense schools close to the bottom; if the discrimination of the sea-bed is poor, there will be large errors in the estimated fish density. Electronic circuits may be used to detect the sea-bed echo, but their adjustment is critical and failures may occur when the sea-bed suddenly rises or deepens. Again, the digital output provided by some modern echosounders allows the sea-bed discrimination to be done more accurately by software in a microprocessor or computer (Simmonds and MacLennan, 1990).

Netsondes

The netsonde is similar to the conventional echosounder, but the transducer is remote from the receiver. As used in trawl fishing, the transducer is attached to the headline of the net, and echoes are transmitted to the receiver in the towing vessel through an electrical cable or by means of an acoustic link. In the latter case, the netsonde on the trawl contains its own power supply and two transmitters, one to pulse the transducer and the other to drive the acoustic link, transmitting signals to a hydrophone in the water near the ship. The frequencies of the echosounder and the acoustic link must be sufficiently different to avoid interfering with each other. On the ship, the received signals are displayed as an echogram in the normal way. The beam width of the headline transducer is typically 15–30 degrees. This ensures that a good proportion of the netmouth can be observed, and the footrope of the trawl may be detected even when it is not directly below the headline.

Figure 3.5 illustrates how the headline transducer is used on a pelagic trawl to detect fish traces and the sea-bed. An example of the echogram as recorded on the ship is shown in Fig. 3.6. The transmission line at the top of the echogram marks the transducer position on the headline. Thus the

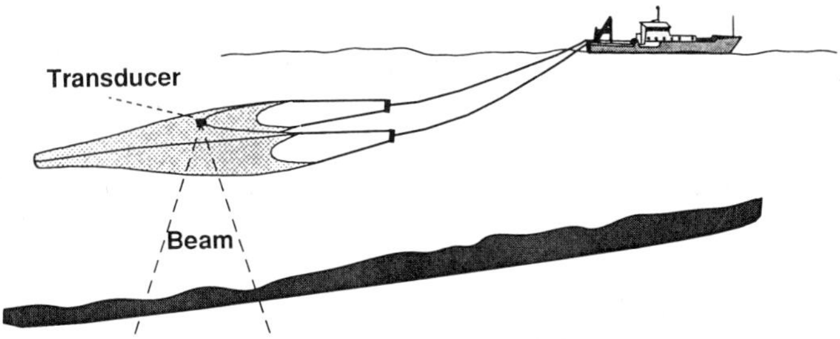

Fig. 3.5 Netsonde attached to a pelagic trawl. The transducer is on the centre of the headline. Echoes received by the transducer are transmitted to the ship via cable or acoustic link (not shown).

'depth' of a mark indicates the distance below the headline, not the distance below the surface as would be the case if the transducer were on the ship. On the left we see the footrope rising towards the headline at the beginning of the haul, as the net accelerates to the fishing speed. In the normal fishing condition, the vertical opening of the net is 12–16 m depending on the towing speed. The higher the speed, the smaller is the opening. The net descends initially and the sea-bed soon appears as the strong trace rising from the foot of the echogram. The distance between the footrope and sea-bed traces indicates the height of the gear above the sea-bed. Thus the

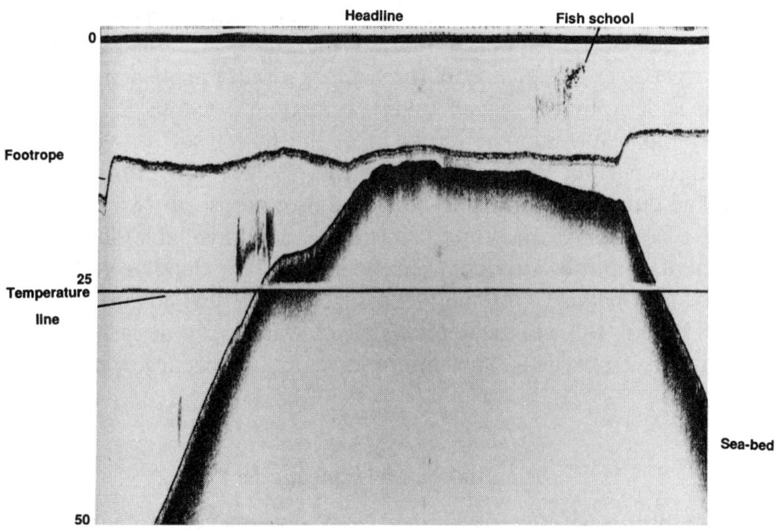

Fig. 3.6 Echogram produced by a netsonde on a pelagic trawl. Scale at left denotes distance (in m) below the headline, not true depth: see text for explanation.

netsonde allows the fishing skipper to control the depth of the gear with great precision, to place the net within one or two metres of the sea-bed while avoiding damage through contact with the ground. The rise and fall of the sea-bed trace in Fig. 3.6 mainly indicates the vertical movement of the net rather than any real change of water depth. A number of fish schools can be seen just above the sea-bed as the net descends, and on the right of the figure, fish are observed entering the net between the headline and the footrope. At the end of the haul, the towing warps are wound in, accelerating the net (and thereby reducing the vertical opening) and at the same time lifting the gear away from the sea-bed. This particular netsonde includes a sensor to measure the water temperature, which is indicated by the position of artificial marks appearing as a line on the echogram. This facility might be used to position the net at the thermocline (a region where the temperature changes suddenly with depth) if it is expected that fish might aggregate there.

Some netsondes have two transducers whose beams are directed upwards and downwards respectively, and the combined echogram then covers the water column from the surface to the sea-bed. As for any echosounder, the netsonde echogram may be recorded on paper or displayed in colour on a monitor screen.

3.3 THE MEASUREMENT OF TARGET STRENGTH

The signal produced by the single-beam echosounder depends upon the direction of the target as well as the acoustic cross-section. The measurement of target strength is an important application of acoustic instruments. When this is done by the so-called 'direct' method (sections 5.3 and 6.3) it is necessary to determine the direction of the target so that the observed echo strength may be compensated for the effect of the transducer beam pattern. The dual-beam and split-beam echosounders are two instruments which have been developed (and are commercially available) for the direct measurement of target strength (Ehrenberg, 1974b; Carlson and Jackson, 1980). The split-beam technique has a slight theoretical advantage (Ehrenberg, 1983), but the practical problems of transducer design and signal processing are more important limitations on the accuracy of these instruments.

The dual-beam echosounder

Figure 3.7 illustrates the transducer and the two beam patterns of a dual-beam echosounder. The transducer consists of 73 elements arranged in four concentric circles around one element at the centre. If all the elements are used as one large transducer, they form a narrow beam. The

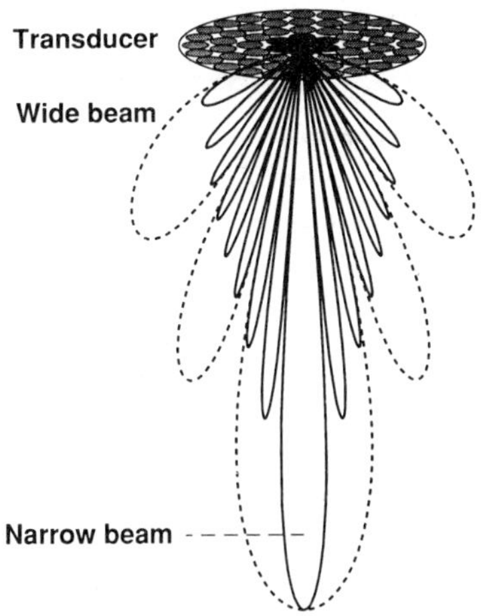

Fig. 3.7 Dual-beam transducer and beam patterns. The narrow and wide beam patterns are the solid and broken lines respectively. The logarithmic radial scale emphasizes the side-lobe structure.

seven central elements (the black ones in Fig. 3.7) may be controlled independently of the others. When used alone as a small transducer, the central elements form a wide beam. The sizes of the two transducers are such that the narrow beam has three side lobes to each one in the wide beam, and the nulls coincide.

The transmitter pulse is applied to all the elements, so the transmission is on the narrow beam only. The signals received by the elements are separately processed to form two output signals, of intensities $\boldsymbol{I}_n$ and $\boldsymbol{I}_w$ corresponding to the narrow and wide beams respectively.

Assuming that the received signals come from one target, the ratio $\boldsymbol{I}_n/\boldsymbol{I}_w$ (called the beam factor) depends upon the direction of the target. Suppose that $\mathbf{b}_n$ and $\mathbf{b}_w$ are the one-way sensitivities in the target direction, for the narrow and wide beams respectively, and that $\mathbf{b}_n$ and $\mathbf{b}_w$ have been determined as functions of direction from the earlier calibration of the instrument. If σ is the acoustic cross-section of the target, and remembering that the transmission is on the narrow beam only, the observed intensities are

$$\boldsymbol{I}_n = \sigma \mathbf{b}_n^2 \tag{3.1}$$

$$\boldsymbol{I}_w = \sigma \mathbf{b}_n \mathbf{b}_w \tag{3.2}$$

Solving these equations for σ and the beam factor, we get

$$\boldsymbol{I}_n/\boldsymbol{I}_w = \mathbf{b}_n/\mathbf{b}_w \tag{3.3}$$

$$\sigma = (1/\mathbf{b}_w^2)\boldsymbol{I}_w^2/\boldsymbol{I}_n \tag{3.4}$$

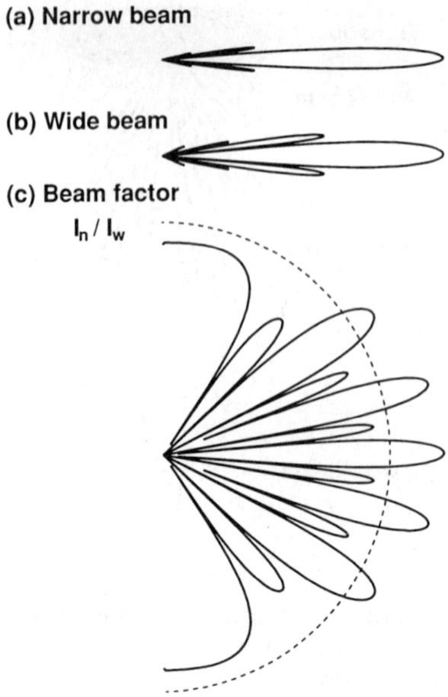

Fig. 3.8 Directional properties of the dual-beam echosounder (a) narrow beam pattern; (b) wide beam pattern; (c) the beam factor. The arc in (c) is a threshold which excludes the smaller lobes of the beam factor.

In principle, the target direction and therefore $\mathbf{b}_w$ are functions of I_n/I_w, and an estimate of σ follows from Equation 3.4. Unfortunately, the indicated direction is ambiguous because it is not a single-valued function of I_n/I_w. The same beam factor might result from targets on the main lobe or the side lobes. This problem is illustrated in Fig. 3.8.

The ambiguities of direction may be greatly reduced by the following techniques. Firstly, we should accept only those echoes having a beam factor greater than some limit. Secondly, if I_n or I_w is less than some threshold, the echo is again ignored. Suppose the limit on I_n/I_w is the arc shown in Fig. 3.8(c). The ambiguities attributable to the smaller lobes of the beam factor (which must not be confused with the different lobes of the transducer beam pattern) are thus removed, since targets in the relevant directions are not detected. In this case the larger lobes of the beam factor (Fig. 3.8c) correspond to nulls of the beam patterns (Fig. 3.8(a) and (b)). In directions around these nulls, the sensitivity of the transducer is relatively low, and most targets in these directions will be ignored because I_n or I_w does not exceed the threshold. Almost all the detected targets are now in directions within the central lobe of the beam factor and the main lobe of the narrow beam. A few strong targets elsewhere may still be wrongly located by these techniques, but the error rate will be acceptably small if the thresholds are well chosen.

The split-beam echosounder

The split-beam echosounder has a transducer which is divided into four quadrants (Fig. 3.9). The target direction is determined by comparing the signals received by each quadrant. In section 2.3 we explained how the phase difference between the signals received by two hydrophones depended on the direction of the source. The same principle is applied in the split-beam technique. The transmission pulse is applied to the whole transducer but the signals received by each quadrant are processed separately. Suppose the four quadrants are labelled 'a' to 'd' as in Fig. 3.9. The angle θ of the target in one plane is determined by the phase differences $(a-b)$ and $(c-d)$, which should be the same. In practice, the summed signal $(a+c)$ is compared with $(b+d)$. The angle ϕ in the plane perpendicular to the first is similarly determined by the phase difference between $(a+b)$ and $(c+d)$. The two angles define the target direction uniquely. The target strength is estimated from the transducer sensitivity in the relevant direction, namely the beam pattern which is determined by calibration as described on p. 85.

The directional ambiguity noted earlier (p. 56) occurs also in the split-beam echosounder. Suppose the difference in the path lengths from quadrants 'a' and 'b' to a particular target is D. If the path–length difference

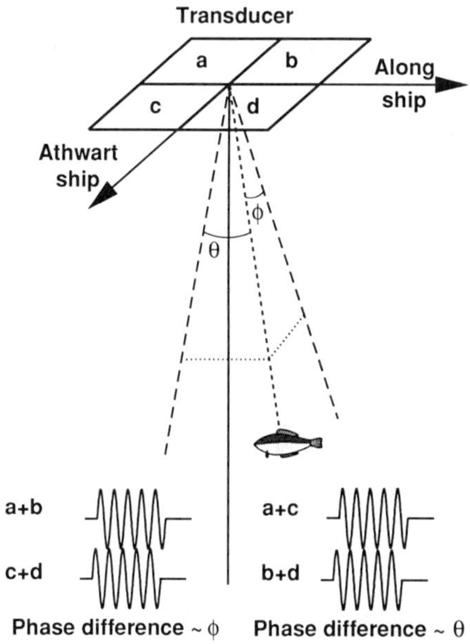

Fig. 3.9 Principles of the split-beam echosounder. Signals from the four transducer quadrants a–d have phase differences which determine the angles θ and ϕ of the target direction.

for another target were $D+\lambda$, the relative phase of the two signals would be the same. Again, this problem can be avoided to a large extent by good transducer design and the application of thresholds which the detected signals must exceed.

Resolution of single targets

The dual-beam and split-beam echosounders measure the target strength directly. There is also an indirect method which works with single-beam instruments (sections 5.3, 6.3). In either case, an important requirement is to identify those echoes that have come from one fish. If the echoes from two or more fish are received at the same time, it is not possible to determine a useful target strength and such echoes must be ignored.

The presence of overlapping echoes may be detected by examining the shape (envelope) and/or the phase coherence of the received signal. Both methods are used together in present designs of target-strength measuring instruments. Figure 3.10 illustrates how these criteria are used to distinguish single and multiple targets. The reader is referred to sections 2.5 and 5.3 for further discussion of this problem.

It is not possible by acoustic measurement alone to say whether a particular echo (satisfying the shape and phase criteria) has come from one fish, or from two fish very close together which happen to reflect echoes which are nearly coincident. When many echoes have been detected and measured, some proportion of the estimated target strengths will be wrong because of coincident multiple targets. Nevertheless, good target-strength estimates are obtained if the mean density of the observed fish is low enough to ensure that the proportion of such errors is acceptably small.

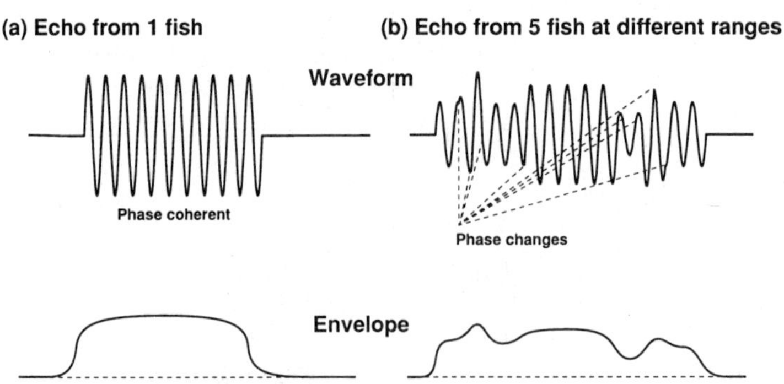

Fig. 3.10 Resolution of single-target echoes based on the pulse shape and phase coherence of the received signal. (a) Echo from one fish; regular envelope and coherent phase. (b) Echo from several fish; irregular envelope and incoherent phase.

3.4 SONARS

There are two kinds of sonar: active and passive. The active sonar transmits an acoustic signal and detects the reflections from objects in the surrounding water. The passive sonar transmits no signal, but detects the sounds or noise produced elsewhere, by ships or submarines for example. There are few applications of passive sonar in fisheries research, and here we shall confine attention to the active instruments which are most useful for the remote observation of fish.

Searchlight sonar

This instrument is a single-beam device used mainly by fishermen when searching for schools of fish. It works on the same principles as the echosounders described in section 3.2, and employs similar sizes of transducer and beam width. However, the transducer is mounted on a stalk which extends about 1 m below the hull of the vessel. The transducer may be rotated and tilted to point the beam in any direction (Fig. 3.11).

Most searchlight sonars allow manual or automatic control of the beam direction. For example, with the transducer tilted slightly below the horizontal, it might be rotated in steps through 360 degrees, transmitting once at each step so that near-surface schools are detected in any direction around the ship. The display may be an echogram, or a plan view presented on a monitor screen with targets shown both in range and direction relative to the ship's position at the centre of the display. The mechanical movement of the transducer is slow, and it takes tens of seconds to construct the complete picture of targets around the ship.

Side-scan sonar

The side-scan sonar is a single-beam instrument with the transducer in a towed body, mounted so that the beam is directed sideways, perpendicular to the cruise track. The transducer has a highly asymmetrical beam pattern

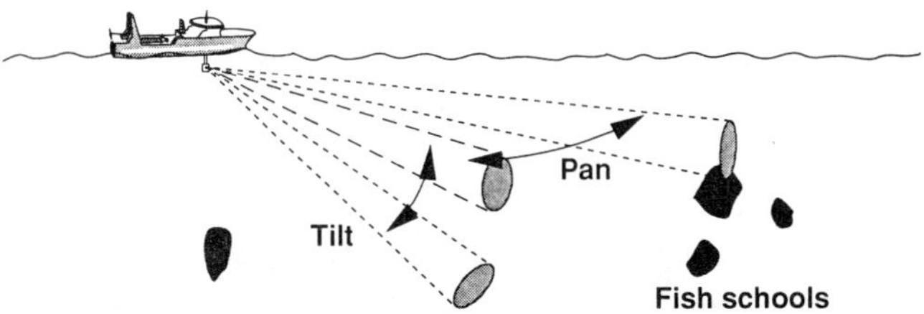

Fig. 3.11 Searchlight sonar. The transducer moves mechanically to pan or tilt the beam in any direction.

which is narrow in the fore–aft direction and wide in the vertical plane perpendicular to the cruise track, typically 1 degree by 40 degrees. Side-scan sonars are most often used to map features of the sea-bed (Fig. 3.12). In this application, the pulse duration is short and the towed body is on a long cable so that the transducer is close to the sea-bed. After each transmission, the marks on the echogram show rising features of the bottom topography, which reflect the transmission strongly, whereas the valleys behind these features are shielded from the transmission and therefore produce no marks on the echogram. As the ship proceeds along the cruise track, successive transmissions are displayed on the echogram, which shows a two-dimensional picture of the sea-bed to one side of the ship.

An example of a side-scan echogram is shown in Fig. 3.13. At the top of this figure is the depth profile recorded simultaneously by an echosounder on the ship. The map of the sea-bed thus produced can reveal the type of bottom (hard or soft) as well as obstacles such as wrecks.

One fishery application of the side-scan sonar is the mapping and counting of fish schools (section 5.2). More generally, it is useful to know the bottom topography in studies of demersal fish and shellfish, whose behaviour and distribution may depend upon the type of habitat which is available.

Sector scanners

In the searchlight and side-scan sonars, the transducer moves so that each transmission insonifies different volumes of water, and the eventual result is a two-dimensional display of targets in the vicinity. The sector scanner does this much more rapidly. A narrow beam is steered or scanned electronically, covering a sector in one plane within the pulse duration. Thus the range and direction of targets in this sector are both determined concurrently.

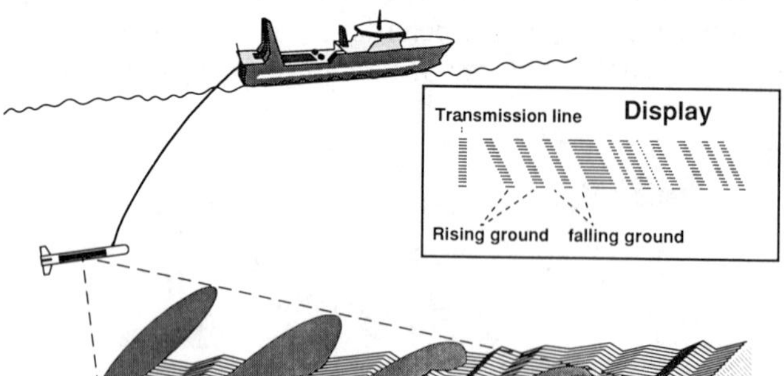

Fig. 3.12 Side-scan sonar. The transducer is in a towed body not far above the sea-bed. Marks on the display indicate topographical features of the sea-bed.

Fig. 3.13 Example of sea-bed features revealed by side-scan sonar. Composite picture built up from nine consecutive passes. The cruise track is from left to right. The boxed echogram at the top is the depth profile along the cruise track. Reproduced with permission from Kline Associates Inc.

The transducer is an array of many elements in a line (Fig. 3.14). The transmitter pulse is applied to one central element, generating a wide beam on transmission. Suppose for the present that the line of elements is horizontal and the transducer face is vertical. The sector is the horizontal cross-section of the transmitted beam. The signals received by the elements are processed in such a way that the receiving (or scanning) beam is a thin vertical section of the transmitted beam. Furthermore, the scanning beam is moved under electronic control and it sweeps from one side of the sector to the other within the pulse duration.

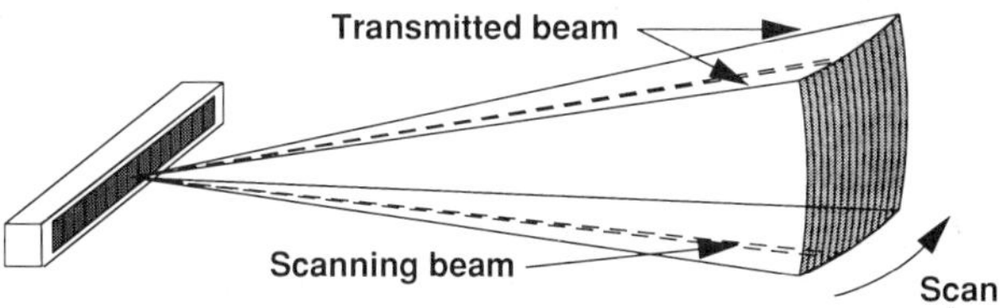

Fig. 3.14 The sector scanner. The transmission beam covers a wide sector within which the narrow scanning beam sweeps to locate targets in any direction.

The scanning technique makes use of the phase differences between the echoes detected by the elements. As discussed earlier (p. 57), the phase differences depend upon the direction of targets within the sector. More generally, the ranges and angles of all targets in the sector are determined uniquely by reference to the time dependence of the received signals (Tucker and Gazey, 1966). Unlike the split-beam echosounder, the sector scanner detects and correctly locates many targets in different directions, even when they are at the same range.

There are two kinds of electronic scanner, which differ in the method of processing the received signals, by modulation or time delay, to steer the beam across the scanned sector (Mitson, 1984b). The modulation scanner (Fig. 3.15) depends upon the fact that when a sine wave of frequency ω is multiplied (modulated) by another of frequency ω_1, the result is the sum of two signals whose frequencies are $(\omega+\omega_1)$ and $(\omega-\omega_1)$ respectively. The lower-frequency component is removed by means of a high-pass filter, leaving one signal which has been shifted to the higher frequency $(\omega+\omega_1)$. The frequency shift ω_1 is incremented by a constant amount $\Delta\omega$ in modulating the signals from successive transducer elements. This introduces a phase difference ϕ between the signals from successive elements; ϕ is proportional to $\Delta\omega$ and it increases linearly with time. The direction of the receiving beam depends upon ϕ, so the beam rotates to scan the sector. However, the modulation scanner has limited power to resolve targets in range, because it requires a pulse long enough to ensure that for targets at

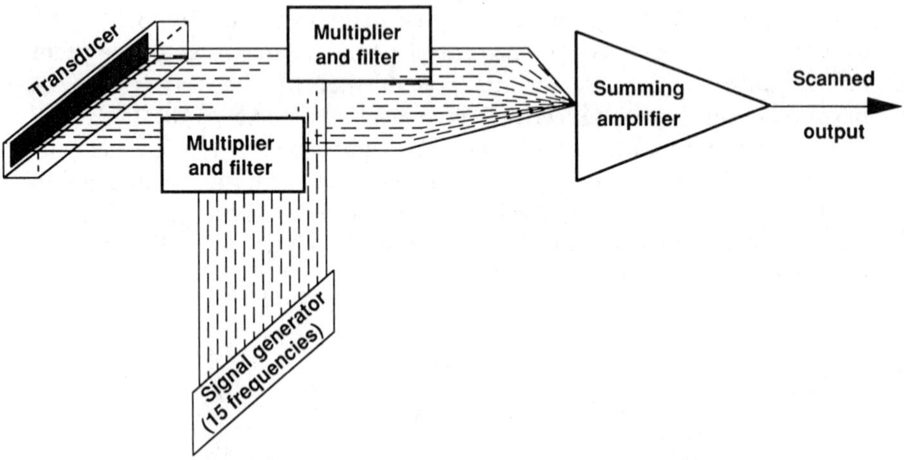

Fig. 3.15 Modulation scanning technique. The received signals are multiplied (modulated) by internally generated signals whose frequency increments in fixed steps $\Delta\omega$ between successive transducer elements. After the modulated signals are filtered to select the higher frequency component, the summed output corresponds to targets in a particular direction which changes with time.

the edge of the sector, the echoes received by all the transducer elements overlap to some extent. If the pulse duration is reduced to improve the range resolution, the discrimination of targets at extreme angles will deteriorate.

The time-delay scanning technique is illustrated in Fig. 3.16. In this case the received signals are delayed for variable periods before they are summed to form the output signal. The delays are incremented in steps Δt between successive transducer elements. This is equivalent to introducing a phase shift ($\omega\ \Delta t$). Thus by changing Δt, the receiving beam is made to move across the sector. Although the time-delay processor is more difficult to implement in hardware, it correctly determines the angle of any target and the same resolution can be achieved with shorter pulses, compared to the modulation scanner. The performance of both types of scanner is limited by noise and the need for a large bandwidth to include the full frequency spectrum of echoes, and in these respects the two methods are similar.

Tucker and Gazey (1966) have described the theory of sector scanning which we need not repeat here. Mitson (1984a,b) provides a good description of the design and operation of scanners. These instruments have been applied in studies of fish behaviour and migration (Cushing, 1977; Arnold *et al.*, 1990).

Scanning netsonde

Ona and Eger (1987) have described this interesting device which works on the same principles as the searchlight sonar (p. 59), but the transducer is

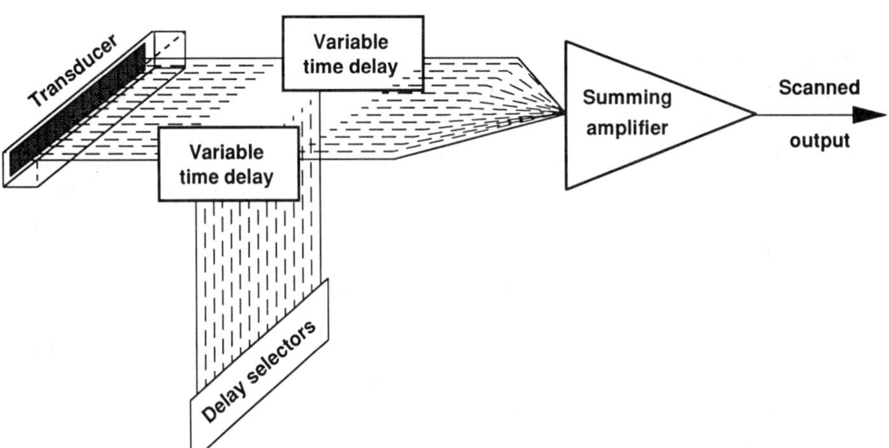

Fig. 3.16 Time-delay scanning technique. The received signals are delayed by an amount which increments in steps Δt between successive transducer elements. The summed output corresponds to targets in a particular direction determined by Δt. The receiving beam pattern is steered by varying Δt with time.

located on the headline of the trawl, connected by an electrical cable to the other parts of the instrument on the towing vessel, as is done with conventional netsondes. The transducer rotates like a radar aerial so that the acoustic beam sweeps around the vertical plane perpendicular to the towing direction. After it has swept through a full circle, the detected targets are displayed on a plan view of the netmouth and the surrounding water. Plate 4 illustrates the type of display produced by the scanning netsonde on a pelagic trawl. The picture shows the outline of the netmouth and fish traces, some in the netmouth but stronger traces beyond the path of the trawl. The white line in the upper part of Plate 4 marks the sea surface. The trawl has been lowered to the depth of a fish layer detected on the ship's echosounder, but the lack of traces near the net shows that many fish have avoided the gear.

This illustrates the difficulty of sampling echo traces with a pelagic trawl, in that some fish are more easily caught than others.

Omni-sonar

The omni-sonar is a versatile instrument which uses electronic scanning to locate targets in two dimensions within various sections of the hemisphere below the ship. The transmitted beam insonifies one-half of a thin conical shell (Fig. 3.17). The receiving beam scans the 180° sector formed by the half-shell to determine both the direction and the range of any targets within the sector. The shape (apex angle of the cone) and the orientation of the shell may be altered to inspect any part of the downward hemisphere. Two particular shells are shown in Fig. 3.17, one with a forward-looking sector and tilted slightly below the horizontal, and the second in a vertical plane, but many others may be selected by the sonar operator.

By connecting navigational instruments to the sonar, it is possible to view the ship's track and the movement of schools over a period of time, on what is called a 'true motion' display. Another useful facility is the automatic tracking of schools. In this mode, the omni-sonar adjusts the inclination of the shell automatically as the ship moves so that a particular target (selected by the operator) is always within the transmitted beam.

Plate 5 shows the display of an omni-sonar, recorded on a vessel which is fishing with a purse seine. The upper display is a near-horizontal plan view with the vessel at the centre. There is a school of sardines to the right of the vessel, marked by the white octagon. The white curve is the track of the vessel, which has circled the school while shooting the purse seine. The lower display is a conventional echogram recorded at the same time. As well as the display of echoes, various markers and numerical data are included, providing additional information from navigational and other instruments. The comprehensive facilities of the omni-sonar are achieved by means of microprocessors which access the acoustic and other data, perform the

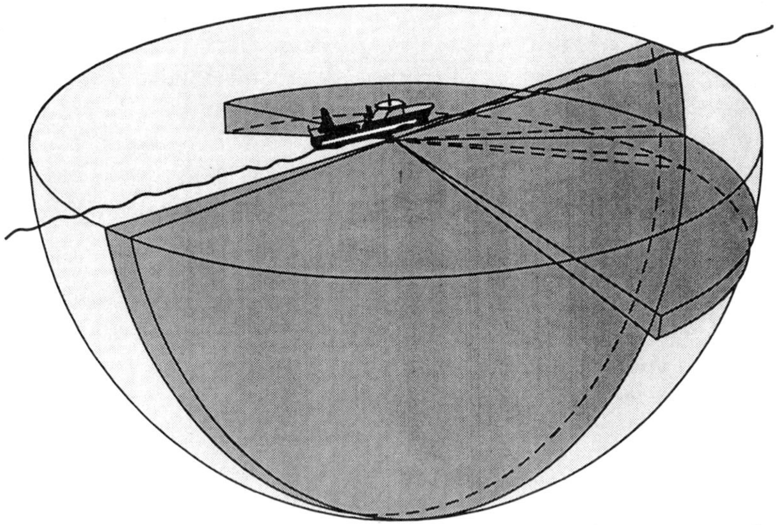

Fig. 3.17 The omni-sonar, showing two examples of transmitted beams formed as conical shells (heavy shading), one vertical in the fore–aft plane, the other directed forward and tilted slightly downwards. The direction of targets within the shell is determined by scanning a narrow receiving beam through a 180° sector. The beams may be rotated and/or tilted to inspect the complete hemisphere below the ship (light shading).

necessary calculations, control the transmitted beam and generate the display.

The Doppler effect

Suppose a sonar transmits sound of frequency f_0, and an echo is received from a moving target whose range is increasing at speed v. The frequency of the received signal is not f_0 but is given by the formula

$$f_r = f_0(1 - 2v/c) \tag{3.5}$$

The change of frequency is caused by the Doppler effect (Kinsler and Frey, 1950). For example, if $c = 1500\ \mathrm{m\ s^{-1}}$, $v = 5\ \mathrm{m\ s^{-1}}$ and $f_0 = 100$ kHz, then the difference is $(f_0 - f_r) = 660$ Hz. Conversely, v may be estimated by measuring f_0 and f_r, provided that the pulse duration is long enough to ensure that the bandwidth of the echo is less than $(f_0 - f_r)$. More generally, the change of frequency may be determined to an accuracy of about $(1/T)$, where T is the pulse duration. Rearranging Equation 3.5, we see that v may be estimated from the formula

$$v = c(f_0 - f_r)/(2f_0) \tag{3.6}$$

Note that v is the radial component of the relative velocity. Sideways components of the velocity do not change the frequency. One application is the Doppler log which measures the speed of the ship relative to the sea-bed or to the water at some depth. The log may have four transducers, whose beams are inclined to the vertical in the fore, aft, port and starboard directions respectively. The speeds in the forward and sideways directions are determined by resolving the radial (along-beam) speeds indicated by the changes in frequency.

A Doppler sonar is one that measures the frequency of the received signal, but otherwise it operates in much the same way as any other sonar. Doppler sonars have been used to study the movement of schooling fish (Holliday, 1977b) and the migration of salmon in rivers (Johnston and Hopelain, 1990).

3.5 WIDE-BAND SYSTEMS

A major problem in fisheries acoustics is how to determine the species and size of targets whose echoes have been detected. Early attempts to do this by acoustic means had limited success. Holliday (1977a) suggested that the size of fish relates to the resonance frequency of the swimbladder, and this frequency may be determined from the spectrum of echoes generated by a broad-band source such as an explosive charge. There are considerable difficulties in this approach (section 4.4). Rose and Leggett (1988) reported that the probability distribution of the echo amplitude from a school depended on the species, perhaps because of different packing densities in the schools. However, Denbigh and Smith (1987) suggest that the echo amplitude distribution is largely independent of the packing density if there are more than five fish per sampling volume.

In principle, the more frequencies there are in a signal (or the wider the spectrum) the more information is conveyed from the source to the receiver. This leads to the concept of the wide-band sonar, which has a much larger bandwidth than that of conventional sonars and echosounders. Simmonds and Copland (1986) have described such an instrument whose bandwidth covers a factor of two in frequency, from 27 to 54 kHz. An important requirement for the measurement of fish is that the transducer beam width should be the same at all frequencies. This is achieved by means of the spherical cap transducer first described by Rogers and van Buren (1978). The transducer consists of a number of independent elements on a spherical surface, and the beam width is controlled by 'shading' the signals from the individual elements (section 2.3).

The spectrum of the echoes received by the wide-band sonar is a useful indicator of the target identity, provided that the target strength of each species that could be present changes with frequency in a characteristic way. In one species, the target strength might increase between 27 and 54 kHz, and in another the change might be in the opposite direction. Simmonds and Copland (1989) and Simmonds and Armstrong (1990) conducted experiments with cod, herring and mackerel in cages. Having first measured the frequency dependence of target strength for each species, they derived algorithms for deciding which species had produced any spectrum of echoes observed with the wide-band sonar. Simmonds and Copland found that the success rate in identifying the species depended on the number of independent samples on which the decision was based. With 100 and 1000 samples, the decision was correct 90% and 95% of the time respectively. The method could also identify mixtures of two species, although the success rate in this case was lower, 75% and 90% respectively for 100 and 1000 samples. Other wide-band experiments have been reported by Zakharia (1990). He used tethered individual fish and showed that the spectrum of echoes depended on both the species and the activity of the fish which might be swimming or stationary.

Wide-band sonar is still being developed. It has not yet been shown to be useful for the identification of wild fish in the sea, but the technique offers considerable promise for the solution of this difficult problem.

3.6 PINGERS AND TRANSPONDERS

All the instruments discussed above are active sonars, which rely upon the reflection of their own transmission to detect targets. Echo detection may be unreliable if the targets are very small or close to a strong reflector such as the sea-bed. The detection is much improved by attaching a transponder to the target. This is a device which transmits a signal when it receives one. The transponded signal is much stronger than the echo, but both arrive simultaneously at the sonar receiver and the target may be located in the usual way. It is possible to detect bottom-living fish that have been tagged with transponders, and this technique has been widely used in the study of fish behaviour (e.g. Harden Jones *et al.*, 1977; Arnold *et al.*, 1990).

The pinger has no receiver but it transmits a pulse of sound at regular intervals. The pulses are received by one or more hydrophones located in the vicinity. If there are three receiving hydrophones in a triangular array, the pinger can be located in two dimensions (in the plane of the array). This is done by measuring the arrival times of the pinger pulses at the hydrophones (Hawkins *et al.*, 1974). If there are four hydrophones at the apexes of a tetrahedron, the pinger can be located in three dimensions. Furthermore, the frequency of the pinger transmission or the pulse interval may be changed in response to some variable, such as the water depth or the temperature, to transmit additional information.

Pingers and transponders used to study the behaviour of fish must be small to avoid disturbing the natural behaviour of the animal. Modern microelectronic circuits are very compact; most of the volume is required for the transducer and the battery which powers the device. The size of battery is determined by the transmission power, the pulse duration and the period of time for which the device is required to continue working. The device may be attached externally in the same way as a Petersen tag, or it may be inside the fish. Plate 6 shows a pinger designed to transmit 76 kHz pulses being inserted into the gut of an anaesthetized cod. After this operation, the fish is released into the sea and its movements may be followed for several days (Hawkins *et al.*, 1974).

3.7 CALIBRATION

The need for careful and accurate calibration of the acoustic instruments cannot be emphasized too strongly. In the early days of acoustic surveys, before 1980 perhaps, calibration was a major source of error in the fish abundance estimate, amounting to a factor of two or worse (Blue, 1984). However, calibration techniques have improved considerably in recent times. Error arising from uncertainty about equipment performance is now of minor significance, provided the calibration is performed in accordance with international practice (Simmonds *et al.*, 1984; Foote *et al.*, 1987). The recommended calibration procedure is that described by Foote *et al.* (1987) and much of the information presented here has been drawn from their report.

In this section we restrict attention to the physical calibration of acoustic instruments, that is to say the calibration of the transducer and electronic equipment as independent pieces of hardware, but excluding the acoustic properties of particular targets, which are discussed in Chapter 6.

What is a calibration? In traditional physics or engineering, it is an experiment conducted to determine the correct value of the scale reading of an instrument, by measurement or comparison with a standard. In the case of the acoustic instruments used for echo integration, the 'correct value' is the density of fish in the transducer beam, and the 'scale reading' is the corresponding output of the echo integrator. Other definitions apply to the dual-beam or split-beam instruments, which are used to measure target strengths expressed in terms of the response to individual targets, but we shall consider first the calibration of the single-beam echosounder. It is perhaps the most difficult instrument to calibrate, because it does not measure the target direction. The calibration must take account of the distribution of targets across the acoustic beam as well as the physical response of the instrument.

The sensitivity of the single-beam echosounder is determined by reference to the echo from a target in a particular direction, normally the acoustic axis of the transducer, which is the direction of maximum sensitivity. Targets in this direction are said to be on-axis. The receiver incorporates time-varied gain (TVG), which is intended to ensure that the same fish density will produce the same signal at any range. The gain changes with time after the transmitted pulse; this dependence is described by the TVG function. Another factor to be considered is the change in sensitivity with direction, known as the beam pattern. Thus the physical calibration of the single-beam echosounder involves three separate measurements, to determine (1) the on-axis sensitivity, (2) the TVG function and (3) the beam pattern.

The echo integrator output, E, is assumed to be proportional to the number of fish per unit area, F, according to the equation

$$F = [C\bar{g}/(\psi\langle\sigma\rangle)]E \quad (3.7)$$

This is the echo integrator equation which is discussed further in section 5.4. To apply it, we need to determine the values of the several factors within the square brackets. $\langle\sigma\rangle$ is the expected value of the acoustic cross-section of the targets, which are assumed to be randomly distributed over the acoustic beam. This is a major topic in its own right, to which we return in Chapter 6. For the present, it is the three other parameters which have to be measured in the physical calibration of the equipment, corresponding to the three aspects of performance noted above. They are C, the on-axis sensitivity; $\bar{g}$, the TVG correction factor; and ψ, the equivalent beam angle of the transducer.

On-axis sensitivity

The acoustic axis of the transducer is the direction in which the transmitted energy is greatest, and from which the largest echo is returned for a given target at constant range. The calibration is performed to determine the combined transmit and receive sensitivity in this direction, from which the value of C is calculated. Three methods might be used to measure the on-axis sensitivity: (1) the reciprocity technique, (2) the calibrated hydrophone or (3) the standard target. The first two are described below for the sake of completeness. In modern practice, the standard target method is preferred as the one that gives the most accurate results (Robinson, 1984).

The reciprocity technique

This method of measuring the on-axis sensitivity employs two additional transducers to transmit and receive signals (Foldy and Primakoff, 1945, 1947). The three transducers are at fixed locations but they can be rotated to align any pair. It is assumed that the two additional transducers are 'reciprocal', which means that the energy losses on transmission are the same as those on reception. The first transducer (which is the subject of the calibration) is not required to be reciprocal, and none of the three needs to be calibrated beforehand. Each transducer is used in turn as a transmitter, and the other two as receivers. The voltages and currents at all the transducer terminals are measured, and also the distances between each pair of transducers. These electrical and geometrical measurements are sufficient to determine the sensitivities of all three transducers, without having to measure the acoustic field directly.

Variants of the reciprocity technique can be applied with fewer than three transducers. When the calibration is done with two transducers, both must be reciprocal. Then there is the 'self reciprocity' technique, in which the sensitivity is determined for one transducer on its own, by reference to the echo from a near-perfect reflector, usually the surface of still water (Widener, 1980). In effect, the reflection is equivalent to the signal from a second transducer transmitting from the image above the surface. Again, electrical measurements at the transducer terminals and knowledge of the geometry are sufficient to determine the sensitivity, provided the transducer is reciprocal.

The reciprocity technique is best suited to calibrations performed in a tank of water where there is little external disturbance, but it is difficult to achieve good accuracy. The transducers must be separated by more than the near-field distance, which is typically 5 m for those used on acoustic surveys. Thus a large body of still water is required. For each pair of transmit–receive measurements, the transducers have to be rotated so that their acoustic axes coincide. Small errors in the alignment can severely degrade the calibration. The accuracy of the reciprocity technique is unlikely to be better than 0.5 dB, or $\pm 13\%$ in the value of C (Robinson and Hood, 1983).

The calibrated hydrophone

The hydrophone is a small transducer which operates over a wide range of frequencies. It is primarily a receiver of broad-band sounds, but most hydrophones are reciprocal devices which can also function as projectors. When supplied with details of the electro-acoustic sensitivities in both directions, as measured by the manufacturer, the hydrophone is a secondary standard which may be used to calibrate other transducers. The calibration is performed by locating the hydrophone at a known distance from the survey transducer in the direction of the acoustic axis. Figure 3.18 shows how this might be done in the case of a hull-mounted transducer. The transmit and receive sensitivities of the transducer are measured separately, by transmitting pulses firstly from the transducer and secondly from the hydrophone, while the electrical signal at the terminals of the receiving device is recorded. The sensitivities are determined from the electrical measurements and the distance between the two devices. However, the results depend critically on the angular alignment of the hydrophone and

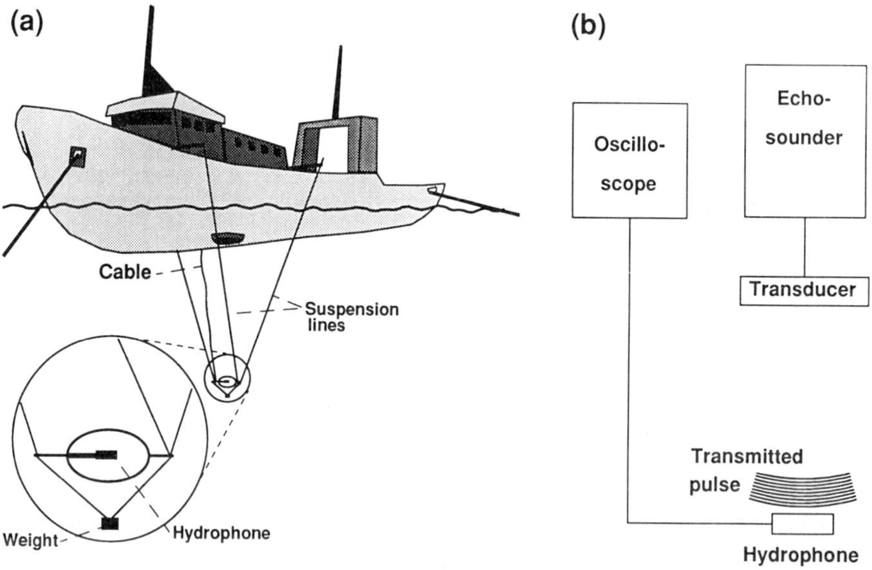

Fig. 3.18 Hydrophone used as a secondary standard to measure the sensitivity of acoustic transducers. (a) Location of the hydrophone below the ship; (b) diagram of the acoustic pulse and electrical connections.

the extent to which its directional sensitivity is known. The accuracy of the technique is generally poor. About ±40% in the value of *C* is as good as should be expected.

The standard target calibration

This method of calibration employs a standard target whose acoustic scattering properties are known. The target is normally a homogeneous solid sphere which is suspended below the transducer. The echosounder is operated in the normal manner with the same pulse length, TVG and power level as would be used during the survey. The echo produced by the target is measured, and also the time delay between the echo and the transmitted pulse. These measurements are sufficient to estimate both the range of the target and the combined transmit–receive sensitivity of the transducer in the direction of the target. By moving the target across the beam, the position at which the echo is strongest for a given range can be found. The target is then on the acoustic axis, and *C* is determined from the corresponding echo measurements.

The standard target technique is both accurate and simple to apply in practice. A particular advantage is that the echosounder transmits and receives signals exactly as it would during the survey, so the one calibration takes account of the transducer sensitivity, the electrical gain and the frequency response of band-pass filters in the receiver. A minor limitation is that because the transmitter and receiver are measured simultaneously, it is less easy to identify the cause of peculiar results, for example if the sensitivities determined from two successive calibrations were greatly different. In that event, however, a few additional electrical measurements of the transmitter power and the receiver gain should quickly reveal the cause of the problem.

Tungsten carbide or copper spheres have been found to give the best results in practice (Foote *et al.*, 1983; Foote and MacLennan, 1984). The acoustic properties of these targets have been discussed in section 2.5, in particular the factors that determine the optimum size of target for calibrating a particular echosounder. The target strength depends upon the sound speed in the water as well as the echosounder frequency, so the environmental conditions must be known to determine the target strength correctly. Table 3.1 gives details of copper and tungsten carbide spheres which are commonly used to calibrate 38 kHz echosounders. Of the two materials, tungsten carbide is the more versatile because it is harder and resonates at higher frequencies (Foote, 1990b). A few examples covering frequencies up to 1 MHz are specified in Table 3.2.

Table 3.1 Target strengths of two spheres used to calibrate 38 kHz echosounders, and their physical properties. Calculated values assuming continuous wave (zero-bandwidth) transmission; c, sound speed in water; c_1 and c_2, longitudinal and transverse (respectively) sound speeds within the spheres; ρ_1, sphere density. Water density 1000–1030 kg m^{-3}

Sphere material	*Diameter (mm)*	ρ_1 *(kg m^{-3})*	c_1 *(m s^{-1})*	c_2 *(m s^{-1})*
Copper	60.0	8945	4760	2288
Tungsten carbide	38.1	14900	6853	4171

c *(m s^{-1})*	*TS (dB) in spheres made of* Copper	*TS (dB) in spheres made of* Tungsten carbide
1430	−34.0	−41.8
1450	−33.8	−42.1
1470	−33.6	−42.3
1490	−33.6	−42.4
1510	−33.6	−42.4
1530	−33.6	−42.4
1550	−33.8	−42.2

Table 3.2 Target strengths of tungsten carbide calibration spheres for sound speeds 1450 m s^{-1} (fresh water) and 1490 m s^{-1} (seawater). Sphere density 14 900 kg m^{-3} and sound speeds $c_1=6853$ m s^{-1}, $c_2=4171$ m s^{-1}. Calculated values for echosounder bandwidth 5 kHz

Frequency (kHz)	*Diameter* (mm)	*Fresh water TS* (dB)	*Seawater TS* (dB)
30	38.1	−39.8	−39.7
38	38.1	−42.1	−42.4
50	36.4	−40.9	−40.4
50	38.1	−41.6	−41.0
70	36.4	−40.9	−40.4
70	38.1	−40.6	−41.0
120	33.2	−40.8	−40.6
120	38.1	−39.8	−39.5
200	36.4	−39.5	−39.8
200	38.1	−39.5	−39.2
420	8.9	−52.3	−52.8
420	21.2	−43.5	−43.7
1000	8.9	−51.0	−51.2

Below 100 kHz, with a sphere that has no resonance close to the echosounder frequency, the calibration should be accurate to better than ±2.5% in the value of C. These targets are best suited to calibrations at low frequencies. At very high frequencies, the target strength is less certain because of the close spacing of the resonances.

Experimental procedure

The standard target is suspended below the transducer, supported by the minimum amount of additional material to avoid unwanted reflections. This is normally done by encasing the target in a web of monofilament nylon (Fig. 3.19). Three suspension lines are attached to the web, and by adjustment of the lengths of the lines, the sphere can be moved to any point in the transducer beam. Before it is put into the water, the sphere should be soaked in a soap solution – a mixture of one part household detergent to four parts fresh water is adequate – to wet the surface thoroughly so that air bubbles on the web are eliminated.

In the case of a hull-mounted transducer, the sphere is suspended by two lines on one side of the ship and one line on the other side (Fig. 3.19). Booms are fitted to the ship's rails to lead the lines clear of the hull. A small winch or line reel on each boom is used to adjust the lengths of the support lines, to move the sphere to any required position.

When the transducer is in a towed body, the sphere is suspended from the three horizontal arms of a frame attached to the top of the body (Fig. 3.20).

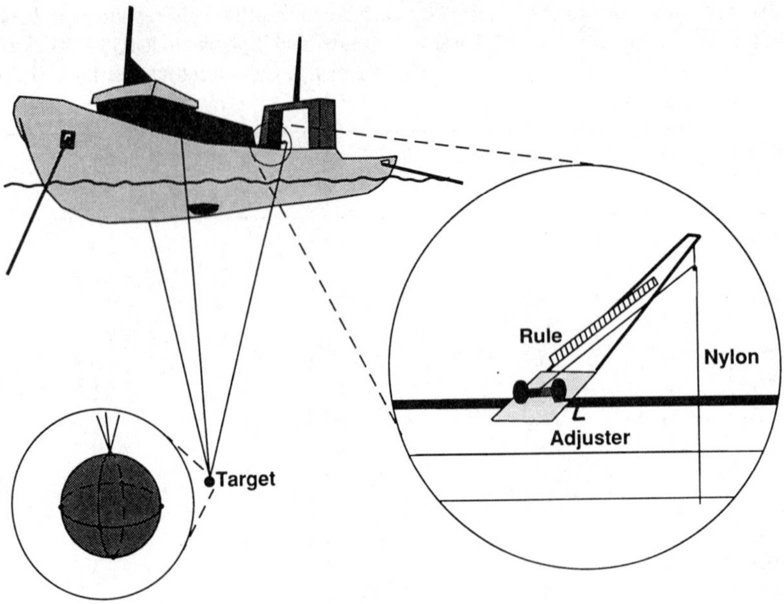

Fig. 3.19 Apparatus for calibrating a hull-mounted transducer. The standard target is suspended below the ship by three lines. The line lengths are adjusted and measured against marks on the rule to locate the target precisely in relation to the transducer.

The entire apparatus (transducer, body, sphere and frame) is lowered into the water, hung from a wire attached to the top of the vertical arm. The apparatus is lowered until the transducer is at its normal depth. The three support lines from the sphere can be led up the vertical arm so that they are accessible for adjustment at the surface. Alternatively, the line lengths may be adjusted by remote control of motors on the frame (Fig. 3.21).

Whether the transducer is in a towed body or on the ship's hull, the conduct of the calibration is much simpler if remotely controlled motors are used to adjust the lengths of the target suspension lines. The support lines must run from the adjusting mechanism to the sphere with the minimum of friction to allow free movement of the lines. Yacht fairleads may be used as a cheap and durable method of leading the support lines with little friction at any point where they have to change direction.

If the calibration is being performed from a vessel anchored in a current or swinging in the wind, the hydrodynamic force on the sphere may cause it to move unpredictably. This is less likely to occur when the support lines are short. On the other hand, the sphere must be far enough from the transducer to be outside the near field. Thus the optimum position of the sphere is just outside the near field, at a distance from the transducer which

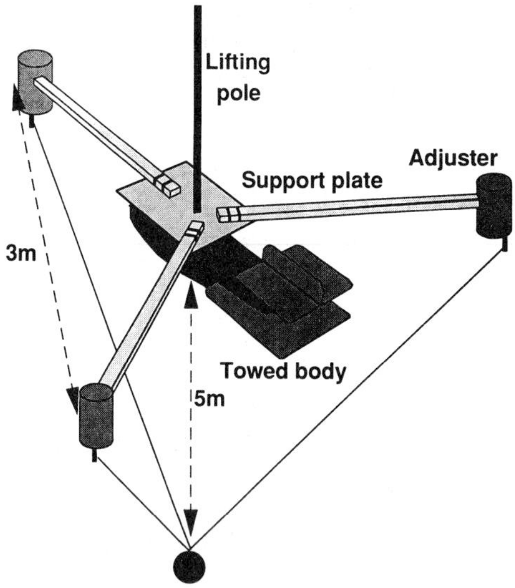

Fig. 3.20 Apparatus for calibrating a transducer in a towed body. The standard target is suspended on three lines from a frame attached to the top of the body. The line lengths are changed by mechanical adjusters to move the target relative to the transducer.

may be estimated from the formula

$$R_{opt} = 2\,d^2 f_0 / c \tag{3.8}$$

where d is the greatest width of the transducer face, f_0 is the echosounder frequency and c is the sound speed in water. For example, suppose that the transducer to be calibrated has a rectangular face of size 45×30 cm. Thus the larger dimension is 0.45 m. If the frequency is 38 kHz and the sound speed is $1500\,\mathrm{m\,s^{-1}}$, R_{opt} is $2 \times (0.45)^2 \times 38\,000/1500$, or 10.26 m. In performing this calculation, remember that all the parameters must be expressed in the same units (metres and seconds in this case). If the range of the sphere is R_{opt} or more, the bias in the on-axis sensitivity measurement resulting from the near-field effect will be much less than 1%.

The echo strength is measured either as the peak amplitude at the echosounder output, or as the energy (integral of the squared amplitude) recorded by the echo integrator if it is included in the calibration. For the same pulse duration, the echo energy is proportional to the square of the peak amplitude. The sphere is moved across the beam to find the position where the echo is strongest for the same range, at which point the sphere is on the acoustic axis. This might be done by adjusting the length of each line in turn until the strongest echo for that movement is observed. With each adjustment the sphere will move progressively closer to the acoustic axis, but this procedure is rather slow. A faster method is to adjust one line

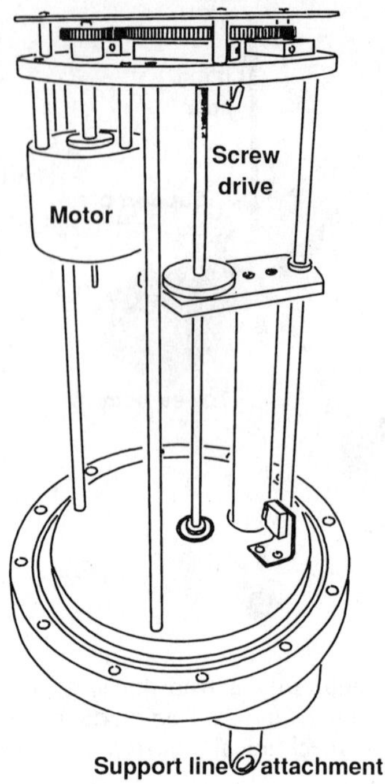

Fig. 3.21 Powered adjuster for moving a standard target under remote control, shown with the outer casing removed. The motor operates a screw drive which raises or lowers the line attachment point.

first, and then the other two by equal amounts in opposite directions (Fig. 3.22). This procedure moves the sphere successively in two directions at right angles to each other. The one-line and two-line adjustments are repeated until the maximum echo energy during each adjustment achieves a steady value. Three or four adjustment cycles are usually sufficient to locate the sphere on the acoustic axis.

It is not essential to move the sphere precisely onto the acoustic axis. An alternative procedure is to measure the echo energy with the sphere at a series of points in a section extending from one side of the beam to the other. The echo increases in amplitude from the first point onwards, then it decreases after the sphere passes the centre of the beam. At each point, the echo energy is measured as the average from several transmissions, perhaps 20 or 40. The length of the moving support line is recorded as a determinant of the sphere position in the section, and the maximum echo energy is estimated by fitting a quadratic curve to the measurements at the several points (Fig. 3.23). As before, the sphere is moved successively across two perpendicular sections of the beam until the overall maximum of the echo energy is determined. For the best results, the quadratic curve is fitted to the

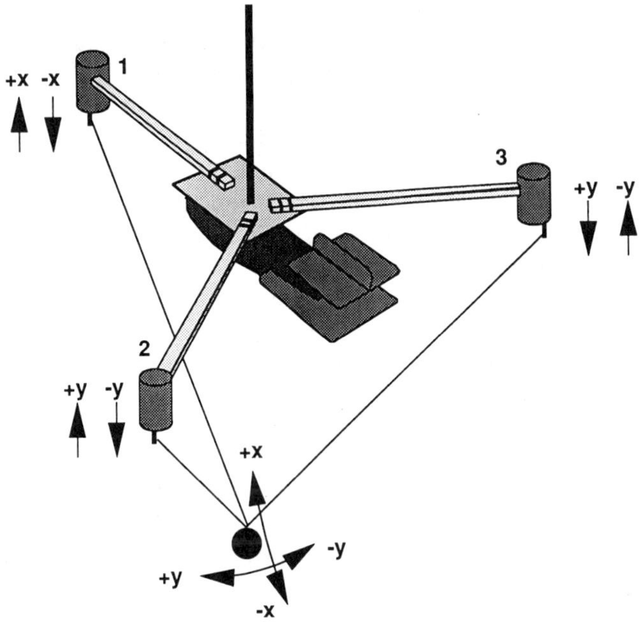

Fig. 3.22 Control of the target position on two axes (x, y). Adjuster (1) controls x; adjusters (2) and (3) work together but in opposite directions to control y.

fourth root of the energy measurements, which matches the fitted curve to the theoretical beam shape, and the sphere is moved over a small portion of the centre of the beam, between points at which the echo energy is within 30% of the maximum.

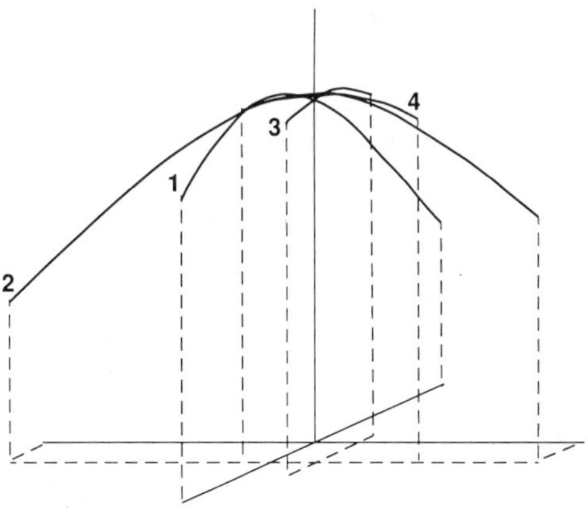

Fig. 3.23 Estimation of the on-axis sensitivity of a transducer. Each curve is a quadratic fitted to the measurement made in one plane. Successive measurement sets (1–4) are in planes at right angles to one another. The maxima of the quadratic curves converge to the on-axis sensitivity.

The range of the target is estimated by measuring the time delay between the transmitter pulse and the echo. This method is easier and more accurate than attempting to measure the distance physically in the water. The arrival time of the echo depends upon the target range and the sound speed in the water, but a small additional delay is introduced by the receiver electronics and must be taken into account in the calculation. Furthermore, the echo has a rounded shape when it is viewed on an oscilloscope, and it is not immediately obvious which point on the waveform should be timed. The recommended procedure is to measure the time between the start of the transmit pulse and the point on the leading edge of the echo at which the amplitude has risen to half the peak value (Fig. 3.24). If this time is t_h, and t_{del} is the additional delay from the receiver electronics, the target range is

$$R_t = c(t_h - t_{del})/2 \tag{3.9}$$

The delay t_{del} may be determined from the theory of echo formation and knowledge of the receiver electronics (Foote *et al.*, 1987). The calculations are not simple, but in some cases the value may be known from published data (Table 3.3).

The echo delay t_h should be measured against an accurate time reference such as a crystal-controlled oscillator. Oscilloscopes with an analogue time base are too inaccurate for this purpose. A digital oscilloscope with a precise time base should be used. Alternatively, an accurate time signal from another instrument may be displayed together with the echo waveform, so that the oscilloscope is only used to make the visual comparison.

Calculation of *C*

When the echosounder is to be used for echo integration, the calibration should be performed with 20 log *R* time-varied gain (p. 78) as would be appropriate during the survey. Since the sphere is a single target, the echo amplitude depends upon the range R_t which therefore has to be measured

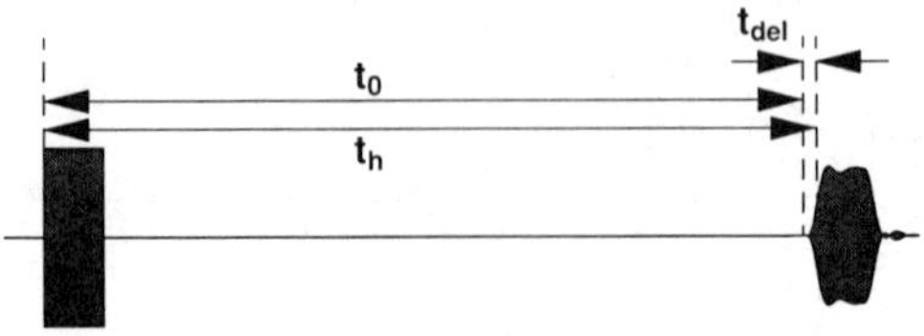

Fig. 3.24 Estimation of the target range. The echo delay t_h is measured at the half-amplitude point on the envelope. The range is $R = ct_0/2$ and $t_0 = t_h - t_{del}$ (Table 3.3).

Table 3.3 Values of the delay parameter t_{del} for standard targets used with the Simrad EK400 echosounder operating at 38 kHz, with 1 ms pulse length, 3 kHz bandwidth and 20 log R time-varied gain

Target material	*Diameter (mm)*	t_{del} *(ms)*
Copper	60.0	0.47
Tungsten carbide	38.1	0.45

independently, as described above. Suppose E_t is the measured echo energy when the sphere is on the acoustic axis. It is also necessary to know σ_t, the acoustic cross-section of the standard target, or TS_t which is the equivalent target strength in decibels (Table 3.1). The sensitivity factor C can now be estimated as

$$C = 4\pi 10^{(TS_t/10)}/(E_t R_t^2)$$
$$= \sigma_t / E_t R_t^2 \qquad (3.10)$$

If a different echo integrator is to be used during the survey, or the echosounder is set to a receiver gain or transmitter power level different from those employed during the calibration, then additional electrical measurements would be necessary to derive the correct value of C from that given by Equation 3.10. It is much preferable to perform the calibration with the same equipment and control settings as will be used during the survey, repeating the calibration for different settings if required. This may not always be possible, for example if the maximum transmitter power level causes the sphere echo to saturate the receiver.

If the echosounder is to be used for counting single-fish echoes, the appropriate time-varied gain function is 40 log R, which removes the range dependence of the standard-target echo. E_t is now the echo integral or the square of the peak echo amplitude, depending on how the echo counter works, and the formula for C reduces to

$$C = \sigma_t / E_t \qquad (3.11)$$

The TVG function

The purpose of the time-varied gain (TVG) function is to compensate the range dependence of the echo. In this section we consider the single-beam echosounder used for echo integration, when the appropriate TVG function is nominally 20 log R. However, the same principles may be applied to the calibration of the 40 log R function which is required for echo counting.

Suppose the voltage gain of the receiver is proportional to $A(t)$ where t is the time after the start of the transmitter pulse. $A(t)$ is the actual TVG function of the echosounder, but in general it will not compensate the range dependence exactly and the purpose of the calibration is to estimate the resulting error. To do this, measurements of $A(t)$ are compared with the ideal TVG function $a(t)$, which does compensate the range dependence exactly. MacLennan (1987) has shown that the following function, although not completely exact, is a good enough representation of 20 log R TVG for all practical purposes.

$$a(t) = c(t - t_0)\exp(\beta ct/2) \qquad (3.12)$$

Here c is the speed of sound, β is the acoustic absorption coefficient and t_0 is the optimum start time of the TVG. t_0 depends upon the pulse duration and the bandwidth of the receiver; it is always more than half the pulse duration. Unfortunately, present echosounders (even those designed for scientific purposes) mostly have a TVG function which is intended to be $ct\exp(\beta ct/2)$. Manufacturers seem to have ignored the need for the start time of the TVG (at $t = t_0$) to be delayed for an interval after the beginning of the transmitter pulse. However, this deficiency is overcome provided that $A(t)$ is compared with the correct ideal function, that given by Equation 3.12. Values of t_0 for a particular echosounder are given in Table 3.4. For the same pulse duration and bandwidth, other echosounders may have slightly different values of t_0 but the differences are unlikely to be important. The calculation of t_0 is rather complicated, (details, MacLennan, 1987).

In performing the calibration, it is necessary to assume a value for c according to the water temperature and salinity likely to be encountered during the survey, to determine the correct ideal function. The actual function $A(t)$ is measured as the ratio of voltage amplitudes at the output and input of the receiver. These measurements have to be made with electronic instruments that respond fast enough to record the change of gain with time over the operating range of the echosounder.

Table 3.4 The optimum start-time of the ideal TVG function (nominally 20 log R) for range compensation of distributed targets. Calculated values for the Simrad EK400 echosounder operating at 38 kHz and targets at long range whose acoustic cross-section is independent of frequency

Bandwidth (kHz)	*Pulse duration (ms)*	*TVG start-time (ms)*
1	1	1.21
1	3	2.18
3	1	0.95
3	3	1.94

The next step is to see how the ratio $A(t)/a(t)$ changes with time. This ratio is the TVG error, and two examples are shown in Fig. 3.25. The upper curve is from measurements on a receiver which generates the TVG function by the analogue technique, using non-linear electronic components in the feedback loop around the amplifier. This error curve varies smoothly with time, but the maximum error is very large. The lower curve shows the TVG error of a receiver whose gain is digitally controlled. In this case the maximum error is smaller, but there are sudden changes due to quantization, because the digital circuits adjust the gain in fixed increments. The sudden changes are most frequent at times corresponding to short ranges. The quantization error is not an important consideration at the range of most fish, but it is relevant to the echo from the standard target during the calibration.

The measurements of the TVG error are used to calculate the TVG correction factor $\bar{g}$ in the echo integrator equation. The error is normalized by supposing that the gain is correct at the range of the standard target. $\bar{g}$ is then calculated as an average of the error over the times corresponding to the depth channel of the echo integration, according to the theory described in Appendix D (Chapter 5). When calculated in this way, the TVG correction is unbiased if the fish targets are distributed evenly with depth. If the fish are unevenly distributed, e.g. concentrated at one end of the depth channel, and the TVG error is greatly different between the top and the bottom of the channel, the echo integrals will be biased. This problem may be overcome by selecting narrower depth intervals so that the TVG error is less variable over the width of each channel. Alternatively, $\bar{g}$ may be calculated as a weighted average with most weight

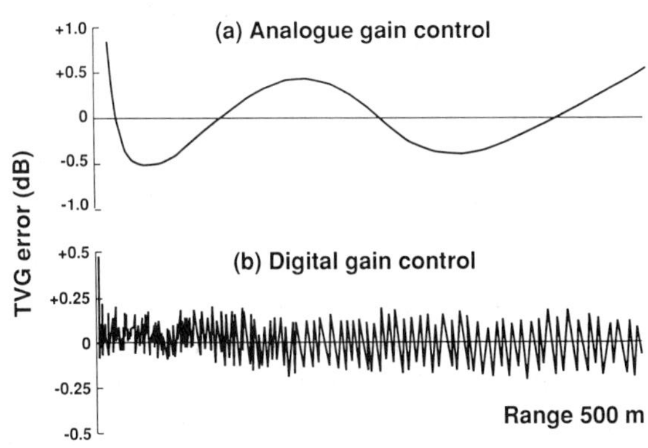

Fig. 3.25 Time-varied-gain (TVG) error v. time after the transmission pulse. Comparison of (a) analogue and (b) digital echo integrators. The horizontal axis covers times corresponding to target ranges 0–500 m; note differing vertical scales.

being given to the TVG error corresponding to the depths of the observed fish.

The equivalent beam angle

The third part of the calibration procedure is the measurement of ψ, the equivalent beam angle of the transducer, which is a measure of the beam width. ψ has been defined in section 2.3 (Equation 2.10).

In most cases, less than 1% of the transmitted energy appears in the side lobes of the transducer beam (Simmonds, 1984a). To determine ψ therefore, it is normally sufficient to measure the beam pattern in detail only within the main lobe, adding a small correction which is calculated from theory to take account of the energy in the side lobes. However, a few measurements should be made in directions beyond the main lobe to check that the side lobes are as expected. Gross changes in the beam pattern might occur, for example, if the transducer had been damaged.

It is possible to calculate ψ from theory and the geometry of the transducer face. However, Simmonds (1984a) has shown that the theoretical predictions can be wrong by more than 20%, and nominally similar transducers produced by the same manufacturer can differ likewise. It is therefore important to determine ψ empirically for any transducer that is to be used for echo integration. Furthermore, the beam pattern is influenced by the mechanical attachment of the transducer to the supporting body (Simmonds, 1984b). Different methods of mounting the transducer have been found to alter ψ by as much as 15%. It is therefore desirable to measure the beam pattern with the transducer in the mounting that will be used when surveying. This is difficult to do in practice because it implies that the calibration apparatus needs to include the bulky towed body, or that the measurements should be performed with the transducer in place on the ship's hull.

Measurements of the same ceramic transducer repeated at intervals over several years have indicated that ψ remains nearly constant, provided the transducer does not suffer damage (Simmonds, 1990). It is not necessary to measure ψ frequently, and one measurement during the working life of a transducer should be sufficient. However, if a sudden change in the on-axis sensitivity were to be observed between two successive calibrations, this might suggest a mechanical failure or deterioration of the transducer, in which case the beam pattern should be measured again. Ideally, the manufacturer should provide a certified empirical value for ψ to go with each transducer, avoiding the need for users to undertake this difficult and time-consuming measurement, but not all manufacturers have the facilities necessary to determine ψ experimentally to the required accuracy, $\pm 2\%$ or better.

The beam pattern is measured by reference to the echo from a standard target suspended below the transducer. Suppose that E_r is the echo energy and R_r is the range when the target is in a particular direction $\hat{\mathbf{r}}$. The combined transmit-receive sensitivity in this direction, equivalent to b^4 in Equation 2.10, is $(E_r R_r^2)/(E_0 R_0^2)$, where the subscript zero refers to the measurements with the target on the acoustic axis. R_r is determined by timing the echo as described earlier, but the direction of the target must be measured independently. The measurements are made with the target at a large number of positions, several thousand perhaps, covering the entire cross-section of the main lobe. Equation 2.10 is evaluated numerically as a weighted sum of the measurements at each position, from the formula

$$\psi = \left(\sum_r E_r R_r^2 \Delta\Omega_r \right) \Big/ (E_0 R_0^2) \tag{3.13}$$

where $\Delta\Omega_r$ is the solid angle containing $\hat{\mathbf{r}}$ and bounded by the directions midway between adjacent measurement positions.

The apparatus is similar to that used to determine the on-axis sensitivity as described earlier. Figure 3.26 shows how the beam pattern can

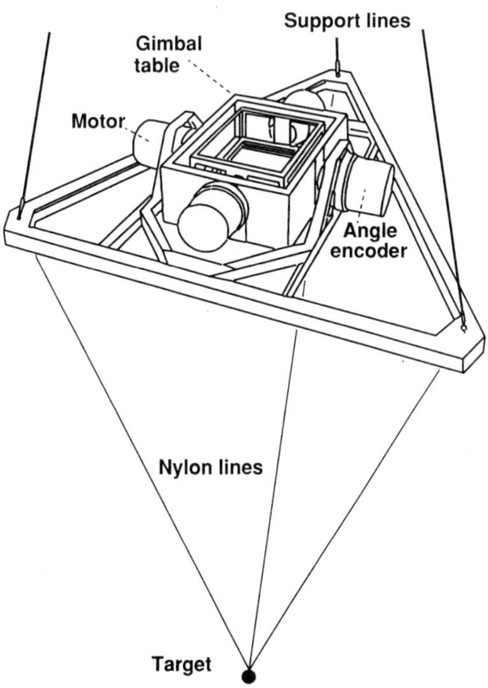

Fig. 3.26 Gimbal table used to measure the beam pattern of transducers. The target is stationary while the transducer rotates on two perpendicular axes. The rotation angles are measured by encoders.

be measured in the case of a transducer installed in a towed body. In this method the target remains stationary, suspended on fixed lengths of nylon line, while the transducer and the supporting body are rotated on a motorized gimbal table. The direction of the target relative to the acoustic axis is determined by angle encoders on the rotation axes of the gimbal table. The results may be displayed as a landscape plot of the range-corrected echo energy (Fig. 3.27). The main lobe is the only feature evident to the eye, which suggests that the side lobes are small enough to be ignored in determining ψ for this particular transducer.

Another technique is required for the *in-situ* calibration of hull-mounted transducers. In this case the transducer must remain stationary, but the lengths of the supporting lines can be adjusted to move the standard target instead (Reynisson, 1990). E_r, R_r and the line lengths are measured with the target at a number of evenly spaced points in the beam cross-section. The direction of the target at each point is calculated by geometry. Reynisson has also suggested an alternative method of analysis, in which the beam pattern is deduced by fitting cubic splines to the measurements (Fig. 3.28). Equation 2.10 is again evaluated numerically, using the fitted splines, to estimate the directivity at points between those measured. This method requires fewer sensitivity measurements and is quicker than the full mapping procedure described in the previous paragraph.

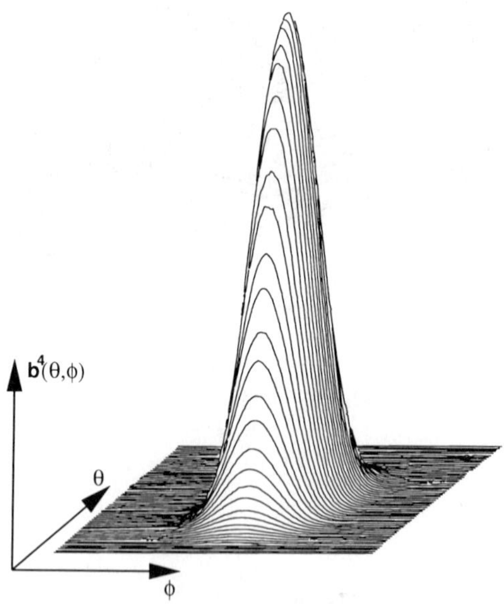

Fig. 3.27 Landscape plot from measurements of the combined transmit–receive sensitivity of a 38 kHz transducer. The beam pattern $\mathbf{b}^4(\theta, \phi)$ is shown as a function of the direction in spherical polar coordinates.

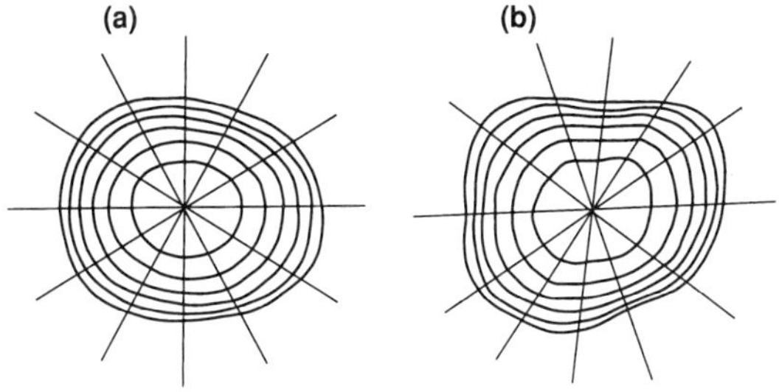

Fig. 3.28 Beam patterns of two transducers, redrawn from Reynisson (1990). Measurements have been made along the radial lines; the closed curves are interpolated contours of the beam pattern from which the equivalent beam angle ψ may be estimated. (a) EK38 transducer, 10 log $\psi = -20.82$ dB; (b) ES transducer, 10 log $\psi = -20.39$ dB.

Direction-sensing echosounders

The signals produced by the split-beam and dual-beam echosounders determine the direction of targets as well as the range and the echo amplitude. When the instrument is to be used for echo counting or the measurement of target strength, the appropriate TVG function is 40 log (R). The TVG is calibrated by the method described above for the single-beam echosounder. The TVG calibration should be performed first, and the results used to correct the subsequent sensitivity measurements for any residual range dependence of the echoes. As before, the sensitivity in various directions may be measured by reference to the echo from a standard target which is placed at a number of representative positions over the active part of the beam. The on-axis direction is not especially important in this case. The usual requirement is to determine the average sensitivity weighted by the area of the beam cross-section, and the additional variation of the output signal caused by inexact compensation for the target direction (Degnbol and Lewy, 1990).

The sensitivity calibration is simply a matter of recording the echo energy while the standard target is placed at a number of positions which are distributed across the beam. It is not necessary to measure the target position independently since the direction of the echo is determined by the instrument itself. Figure 3.29 illustrates the method proposed by MacLennan and Svellingen (1989). The cross-section of the beam is divided into seven regions of equal area, a circle in the centre and six segments on the periphery. The target is moved so that the number of measurements is about the same in each region, ensuring that all parts of the beam receive the same attention.

Suppose that n measurements are made and $\boldsymbol{B}_i$ is the observed sensitivity at the ith position of the target. The mean sensitivity is estimated by

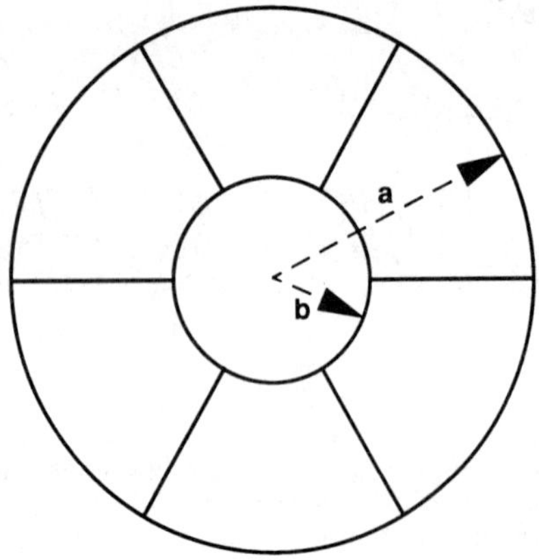

Fig. 3.29 Division of the cross-section of a transducer beam into seven equal sub-areas. The radius of the outer circle (a) is $\sqrt{7}$ times that of the inner circle (b). Redrawn from MacLennan and Svellingen (1989).

weighting each measurement in proportion to the area it represents. To a first approximation, the weighting factor w_i is proportional to the angular distance of the *i*th measurement position from the acoustic axis at the centre of the beam. The mean $\bar{B}$ and variance V_B of the sensitivity are obtained from the formulas

$$\bar{B} = \sum_{i=1}^{n} B_i w_i / \sum_{i=1}^{n} w_i \tag{3.14}$$

$$V_B = \sum_{i=1}^{n} (B_i - \bar{B})^2 w_i^2 / \left\{ (n-1) \left(\sum_{i=1}^{n} w_i \right)^2 \right\} \tag{3.15}$$

The direction-sensing echosounder is normally used to determine the distribution of target strengths in the fish population which is to be surveyed. The results may be considered in terms of the target strength or the equivalent acoustic cross-section of the detected fish. Suppose that $\bar{\sigma}$ is the true mean of the cross-section, and V_σ is the corresponding variance. To estimate these statistics, many observations are required of individual targets as a representative sample of the population. It is also assumed that the detected targets are randomly distributed over the active area of the beam. This implies that if $\bar{E}$ is the mean of the observed echo energy, then $\bar{E}\bar{B}$ is an unbiased estimate of $\bar{\sigma}$. However, any variation of the sensitivity across the beam will increase the spread of the observed distribution. V_B is a measure of the sensitivity variation, and this quantity as determined from the calibration may be used to correct V_s, the sample variance of the observed cross-sections. On the reasonable assumption that the variations of sensitivity and target strength are not correlated, an unbiased estimate of V_σ is given by

$$V_\sigma / \bar{\sigma}^2 = (V_s/\bar{\sigma}^2 - V_B / \bar{\boldsymbol{B}}^2)/(1 - V_B / \bar{\boldsymbol{B}}^2) \qquad (3.16)$$

So far we have considered only the mean and variance of the target strength distribution. Other statistics such as the skewness may also be distorted by the variation of sensitivity across the beam. It is possible to reconstruct the true target strength distribution from the histogram of observations, given a large number of measurements of target strengths and the sensitivity as a function of the target direction. This may be done by the deconvolution method described by Clay (1983). Although originally developed for the treatment of signals from the single-beam echosounder, deconvolution is equally applicable to direction-sensitive instruments. However, it is not necessary to consider the complexities of deconvolution if the compensation for the target direction is good enough, that is to say if $V_s/\bar{\sigma}^2$ is much larger than $V_B/\bar{\boldsymbol{B}}^2$.

Figure 3.30 shows some measurements obtained during the calibration of a dual-beam echosounder. The same target has been moved to different positions across the beam. The TS factor is I_w^2/I_n which is proportional to $\bar{\sigma}$, the acoustic cross-section of the target, and is essentially constant if the

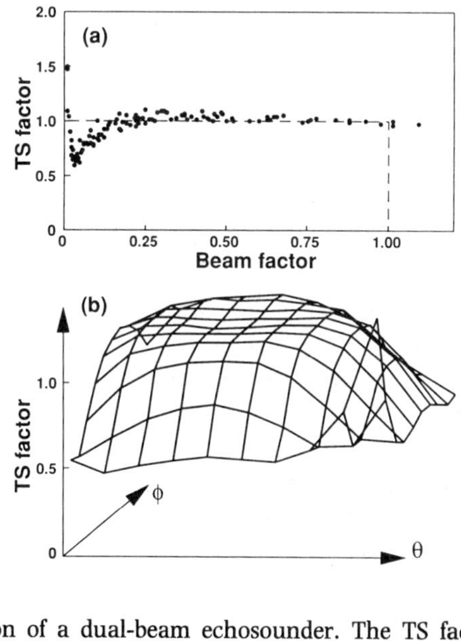

Fig. 3.30 Calibration of a dual-beam echosounder. The TS factor I_w^2/I_n is proportional to the acoustic cross-section. The beam factor is I_n/I_w. The vertical scales are arbitrary. (a) Measurements of the same target in various directions; (b) landscape plot of the TS factor v. direction. θ and ϕ are respectively the fore–aft and athwartship angles of the target direction relative to the acoustic axis.

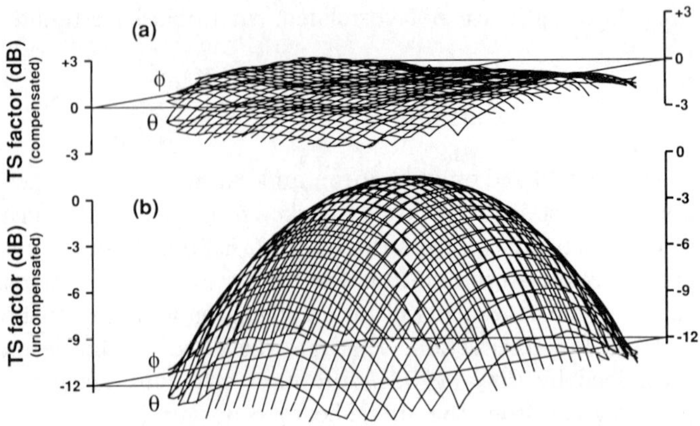

Fig. 3.31 Calibration of a split-beam echosounder presented as landscape plots of the combined transmit–receive sensitivity, (a) with compensation for the target direction and (b) without compensation. θ and ϕ are respectively the fore–aft and athwartship angles of the target direction relative to the acoustic axis.

beam factor I_n/I_w is large enough. A threshold is applied to the beam factor (section 3.3), to reject echoes in the anomalous region for which I_n/I_w is less than about 0.3.

The measured sensitivity of a split-beam transducer is illustrated in Fig. 3.31. The upper plot is a landscape showing the directional sensitivity after the echo amplitude has been compensated using the measured phases of the received signals. There is some residual variation of sensitivity with direction, but it is much less than that of the uncompensated beam pattern shown in Fig. 3.31(b).

Good calibration practice

To achieve satisfactory results, the calibration of acoustic instruments must be performed with careful attention to detail, using equipment that is capable of the required accuracy, and in accordance with the recommended procedures described here. It is all too easy to make mistakes, and they are costly because any error in the calibration will bias the estimate of fish abundance derived from the survey. Good calibration practice should determine the performance of the echo integration system to within $\pm 7\%$ (Simmonds, 1990). The aim should be to develop a routine for calibration which will achieve this accuracy consistently. Many problems arise in the interpretation of acoustic survey results, but there is no need for calibration error to be one of them.

Chapter four

Biological acoustics

4.1 INTRODUCTION

Sound is important for the animals and other creatures that live in water. On the most general level, the propagation of sound waves in space, and their reflection by discontinuities in the medium, provide the means for remote sensing of the environment. Most aquatic animals have organs which are sensitive to acoustic waves, and they may in addition be able to produce sound as a deliberate act. There are several facets of animal behaviour in which sound has a key role. These include communication between individuals, mating behaviour, detection of prey, escape from predators and navigation.

This Chapter is primarily concerned with the significance of sound to living things, including the physiology of sound production and reception, and behaviours which are motivated by hearing. However, we have extended the scope beyond the low-amplitude sounds which are sufficient for communication, to include all physiological effects of sound sources, and in particular the damage caused by underwater explosions.

We begin by describing the many different sounds of biological origin that are produced underwater. Then we discuss the hearing mechanisms of fish, invertebrates and mammals. The acoustic sensitivity of different species is compared, and their ability to detect signals in background noise. This leads to a discussion of the critical bandwidth, which is the frequency range of noise that interferes with a signal through 'masking'. Next, we consider the intriguing ability of some aquatic mammals to navigate and detect prey by echolocation. These animals have evolved a sophisticated form of active sonar. The frequency composition and directional properties of the sonar 'clicks' are discussed in relation to the information content of the returning echoes and the signal processing required to extract this information.

The biological effects of underwater explosions are discussed. The physical properties of the shock waves generated by explosions are described, and equations are presented for estimating the strength of the shock. Published data and theories for predicting the mortality of fish exposed to shock waves are reviewed. Other topics include the phenomenon of swimbladder resonance, which is relevant to the sizing of fish by remote sensing, and the study of plankton by means of multi-frequency sonar.

List of symbols

a	Equilibrium radius of a gas bubble
A	Radius of the sphere having the same volume as the swimbladder
b	Critical bandwidth for noise masking of a pure tone
b_r	Critical ratio, an approximation to b
B	Mortality parameter in Baxter's model
B_{50}	Value of B for 50% mortality
c	Sound speed in water
D	Depth below the surface
E	Energy flux transmitted by shock waves
ESR	Radius of a sphere having the same volume as a zooplankter
f	Frequency of a sound wave
f_n	Frequency of narrow-band noise
f_r	Resonance frequency of the swimbladder or a gas bubble
f_0	Frequency of sound for maximum hearing sensitivity
g	Density of a fluid sphere divided by ρ
h	Sound speed within a fluid sphere divided by c
I	Pressure impulse transmitted by shock waves
I_{50}	Impulse causing 50% mortality of fish
k	Wavenumber, $2\pi/\lambda$
K	Species-dependent constant factor in equation for f_r
L	Fish length
P	Pressure-amplitude of a sound wave
P_0	Ambient (undisturbed) pressure
P_{max}	Peak pressure of an explosive shock
P_{min}	Minimum pressure in the rarefaction after the shock front
P_n	Noise power per unit bandwidth
P_t	Power of a tone signal
Q	Quality factor of the swimbladder resonance
R	Distance between two points
R_{50}	Distance from point of explosion within which at least 50% of fish die
S_v	Volume scattering coefficient
t	Time
T	Time constant for decay of the pressure shock
T_b	Arrival time of the first bubble pulse after the shock front
w_f	Weight of a fish

W Weight of an explosive charge
Z Swimbladder oscillation parameter
γ Ratio of the specific heats of a gas
λ Wavelength of sound
ρ Density of water

4.2 BIOLOGICAL SOUNDS AND HEARING

Compared with light, sound has several advantages as a means of communication in water. It can propagate over hundreds of metres; it is easily produced; and the frequency composition of the signal may convey useful information from the sender to the receiver. In contrast, light waves attenuate very rapidly with distance, especially when the water contains suspended solids as often occurs in fast-flowing rivers, lakes and estuaries. It is not surprising, therefore, to find that many aquatic animals have evolved sensitive hearing organs, and that many of the background sounds heard underwater are biological in origin (Tavolga *et al.*, 1981; Hawkins and Myrberg, 1983).

Sound production

Aquatic lifeforms produce a great variety of sounds, which can be detected by listening to the output from a hydrophone placed judiciously in the water, provided the level of man-made noise (from boat engines for example) is not too great. Three groups of animals are known to produce sounds which have distinctive characteristics. These are (a) some crustaceans, especially shrimp, (b) teleost fish with swimbladders and (c) the aquatic mammals, notably the whales and dolphins.

When the hydrophone is placed close to the bottom, the snaps, clicks and rustles produced by benthic crustaceans may be heard (Fish, 1964). These sounds may be caused simply by the movement of the creatures over the ground, or more deliberately by the rubbing or impact of hard parts of the body against each other, a mechanism called stridulation. When many individuals do this within a local population, the sounds can merge into a continuous noise which has been described variously as crackling, sizzling and frying. The frequency content of this sound is broadband, extending well above 10 kHz. The snapping shrimp (Crangonidae), widespread inhabitants of shallow tropical and semi-tropical waters, are perhaps the best known of the invertebrate sound producers.

Fish are able to produce sounds by oscillation of the swimbladder induced by muscular force. The aptly-named croaker, *Micropogon undulatus*, is well known in this respect (Fish, 1964). It produces a tapping sound like that of the woodpecker, containing frequencies mainly in the range 350 to

1500 Hz. Groups of fish may sound in concert, giving rise to the 'croaker chorus' which occurs in the evening, a diurnal behaviour similar to the evensong of birds. The catfish, *Galeichthys felis*, also produce evening choruses, whose frequency is somewhat lower, like the bubbling of a percolator (Tavolga, 1962). Fish may also generate sounds by mechanisms not involving the swimbladder. They may produce squeaks and rasps by rubbing hard parts of the body together (stridulation), crunching due to feeding, and low-frequency noise by the hydrodynamic turbulence associated with swimming, especially near schools of fast swimmers such as tuna (Fine *et al.*, 1977).

The aquatic mammals produce many different kinds of sound. At the one extreme, there are the high-frequency clicks which whales and dolphins use for echolocation, a topic to which we shall return later. At the other end of the spectrum, there are the powerful sounds of the great whales, narrow-band signals at 20 Hz which can be detected at distances of several hundred kilometres (Schevill *et al.*, 1954; Walker, 1963). Other sounds of the mammals are described as moans, knocks and screams among an extensive repertoire. The 'song' of the humpback whale, *Megaptera novaeangliae*, is worthy of special note in this context. It is perhaps the most fascinating acoustical phenomenon in nature. Each song is characterized by a series of themes and phrases within a unique sequence which is the signature of an individual whale. The song may last for more than 30 minutes, and is repeated exactly (Payne and McVay, 1971).

Hearing

Several organs have been implicated as sensors of sound in aquatic lifeforms. The swimbladder is especially important in this connection for those fish which possess one. Movements of the bladder wall in response to sound waves are transmitted to the earbones (otoliths) whose function has been described by Fay and Popper (1980). There are also the hair cells of the lateral line and the labyrinthine structures which respond to movements in the surrounding fluid relative to the fish body. In the aquatic mammals, the ear is constructed on similar principles to that of their terrestrial cousins, with sound being transmitted to the cochlea via the tympanic membrane and the auditory ossicular mechanism.

The sensitivity of fish and other animals to sound has been the subject of much experimental work (review, Popper *et al.*, 1988). Classical conditioning techniques have generally been employed, in which the animal is first trained to respond to a sudden burst of sound of intensity high enough to be easily detected. The intensity is then reduced in successive trials until the response fails. The auditory threshold is the sound level at which the animal responds in 50% of the trials. The response may be induced by presenting a food reward which the animal associates with the stimulus, and this method is particularly successful with mammals (Johnstone, 1967; Hall and Johnson, 1972; Jacobs and Hall, 1972; Terhune and Ronald, 1975). In the case of fish, more reliable results have been obtained by the cardiac conditioning technique, first proposed by Otis *et al.* (1957) and subsequently applied in experimental studies of many species, for example by Chapman and Hawkins (1973). In this technique, a mild electric shock is applied shortly after the sound burst. This quickly establishes a conditioned change in the cardiac rhythm which can be detected by means of electrodes attached to the body of the fish. When the sound is heard, the heart misses a beat immediately; when it is not heard, the heartrate remains the same until the shock arrives.

By repeating the experiments on the same fish with tone bursts at different frequencies, we can construct the audiogram, which is a graph of the auditory threshold against frequency. Some examples are shown in Fig. 4.1. In most cases the threshold changes slowly at the lower frequencies, but there is a sharp deterioration of hearing performance at the high-frequency end of the audiogram. The maximum sensitivity (or the minimum auditory threshold) and the corresponding frequency are of particular interest in comparing audiograms. The results of hearing experiments are summarized in Table 4.1, from which it will be seen that there is a wide spread of the sensitivity and the frequency range of hearing among species.

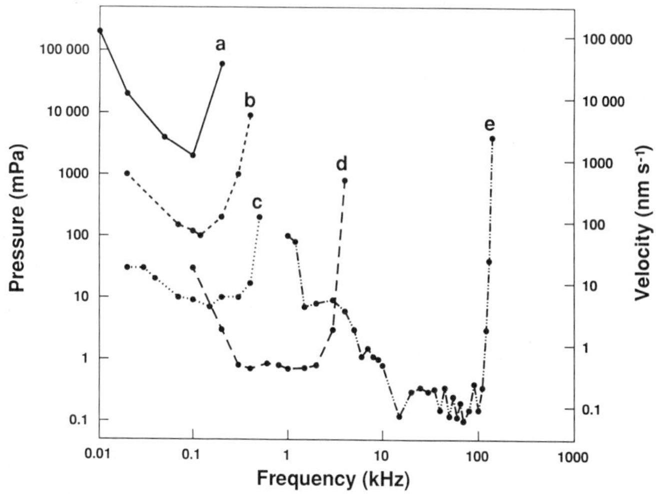

Fig. 4.1 Audiograms of hearing sensitivity for five species: (a) lobster, *Homarus americanus*; (b) Atlantic salmon, *Salmo salar*; (c) cod, *Gadus morhua*; (d) soldier fish, *Myripristis kuntee*; (e) bottlenose dolphin, *Tursiops truncatus* (references, Table 4.1). Vertical axes are (left) sound pressure amplitude at the auditory threshold; (right) corresponding particle velocity for plane-wave free-field transmission.

Table 4.1 Examples of typical auditory thresholds and the frequency range of hearing. Thresholds are given as the sound pressure and the corresponding particle velocity for free-field propagation. f_0 is the frequency of maximum sensitivity; f_{max} is the upper hearing limit at which the threshold is 10 dB above the minimum

Species	*Minimum threshold*		f_0 (kHz)	f_{max} (kHz)	*Reference*
	(mPa)	($nm\,s^{-1}$)			
Invertebrate					
Lobster *Homarus americanus*	1260.0	820.0	0.07	0.1	Offutt (1970)
Non-specialist fish					
Atlantic salmon *Salmo salar*	100.0	65.0	0.16	0.3	Hawkins and Johnstone (1978)
Dab *Limanda limanda*	35.0	23.0	0.1	0.2	Chapman and Sand (1974)
Cod *Gadus morhua*	5.1	3.3	0.16	0.4	Chapman and Hawkins (1973)
Specialist fish					
Carp *Cyprinus carpio*	0.79	0.52	0.46	1.3	Popper (1972)
Goldfish *Carassius auratus*	0.56	0.36	1.5	3.0	Enger (1966)
Soldier fish *Myripristis kuntee*	0.32	0.2	1.3	2.3	Coombs and Popper (1979)
Mammals					
Ringed seal *Pusa hispida*	2.6	1.7	40.0	49.0	Terhune and Ronald (1975)
River dolphin *Inia geoffrensis*	0.32	0.21	30.0	60.0	Jacobs and Hall (1972)
Bottlenose dolphin *Tursiops truncatus*	0.1	0.06	50.0	125.0	Johnstone (1967)
Killer whale *Orcinus orca*	0.02	0.013	15.0	30.0	Hall and Johnson (1972)

Aquatic mammals are particularly sensitive to sounds (Fig. 4.1). The invertebrates have very poor hearing by comparison (Offutt, 1970), while most fish have intermediate capabilities. Some fish such as the carp, *Cyprinus carpio*, are said to be 'hearing specialists' because they are especially sensitive to sounds over a wide frequency range (Popper, 1972). The specialists have an intimate connection between the otoliths and the swimbladder which acts as an efficient aid to hearing. Species in which the connection is made by soft tissue have relatively poor hearing, although the swimbladder still has an auditory role by amplifying the particle velocity of the sound waves near the otoliths. The Atlantic cod, *Gadus morhua*, is one example of fish which are 'non-specialist' as regards their hearing ability (Chapman and Hawkins, 1973).

An important question is whether the hearing organs of fish respond primarily to the pressure or to the particle velocity of sound waves (Hawkins and MacLennan, 1976). The distinction between these variables does not present much difficulty when the animal is remote from the sound source or any boundary, in free-field conditions when the ratio of the particle velocity and pressure amplitudes is constant. However, when the distance to the source is comparable to the sound wavelength, the velocity amplitude is relatively large (section 2.2). The velocity–pressure ratio is also affected by nearby boundaries such as the water surface, where the acoustic impedance is mismatched and the correct ratio is difficult to predict; this problem should be borne in mind when considering studies of sound or hearing in small aquaria or tanks. On the other hand, the near-field effect (which is predictable) can be used experimentally to show the relative importance of pressure and particle velocity in hearing (Hawkins, 1986).

Both the hair cells and the otoliths are sensitive primarily to the particle velocity. However, the gas-filled bladder responds to the incident sound by generating a scattered wave whose velocity amplitude at the otolith may exceed that of the incident wave, at least at frequencies below some limit (de Munck and Schellart, 1987). Thus the bladder functions as a pressure–velocity converter. This explains why some fish whose hearing depends upon the otoliths are nevertheless sensitive to the sound pressure, notably the hearing specialists but also the cod. Other non-specialists have been shown to be sensitive to the particle velocity, which is to be expected in fish that depend upon the unaided otolith for hearing (Chapman and Sand, 1974; Hawkins and Johnstone, 1978).

Fish are also sensitive to the direction of sound waves (Schuijf and Buwalda, 1980). Myrberg *et al.* (1972) have described evidence from field studies that the predatory sharks are particularly able to locate sound sources. Conditioning experiments have shown similar if less acute ability in teleost species such as the cod (Chapman and Johnstone, 1974; Hawkins and Sand, 1977). Sorokin *et al.* (1988) have observed the effect of airgun pulses on schools of the sardine *Sardinops sagax melanosticta*. The airgun releases a volume of compressed air which creates a shock wave, similar to an underwater explosion (see section 4.3). The fish reacted by swimming away from the airgun at ranges up to 150 m. Not surprisingly, Sorokin *et al.* found that the directional sensitivity is strongest for sounds well above the auditory threshold and at frequencies near the middle of the hearing range.

There are two mechanisms that might provide fish with an awareness of direction. Firstly, the otolith vibrates in the direction of propagation of the incident sound wave. Secondly, the phase difference between the signals at two spatially separated receptors partly locates the source. Neither mechanism is sufficient on its own to determine the direction uniquely, but this ability is explained if both are involved, or if the fish is able to compare the signals from several receptors.

Masking and the critical bandwidth

When a tone burst is presented to an animal in the presence of background noise, the threshold for hearing the tone depends upon the intensity of the noise. As the noise increases, so does the threshold, because the noise impairs the ability of the animal to detect the tone (Fig. 4.2). This phenomenon is called masking (Tavolga, 1974; Coombs and Fay, 1989). According to Chapman and Hawkins (1973), noticeable masking

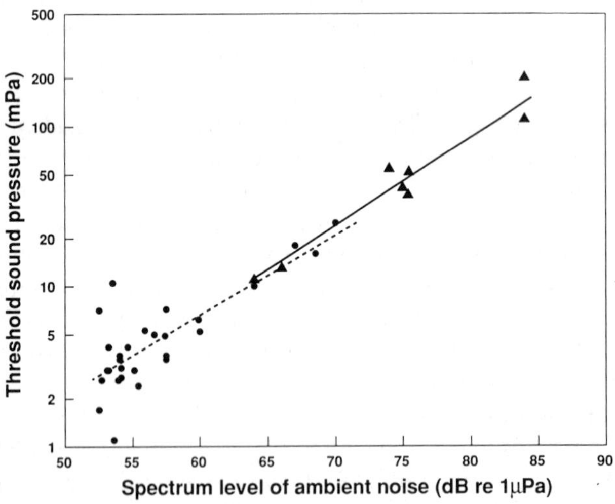

Fig. 4.2 Relationship between the auditory threshold of cod, *Gadus morhua*, at 160 Hz and the spectrum level of background noise *re* 1 Hz bandwidth, redrawn from Chapman and Hawkins (1973). Comparison of thresholds obtained in natural sea noise (●, broken curve) and artificial white noise (▲, solid curve). Each point is the result of one experiment. The curves are linear regressions on the experimental data.

can result from the natural background noise which is always present in the sea.

The frequency components of broad-band noise are not all equally effective in masking a tone signal. The tone is most severely masked by noise at the same and nearby frequencies. In experiments where the noise is produced artificially, permitting the frequency content to be controlled by electronic filtering, it is found that noise frequencies well separated from that of the tone do not mask the signal (Hawkins and Chapman, 1975). This finding leads to the concept of the critical bandwidth b which is the effective range of noise frequencies that mask a tone.

In principle, b may be determined experimentally as follows. Suppose that a tone of frequency f is presented to a fish along with a noise signal which is filtered to include only frequencies in the band $f \pm \Delta f/2$. The noise power per unit frequency is constant within the bandwidth Δf. The auditory threshold is measured in a series of trials with different noise bandwidths. The results should follow the curve illustrated in Fig. 4.3. The threshold increases with Δf at first, up to some limit after which it is constant. b is estimated as the bandwidth at which the threshold stops increasing, the implication being that frequencies beyond $f \pm b/2$ do not mask the tone (Fletcher, 1940). However, this experimental technique has not been successful in the case of fish. In experiments with cod, Hawkins and Chapman (1975) developed a more accurate method, in which the noise is presented in a constant narrow bandwidth centred on different frequencies, f_n. They found that the masking was greatest when $f_n = f$, and they deduced b from the shape of the masking curve, which is the graph of the auditory threshold against f_n (examples, Fig. 4.4). Hawkins and Chapman found that the critical bandwidth increased with the tone frequency (Fig. 4.5).

The ratio b/f is a measure of the hearing ability. The smaller this ratio, the better is the ability to discriminate signals from their frequency composition. Some examples are given in Table 4.2, arranged in decreasing order of b/f.

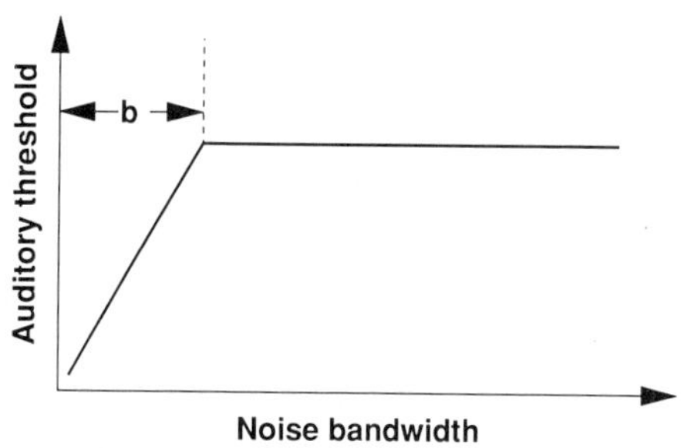

Fig. 4.3 Auditory threshold for a tone presented with noise of various bandwidths centred on the tone frequency. The critical bandwidth b is indicated by the inflection of the threshold curve.

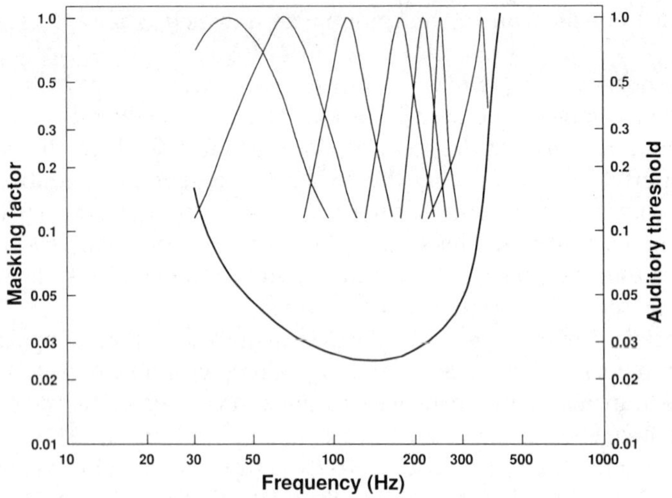

Fig. 4.4 Masking effect of 10-Hz-wide noise bands upon pure tone signals at different frequencies for the cod, *Gadus morhua*, redrawn from Hawkins and Chapman (1975). The auditory threshold is increased in proportion to the masking factor, which is normalized to unity when the noise and tone frequencies coincide (upper curves, masking factor v. noise frequency). The cod audiogram (lower curve, auditory threshold v. tone frequency, arbitrary units) is superimposed to show the hearing range.

A simple but approximate method for estimating b is to measure the auditory threshold for a tone signal in the presence of white noise. Johnson (1968) and Johnson *et al.* (1989) have used this technique in studies of marine mammals. If P_t is the power of the tone signal at the threshold, and

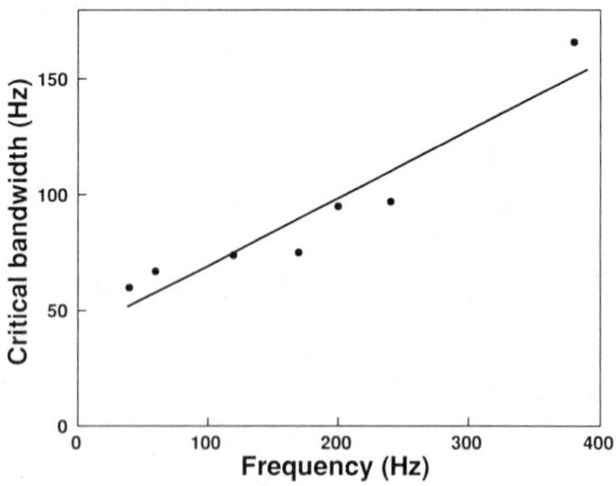

Fig. 4.5 Values of the critical bandwidth for various tone frequencies presented to cod, *Gadus morhua*, redrawn from Hawkins and Chapman (1975). The least-squares fitted line is $b = 38.853 + 0.291f$.

Table 4.2 Experimental value of the critical bandwidth b for noise masking of a tone at frequency f.

Species	f (Hz)	b (Hz)	b/f	*Reference*
Atlantic salmon *Salmo salar*	160	183	1.14	Hawkins and Johnstone (1978)
Cod *Gadus morhua*	160	85	0.53	Hawkins and Chapman (1975)
Bottlenose dolphin *Tursiops truncatus*	60 000	22 800	0.38	Au and Moore (1990)

P_n is the noise power per unit bandwidth, which is the same at all frequencies for white noise, then $b_r = P_t/P_n$ is an estimate of b. It is assumed that when white noise just masks a signal, the noise power in the critical band is equal to the signal power. This assumption is not generally true, however, and b_r is called the critical ratio, to distinguish it from the bandwidth b, which is not necessarily the same. Experimental comparisons have shown that b_r is generally less than b (Hawkins and Chapman, 1975; Au and Moore, 1990).

Biological sonar

Some animals use active sonar as a means of sensing objects in their environment, notably the bats (Microchiroptera) in air and the whales and dolphins (Cetacea) in water. This ability is seldom observed in fish, although Tavolga (1976) has reported a simple form of echo detection in the catfish, *Arius felis*. The aquatic mammals have the most advanced forms of biological sonar. Living in a world dominated by sounds and echoes, where long-range vision is relatively useless, these animals can 'see' their surroundings to a remarkable extent by interpreting the reflections of self-generated sounds (Purves and Pilleri, 1983; Turl *et al.*, 1987).

Cetaceans produce many different sounds over a broad spectrum extending to more than 100 kHz; these are described as whistles, screams or clicks according to their amplitude and frequency modulation. The clicks are one kind of sonar transmission, consisting of short pulses with a frequency content at the upper end of the phonation range (Diercks *et al.*, 1971). Key features of the clicks produced by several species are compared in Table 4.3. Some animals generate superimposed clicks of different frequencies (Pilleri *et al.*, 1976). The whistle is another kind of transmission. As described by Evans (1973), it is a frequency-modulated pulse which can last up to 3 s, sweeping a frequency range of 2–30 kHz. The ability to transmit (and to receive) over a wide frequency range increases the information content of echoes, compared with the man-made sonars described in Chapter 3 which are generally limited to narrow-band transmissions.

Sometimes the clicks are emitted at a low rate, when each one can be distinguished by a human listener with a hydrophone. In this mode, the animal appears to be maintaining a general awareness of its surroundings. The repetition rate may change, however, and at times is so high that the clicks merge into a creaking or crackling sound, presumably when a particular object has excited the animal's interest. Turl and Penner (1989) have shown that the bottlenose dolphin, *Tursiops truncatus*, adjusts the click interval according to the range of the target, transmitting shortly after the receipt of each echo. The beluga, *Delphinapterus leucas*, has a different strategy. In one mode (among several), the clicks occur at a relatively

Table 4.3 Characteristics of sonar clicks: *f*, dominant frequency; *T*, click duration; SPL, sound pressure level in decibels relative to 1 μPa at 1 m

Species		*f** (kHz)	*T* (μs)	*SPL* (dB)	*Reference*
Amazon dolphin *Inia geoffrensis*		65–100	40–60	166	Purves and Pilleri (1983)
Bottlenose dolphin *Tursiops truncatus*		50–60	50–250	170	Evans (1973)
Plumbeous dolphin *Sousa plumbea*		10–30	100–150		Purves and Pilleri (1983)
Indus river dolphin *Platinista indi*	(a)	25–30	50–100		Pilleri *et al.* (1976)
	(b)	50–80			
	(c)	160–200			
Killer whale *Orcinus orca*		14	500–1500	178	Diercks *et al.* (1971)
Beluga *Delphinapterus leucas*	(a)	1.2–1.6	2000–4000	205–225	Turl and Penner (1989)
	(b)	30–40	75–250		
Harbour porpoise *Phocoena phocoena*	(a)	2	500–5000	93–108	Møhl and Andersen (1973)
	(b)	10–150	100	132–149	

*a, b and c denote superimposed clicks of different frequency.

constant rate, even when the echo delay exceeds the click interval. Evidently the beluga can extract useful information from echoes while transmitting and receiving concurrently.

The physiology of sound production by the cetacea is not well understood. Several organs are believed to be involved, notably the larynx, the diverticuli associated with the blowhole, and the muscular plugs that seal the internal nares (Purves and Pilleri, 1983). Simultaneous sound production by different organs could explain the ability to transmit signals with spectrum peaks at widely separated frequencies. Another important factor is the reflection and refraction of the transmitted sound by the bones, air spaces and fatty tissues of the animal's head and rostrum. As a result, the sound is focused into the water as directional beams whose shape may depend upon the frequency. In some species, the sound is projected in a single beam in the forward direction. The Amazon dolphin, *Inia geoffrensis*, is one example (Fig. 4.6(a)), projecting a beam 30° wide in vertical section and a few degrees narrower in the horizontal plane (Pilleri *et al.*, 1979). A few species generate more complicated sonar fields (Purves and Pilleri, 1983). In the Indus river dolphin, *Platinista indi*, transmission in the forward direction is blocked by the cristae maxillares, a pneumatized bony structure in the frontal part of the skull. The sound transmission is thus divided into two beams, one above and one below the head (Fig. 4.6(b)). Yet another mechanism is found in the beluga, which produces superimposed low frequency (LF) and high frequency (HF) clicks, with dominant frequencies at 1.6 and 35 kHz respectively. The HF field is the more powerful in the forward direction. The LF field is angled slightly downwards and has a wider beam. Thus the LF field is the more powerful in the ventral direction. Frequency analysis of the received echoes may therefore provide additional directional clues to the beluga. The sonar field of the harbour porpoise, *Phocoena phocoena*, has a similar characteristic based on dual-frequency clicks (Møhl and Andersen, 1973).

In addition to their remarkable ability to locate objects in space, the cetaceans can interpret the echoes well enough to have some awareness of the size, shape and material composition of targets (Au and Martin, 1989; Au and Pawloski, 1989). For example, Au *et al.* (1980) and Nachtigall (1980) have shown that *Tursiops truncatus* can distinguish (a) spheres and cylinders; (b) aluminium and copper discs, for the same target strength; and (c) objects 1 dB different in target strength but similar in shape. There is a large acoustic nerve which connects the auditory sensors to the cerebral cortex. This suggests that a substantial proportion of the brain is involved in the interpretative function.

The Indus river dolphin, *Platinista indi*, is perhaps the most specialized of all the cetaceans as regards the use of biological sonar. This animal is blind and therefore relies entirely on acoustic information to navigate and

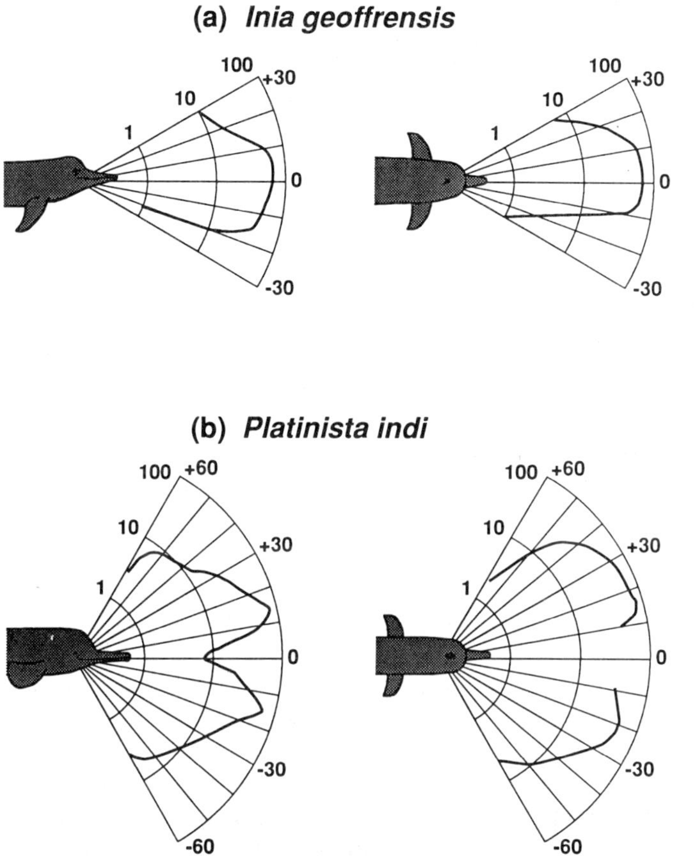

Fig. 4.6 Beam patterns of dolphin sonar transmissions. Each polar diagram shows the sound pressure level at 80 kHz (radial axis, logarithmic scale, arbitrary units) v. direction (azimuthal axis, degrees) in one plane. Left, dorsal–ventral sections; right, lateral sections viewed from the dorsal direction. Redrawn from Pilleri *et al.* (1976).

to capture prey. We have already reviewed some important features of the *P. indi* sonar, notably the broadband clicks and the complex shape of the transmitted beam, which has a null in the direction of the rostrum axis (Fig. 4.6(b)). While this null is present in both sexes, it is more pronounced in the male, which also exhibits greater asymmetry between the dorsal and ventral sectors of the beam. Purves and Pilleri (1983) believe that the larynx is the only source of acoustic energy in *P. indi*. The dorsal field is due to direct transmission in directions constrained by the air spaces and bony tissues of the head, while the ventral field is the same larynx signal reflected downwards by the cristae maxillares. The maximum intensities occur in forward directions about 25° above and below the rostrum axis.

Pilleri *et al.* (1976) describe the behaviours adopted by *P. indi* according to whether the animal is exploring its surroundings, locating an object in space or inspecting an already-located object. *P. indi* normally swims on its side, and when exploring, nods its head back and forth as the animal moves forward. The head movement causes the sonar beam to sweep horizontally, and echoes are returned from all directions, despite the 'blind spot' on the rostrum axis. Moreover, the changing echo strength as the beam sweeps, and the comparison of signals from the dorsal and ventral sectors, will allow the direction of targets to be determined precisely. When an echo is detected that the dolphin wishes to locate, it swims in an arc until the position of the target is established. Finally, if the target is considered worthy of further investigation, the dolphin stops swimming but remains on its side with the head bent and the throat directed at the target one or two metres distant. The ventral sector of the sonar beam is used exclusively during this inspection behaviour. In figure 130 of Purves and Pilleri (1983), there is a remarkable photograph of *P. indi* inspecting a piece of lead shot which is only 3 mm in diameter.

4.3 UNDERWATER EXPLOSIONS

There are a number of applications, in fields other than fisheries, which involve the sudden release of large amounts of energy into the water (Cole, 1948; Greene *et al.*, 1985). In seismic surveying, for example, the transmitted pulse of acoustic energy must be sufficient to penetrate the deep subterranean structures where oil or gas may be found. Then there is the use of explosives to excavate soil and rock during construction work. Even when the excavations are on land, shock waves can be transmitted through the ground into nearby bodies of water.

It is important to understand the biological effects of underwater explosions, so that fishery scientists can give an informed view on the consequences for aquatic animals, and advice on how the work should be done to minimize the risk of damage or even mortality. As a first step, it is necessary to understand the physical nature of the shock waves produced by explosives, or devices such as the airgun which is commonly used in seismic surveys (Duncan, 1985). The damage which exposed animals may suffer depends upon the nature of the shock wave and the distance from the explosion site. Even at long range, however, beyond the area where physical damage might occur, continuous exposure to high-energy pulses might have undesirable effects, perhaps in relation to fish reactions. For example, fishermen often complain that the disturbance created by seismic surveying 'drives the fish away'. Again, the opinion of scientists may be sought in resolving disputes between fishermen and seismic surveyors who compete for use of the sea.

Characteristics of shock waves

When a large amount of energy is suddenly released within a small volume of water, the immediate effect is the generation of a shock front, which travels outwards initially at a speed greater than that of low-amplitude sound waves. As the front passes, the pressure at a particular location rises sharply to a peak value, then decays exponentially with time (Fig. 4.7). At the explosion point, there is a bubble of gas which first expands, creating the shock front, then contracts and continues to oscillate in size. During the first contraction, the pressure at the bubble surface falls below the ambient (undisturbed) level. Thus the shock front is followed by a rarefaction, and then by a series of pulses caused by the bubble oscillations. The bubble pulses are relatively small and their speed is close to that of low-amplitude sound. Moreover, the shock front attenuates more rapidly as it travels than do the bubble pulses. This means that the shape of the pressure–time waveform, which is called the 'signature' of the explosion, changes with distance from the explosion point. As the range increases, the time-delay between the shock front and the first bubble pulse lengthens and a larger proportion of the transmitted energy is contained in the bubble pulses.

The signature illustrated in Fig. 4.7 applies when the shock wave is propagating in free-field conditions, at any range R more than a few times the bubble radius but less than the distance to the water surface or the bottom. The qualitative features of the signature are similar for all high-energy sources which produce gas bubbles – sparkers, airguns or explosive

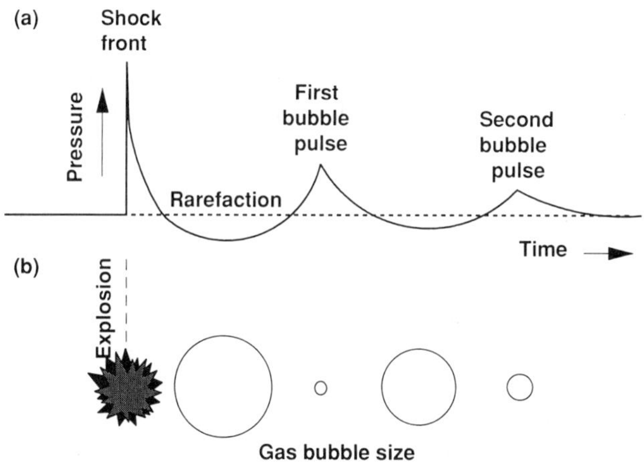

Fig. 4.7 Sequence of events following an underwater explosion. (a) Pressure waveform at distance R from an underwater explosion. Free-field propagation. The peaks and troughs about the ambient level are produced by oscillations of the gas bubble at the explosion site. (b) The bubble size is shown at the corresponding time on the waveform.

charges (details, Duncan, 1985). These sources differ in the amount of energy which can be released in practical applications (Table 4.4), and in the rate of energy release, which determines the amplitude of the shock front. For example, there are two types of chemical explosive in common use – propellants such as gunpowder, and high explosives such as TNT. The high explosives are more energetic than the propellants and they detonate more rapidly. Of all the high-energy sources mentioned above, the high explosives produce the strongest shocks, which are most likely to have adverse biological consequences.

In free-field conditions, when the shock front propagates far from the surface or other boundaries, the physical effects of detonating a charge of high explosive can be predicted by empirical equations (Cole, 1948; Staal, 1985). Of particular interest from the biological point of view are: P, the pressure relative to the ambient level (P_0) and how it changes with time; E, the energy flux transmitted by the shock waves; and I, the impulse, which is the integral of the pressure with time over the duration of the initial shock. When the shock front arrives at a given point, the pressure rises almost instantaneously to P_{max}, the peak value, then it decays more slowly. The time constant T is a measure of the width of the shock pulse. T is the time at which the pressure has dropped to $\exp(-1)=0.37$ of the peak value. These variables are illustrated in Fig. 4.8. Predictive equations are given in Table 4.5 for TNT and Pentolite, two types of high explosive in common use. To illustrate the scale of Fig. 4.8, suppose 10 kg of TNT explodes at 100 m depth. At 30 m distance, P_{max} is 2.7 MPA, P_{min} is 53 kPa below ambient, T is 0.36 ms and the first bubble pulse arrives 90 ms after the shock front.

The rarefaction which follows the initial shock is much smaller than P_{max}. However, fish are more resistant to the compressive forces implied by positive pressures. The expansion caused by a sudden reduction in pressure can be particularly damaging, especially to gas-filled organs such as the swimbladder. This effect depends upon P_{min}, the lowest pressure relative to the ambient level during the rarefaction, and upon the time scale of the pressure changes. Equations for predicting P_{min} are included in Table 4.5.

When the distance of the observation point from the explosion, R, is more than the charge depth below the surface or height above the sea-bed, the pressure waveform is modified by reflections from the two boundaries. Since the shock front precedes the reflections, which have longer path lengths, the peak pressure is unaffected. However, the free-field equations cannot be used to predict the energy flux or the impulse in the presence of boundary reflections.

The changes in pressure in the surface reflection are the reverse of those in the direct wave, because the sum of the two must be zero at the surface.

Table 4.4 Properties of high-energy sources. The stated values refer to the shockfront observed at distance R from the explosion centre, propagating in free-field conditions; calculations based on data from Cole (1948) and Duncan (1985)

Source	*Specification*	R (m)	*Maximum pressure* (kPa)	*Time constant* (μs)	*Energy flux* ($J\,m^{-2}$)	*Impulse* (Pas)
Pentolite	10 kg charge at depth $>R$	10	135000	258	8730	3190
TNT	10 kg charge at depth $>R$	10	9250	284	7600	3160
Black powder	40 kg charge at depth $>R$	10	190	–	70	1100
Airgun	82 litres at 14 MPa 8 m depth 32 gun array size 100 m × 21 m	100	31	–	9	90
Airgun	37 litres at 14 MPa 7 m depth 16 gun array	10	225	–	40	570
Airgun	1.6 litres at 14 MPa 1 gun at 8 m depth	10	17	–	<1	90
Sparker	3 kJ discharge	10	5	–	<1	<10

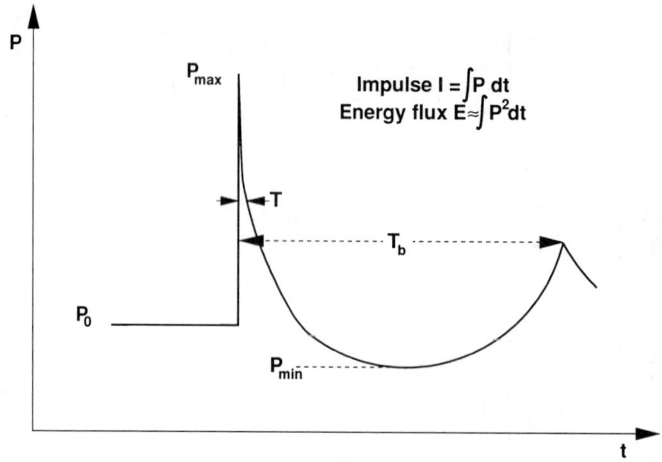

Fig. 4.8 Physical variables of the pressure signature (not drawn to scale). *P*, pressure at time *t* at the point where the signature is observed; P_{max} peak pressure, which coincides with the shock front; P_{min}, greatest rarefaction; P_0, ambient pressure; *T*, time constant of the decaying shock; T_b, time delay of the first bubble pulse.

Table 4.5 Equations for predicting the characteristics of explosive shockfronts propagating in free-field conditions. W is the charge weight (kg); R is the distance of the observation point from the explosion centre (m). P_1 and P_0 are respectively the ambient undisturbed pressures at the observation point and the explosion centre (Pa)

Parameter	*Explosive material*	
	TNT	*Pentolite*
Peak pressure (P_{max}, MPa)	$52.4(W^{1/3}/R)^{1.13}$	$56.3(W^{1/3}/R)^{1.14}$
Energy flux (E, kJ m^{-2})	$82.1W^{1/3}(W^{1/3}/R)^{2.05}$	$105W^{1/3}(W^{1/3}/R)^{2.12}$
Impulse (I, Pas)	$5750W^{1/3}(W^{1/3}/R)^{0.89}$	$7410W^{1/3}(W^{1/3}/R)^{1.05}$
Time constant (T, μs)	$92.5W^{1/3}(W^{1/3}/R)^{-0.22}$	$84.0W^{1/3}(W^{1/3}/R)^{-0.23}$
Min. pressure (P_{min}, Pa)	$P_1 - 69W^{1/3}P_0^{2/3}/R$	$P_1 - 74W^{1/3}P_0^{2/3}/R$

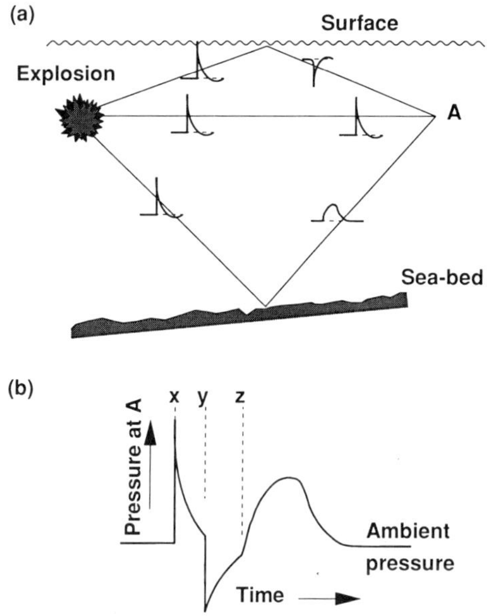

Fig. 4.9 Effect of boundary reflections. (a) Direct and reflected transmission paths to the observation point A, showing the individual pressure waveforms. (b) The pressure signature observed at A: *x*, the shock front arrives first by the direct path; *y*, the surface reflection reverses the pressure change; *z*, the bottom reflection arrives last.

On the other hand, the pressures in the bottom reflection and the direct waves change in the same way (Fig. 4.9). Note that the inverted reflection from the surface can result in a substantial rarefaction when it combines with the direct wave. In the absence of any other boundaries, the resultant pressure is the sum of contributions from the direct wave and the two reflections. Rather complicated waveforms can arise from these interactions. Nevertheless, it is possible to calculate important parameters, such as the impulse and the minimum pressure, if a good assumption can be made about the magnitude of the bottom reflection. An alternative practical approach is to experiment with the detonation of a very small charge at the intended explosion site, and to record the resulting shock by means of hydrophones placed at several locations. The measurements may be scaled to predict what will happen when the 'big bang' occurs, by reference to the similarity principle, which states that the parameters of the shock wave are the same when distances are measured as the number of charge radii (Cole, 1948). In the case of spherical charges, the radius is proportional to the cube root of W, the weight of the explosive material. Thus the maximum pressure, for example, is a function of the ratio $(W^{1/3}/R)$ and does not otherwise depend upon W or R.

Biological effects of explosions

The damage to aquatic life caused by underwater explosions has been the subject of many investigations over the past 50 years (e.g. Aplin, 1947; Hubbs and Rechnitzer, 1952; Teleki and Chamberlain, 1978). The shock fronts described above may injure nearby fish. The damage depends upon a number of variables apart from the size of the explosion, notably the relative positions of the source, the fish, the water surface and the sea-bed. Fish close to the blast may be killed instantly by the shock wave, while those at greater distances suffer less damage. Depending on such factors, death is not necessarily immediate but may still occur a day or two later.

The damage to exposed animals may include scale loss, the tearing of muscle tissue, haemorrhaging, and rupturing of internal organs such as the liver or spleen. The swimbladder, in those fish which possess one, is particularly responsive to pressure changes. Fish that have no swimbladder, and shellfish, are much more resistant to the stress of explosive shock, size for size. For the same species, large animals are more susceptible to damage than small ones.

Mortality in the vicinity of an explosion is revealed by the dead fish which float to the surface following the blast. However, counting them provides no indication of the mortality rate: the number of live fish will be uncertain, and some of the corpses may sink. A more scientific technique is to place fish in cages at different distances from the centre of the explosion. The proportion of mortalities in each cage can then be related to the physical parameters of the shock wave, to describe the biological effect in quantitative terms.

The proportion of mortalities is one measure of the biological effect of the explosion. For example, we may describe an explosion in terms of R_{50}, the range at which 50% of the exposed fish are expected to die, or I_{50}, the corresponding size of the impulse. Yelverton *et al.* (1975) have conducted experiments on several different species and sizes of freshwater fish, including both physostomes and physoclists, held in cages exposed to the detonation of 0.45 kg Pentolite charges. They found that the mortality correlated well with the impulse but not with the peak pressure. When the experimental results are plotted on logarithmic graph paper, the data lie close to a straight line (Fig. 4.10). This means that I_{50} is proportional to a power of w_f, the weight of one fish, and a simple empirical formula can be derived by log–log regression. If w_f is in kilograms, the Yelverton formula is

$$I_{50} = 371 w_f^{0.32} \qquad (\mathrm{Pa-s}) \tag{4.1}$$

However, Yelverton *et al.* worked with fish close to the surface. Following the initial shock, the pressure reversal in the surface reflection results in a strong rarefaction which is particularly damaging to fish. As the depth

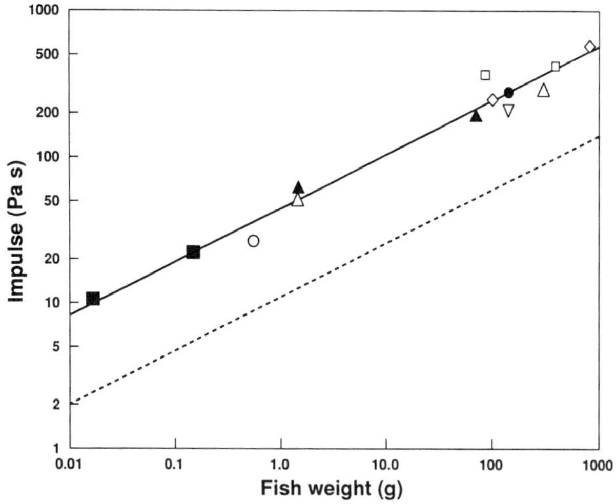

Fig. 4.10 Mortality of fish exposed to underwater explosions as a function of the explosive impulse and the fish weight, redrawn from Yelverton *et al.* (1975). Points are observed 50% mortalities. Physoclists: ■ guppy, *Lebistes reticulatus*; ▲ bluegill, *Lepomis macrochirus*; ●, black bass, *Micropterus salmoides*. Physostomes: ○, minnow, *Gambusia affinis*; △, goldfish, *Carrasius auratus*; ▽, rainbow trout, *Oncorhynchus mykiss*; □, channel catfish, *Ictalurus punctatus*; ◇, carp, *Cyprinus carpio*. Solid curve log–log regression of 50% mortality data (Equation 4.1); broken curve, impulse for no injury.

increases, the surface rarefaction is reduced by the ambient pressure; it is unimportant at depths greater than 10 m. Furthermore, fish near the surface are exposed in ventral aspect, from which direction they are most vulnerable to damage from shock waves (Sakaguchi *et al.*, 1976). It is therefore considered that the Yelverton model can only be applied to fish within a few metres of the surface. For deeper fish, Equation 4.1 greatly underestimates I_{50}.

Wiley *et al.* (1981) suggest that the mortality of fish at any depth depends primarily on the expansion and contraction of the swimbladder in response to the explosive forces. They define the 'bladder oscillation parameter' to be

$$Z = 100 \ln(A_{max}/A_{min}) \tag{4.2}$$

where A_{max} and A_{min} are respectively the maximum and minimum radii of a sphere having the same volume as the expanded and contracted swimbladder respectively. From experiments with spot, *Leiostomous xanthurus*, and white perch, *Morone americana*, Wiley *et al.* find that for fish at different depths, the mortality correlates well with Z but not with *I*. Goertner (1978) has shown how Z may be calculated from the observed pressure waveform, leading to a model which predicts the mortality for any positions of the charge and the fish relative to the surface. Goertner suggests that the mortality is 50% when Z=125. However, his model shows that R_{50} depends upon the fish size and depth in a complicated way. In certain circumstances, perhaps owing to resonance between the bladder oscillation and the pressure changes, the mortality is more than that expected from simple theory. When the shock front is propagating in free-field conditions, distant from the surface, the Goertner model predicts that I_{50} is more than 10 times the value suggested by Yelverton *et al.* (1975).

There is little doubt that the Goertner model is the most accurate of those developed to date. However, it is complicated and difficult to apply in practice, since it requires detailed knowledge of the pressure signature throughout the volume affected by the explosion. Baxter (1985) has proposed a simpler method for predicting the fish kill, which nevertheless takes account of the fish weight and depth. He suggests that the energy flux rather than the impulse provides the best correlation with the mortality. The Baxter model determines the mortality as a function of the parameter B which is defined by the formula

$$B = E/[w_f^{1/3}(1.0 + 0.1D)] \qquad (4.3)$$

where D is the depth of the fish in metres. When E and w_f are expressed in SI units as previously, comparison with experimental data shows that 50% mortality is likely when $B = B_{50} = 400$. Also, the values of B corresponding to 0% and 100% mortality are $B_0 = 49$ and $B_{100} = 1860$ respectively. These criteria can be used to draw contour maps of the region where fish are likely to be killed. An example is shown in Fig. 4.11, using the relevant formula from Table 4.5 to estimate the energy flux. This calculation does not take

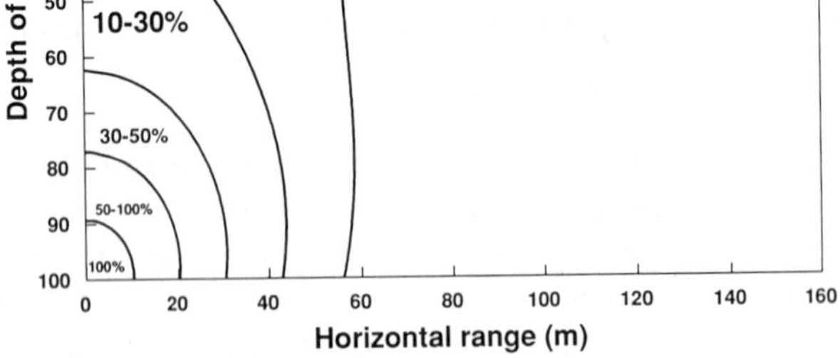

Fig. 4.11 Predicted mortalities according to the Baxter model for a 90 g fish and 11 kg charge of TNT exploded at 100 m depth. Owing to pressure release at the surface, the mortality above 5 m (broken curves) will be less than predicted here.

account of the pressure release at the surface, so the Baxter model should only be used to predict the fish kill at depths of several metres or more.

So far we have only considered fish with swimbladders. The fate of marine mammals is also a matter for concern. O'Keefe (1985) has extended the Goertner model to mammals on the assumption that changes in volume in the lungs will be a major factor, and including a margin for intestinal injury which is caused primarily by the peak pressure. For example, suppose a 4500 kg charge of TNT is detonated at 400 m depth. The model predicts that a 6 m whale at the same depth would be uninjured at a range of 330 m or more. For the same whale at 100 m depth, however, the safe range is 820 m. Again, contours of the damage risk may be drawn to show how the danger zone depends upon the depth and size of exposed animals.

High-energy sources other than the high explosives appear to have little biological effect, although Kostyuchenko (1971) reports that airguns are capable of destroying fish eggs within a few metres' range. It is possible that airguns might scare fish sufficiently to drive them from the vicinity. If this did happen, the fish would probably return within a short time after the seismic operations had ended.

Damage limitation

Given that underwater explosions are necessary for various purposes, a few precautions may be suggested to minimize the risk of biological damage. The work might be done in a season when fish are least likely to be present. A scouting ship could be used to search the area by sonar, to ensure that no large schools were nearby. In the case of sea-bed excavations, the deeper the explosive is buried the better, and the burial hole should be properly backfilled to reduce the energy transmitted into the water. Much of the biological damage appears to be due to the steep rise of the shock front which is characteristic of high explosives. One method of reducing the peak pressure, for the same total release of energy, is to divide the explosive into several relatively small charges which are detonated in succession. The result is a number of shock fronts separated in time, but the amplitude of each one is much less than that from a single charge of the same total weight. For the same reason, the use of linear explosions should be encouraged as an alternative to the single charge. A linear explosive is constructed as a cord which is detonated at one end. The detonation travels along the cord, thus extending the time over which the explosion occurs.

4.4 SWIMBLADDER RESONANCE

Writers who have studied the broad-band acoustic properties of the swimbladder often assume, as a first approximation, that the organ behaves like a free gas bubble of the same volume. A bubble surrounded only by water is nearly spherical, and a well-established theory describes how it responds to acoustic waves from a distant source (Weston, 1967). When undisturbed, the gas pressure inside the bubble equals the ambient pressure in the water. If the size of the bubble is changed from this equilibrium by an external force such as acoustic waves, the gas pressure will also change, in accordance with Boyle's Law. If the temperature of the gas does not change, the pressure is inversely proportional to the volume or the cube of the bubble radius. The disturbed bubble has a natural oscillation frequency as it alternately expands and contracts about the equilibrium. This frequency is determined by the compressibility of the gas (the bubble acts like a spring) and the inertia of the water, which must move to accommodate the changing volume of gas. If the frequency of the external force is close to the natural oscillation frequency, the bubble is said to resonate. The motion of the bubble wall is then greater, and the radiated acoustic waves are stronger than at other frequencies (Fig. 4.12).

The resonance frequency of the free bubble depends upon the ambient pressure P_0, the water density ρ and the equilibrium bubble radius a, according to the equation

$$f_r = \frac{1}{2\pi a}\sqrt{\left(\frac{3\gamma P_0}{\rho}\right)} \tag{4.4}$$

where γ is the ratio of the specific heats of the gas (1.4 for air). To apply this equation to fish, we must first decide on the appropriate value for a. There is likely to be much variation among species, and even between individual fish of the same size and species, but for rough calculations it can be assumed that a is typically 5% of the total fish length. For example, consider a 10 cm fish near the surface where P_0 is 1 atmosphere ($10^5\,\mathrm{N\,m^{-2}}$). The

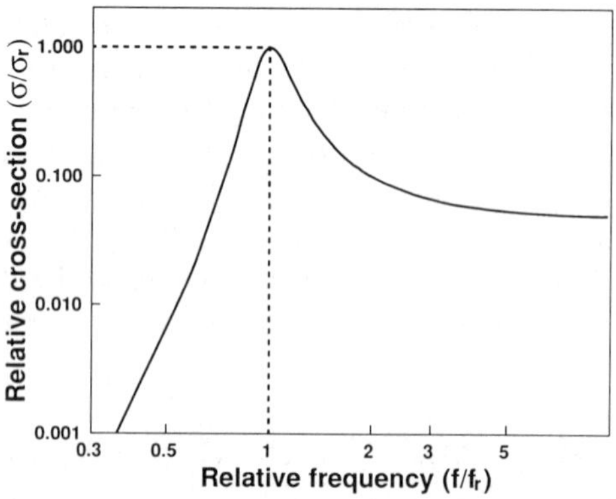

Fig. 4.12 Frequency dependence of acoustic scattering by a free gas bubble. σ_r is the acoustic backscattering cross-section at the resonance frequency f_r.

density of seawater is 1029 kg m^{-3} and a is 0.4 cm. Remembering that all quantities must be expressed in compatible units, MKS (m, kg, s) in this instance, f_r is calculated to be 640 Hz. This is much lower than the frequencies typical of fishery sonars. However, f_r increases with depth and is greater for smaller fish. In the case of fish larvae, f_r could be as much as 40 kHz.

The resonance frequency changes with depth in a manner which depends upon the mass of gas in the swimbladder. If this mass remains constant, as is likely for physostomes such as the clupeoids, which are unable to secrete gas into the swimbladder, then the equivalent bubble radius is proportional to $P_0^{-1/3}$ and f_r will vary as $P_0^{5/6}$. However, some fish adjust the amount of gas in the swimbladder, for example physoclists such as the cod, *Gadus morhua*, in order to maintain the same volume (and the same buoyancy) against changes in the ambient pressure. In that case, the equivalent bubble radius is constant and f_r is expected to vary as $P_0^{1/2}$. Both the $P_0^{1/2}$ and $P_0^{5/6}$ dependences have been observed in measurements of the resonance frequency of species which undergo vertical migrations (Hersey *et al.*, 1962; Mozgovoy, 1986). These migrations usually occur as a diurnal rhythm, in response to changes in the light level between night and day.

The swimbladder in many fish species has an elongated shape with an irregular surface (Whitehead and Blaxter, 1989). The natural tendency of the contained gas to form a sphere is resisted by the bladder wall and the surrounding tissues. Thus the acoustic properties of the swimbladder are unlikely to match those of the equivalent free bubble. In experiments where cod are moved rapidly from one depth to another, so that the fish do not have time to adapt by secreting or absorbing gas, it can be shown that the $P_0^{5/6}$ law is followed while the swimbladder is compressed below its normal volume (Sand and Hawkins, 1973). However, the resonance frequency is about 25% higher than that predicted by Equation 4.4. For fish close to the adaptation depth, with fully inflated swimbladders, the difference is much greater. The observed resonance frequency can be as much as six times the predicted value. These results are explained by the elasticity of the bladder wall and the surrounding tissues, which have a greater effect on the resonance frequency than the non-spherical shape of the organ.

Fish sizing

The spectrum of echoes might be used to determine the size of fish targets, on the assumption that peaks in the spectrum correspond to the resonance frequencies of swimbladders (Hawkins, 1977). Løvik and Hovem (1979) have shown that for several species of fish adapted to surface pressure, including both physoclists and physostomes, the resonance frequency is determined approximately by the empirical formula

$$f_r = 120/L \tag{4.5}$$

where L is the fish length in metres and f_r is in hertz. The frequency predicted by this formula is too low for herring and too high for cod (Fig. 4.13). This is perhaps not surprising, since the swimbladders of the two species are dissimilar in structure and function (Blaxter and Batty, 1990). It would be more accurate to apply Equation 4.5 with a different constant multiplier for each species, 170 for herring and 80 for cod according to Løvik and Hovem (1979).

The variation with depth complicates the calculation of fish size from the observed resonance frequency. For fish that are adapted to surface pressure and then move quickly downwards, without changing the amount of gas in the swimbladder, the 5/6 pressure law applies and we suppose that

$$f_r = K(1 + D/10)^{5/6}/L \qquad (4.6)$$

where K is a constant for each species, typically 120, and D is the depth in metres. This equation should apply to physostomes in the course of vertical migration as a natural behaviour. However, in the case of physoclists which have been at depth for some time, the resonance frequency is expected to be less than that predicted by Equation 4.6. Depending on how much gas the fish has secreted to reinflate the bladder, f_r could be reduced by as much as $(1 + D/10)^{1/3}$ times.

In practice, the received signal will normally consist of the superimposed echoes from many fish or other scatterers at similar ranges. The ability to distinguish targets of different size depends upon the sharpness of the

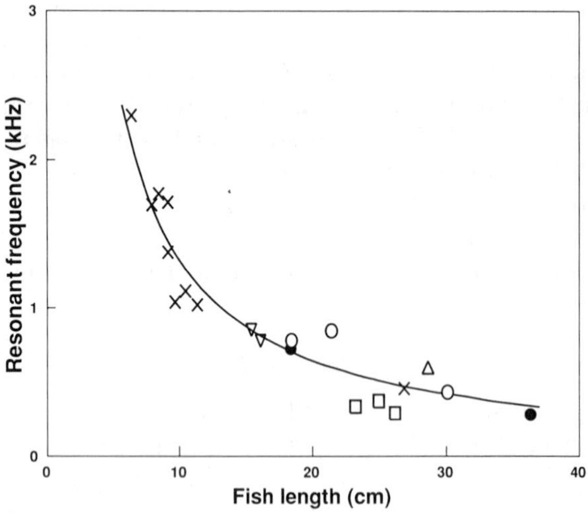

Fig. 4.13 Relationship between swimbladder resonance frequency and length of surface-adapted fish, redrawn from Løvik and Hovem (1979). ×, saithe, *Pollachius virens*; ○, pollack, *Pollachius pollachius*; □, cod, *Gadus morhua*; ●, trout *Salmo trutta*; △, herring, *Clupea harengus*; ▽, sprat, *Sprattus sprattus*. The fitted curve is $f_r = 120/L$.

resonance peak, which is described by the Q factor. If f_1 and f_2 are the frequencies on either side of the peak at which the echo intensity is half the peak value, Q is defined as the ratio $f_r/(f_2-f_1)$. For fish swimbladders, Q is generally in the range 2–5. When fish move rapidly downwards from the adaptation depth, so that the swimbladder is compressed, Løvik and Hovem report that Q increases linearly with depth. However, in the case of physoclists that are left at the new depth for some time and are able to restore the bladder volume, Q gradually decreases to the original value. This implies that for a fully inflated swimbladder, Q is independent of the ambient pressure.

As a general rule, it should be possible to distinguish two peaks in the echo spectrum which are more than f_r/Q apart. Figure 4.14 shows an echo spectrum which Holliday (1977a) obtained by detonating small explosive charges near the surface. Two peaks are evident at the low-frequency end of the spectrum, and can be associated with particular species by comparison with catches of schooling fish in the same area. The lower resonance is due to jack mackerel, *Trachurus symmetricus*, of mean length 180 mm, and the other is northern anchovy, *Engraulis mordax*, of lengths (from a small sample) in the range 97–146 mm. Holliday confirmed that the higher resonance is associated with the anchovy by means of other experiments when no other species was present.

The resonance technique for fish sizing is difficult to apply in practice because of the strong dependence of the resonance frequency on the depth

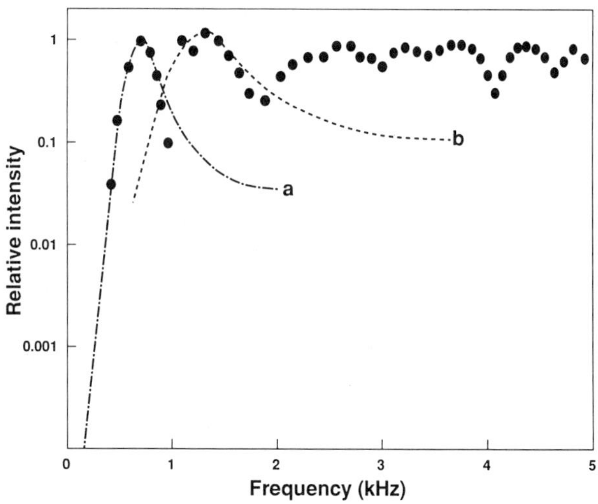

Fig. 4.14 Spectrum of echoes from schooling fish, redrawn from Holliday (1977a). Points are measured intensity per unit bandwidth, relative to the maximum spectrum level. Theoretical curves fitted to the peaks: (a) jack mackerel, *Trachurus symmetricus*; (b) northern anchovy, *Engraulis mordax*.

as well as the size of targets. To obtain useful results, it would be necessary to know the history of vertical movements of the observed fish, but this is unlikely to be known in sufficient detail.

4.5 PLANKTON ACOUSTICS

Diffuse background traces of the kind shown in Fig. 4.15 are often seen on the echogram. They are caused by the echoes from dense concentrations of small creatures, the plankton and the micronekton. The planktonic background is one factor which limits the performance of sonars in detecting fish or other large targets. Although the individual plankters are small or even microscopic, the overlapping echoes from a dense concentration combine to produce signals comparable to those from isolated fish. However, the plankton deserve study in their own right as an important component of aquatic ecosystems. Acoustic methods can be used to investigate plankton in much the same way as they are applied to fish.

There is some evidence to suggest that phytoplankton can be detected by high-frequency sonar (Castile, 1975), but here we restrict attention to the zooplankton and micronekton, whose stronger echoes dominate the planktonic background in most cases.

Early investigators studied the volume scattering coefficient S_v, measured by echo integration at one frequency (e.g. Bary, 1966; Pieper, 1979; Sameoto, 1982). They found that S_v correlated well with zooplankton biomass determined by direct sampling. Thus echo integration reveals the spatial distribution of the biomass, and a large volume of water may be surveyed acoustically much more rapidly than by mechanical samplers.

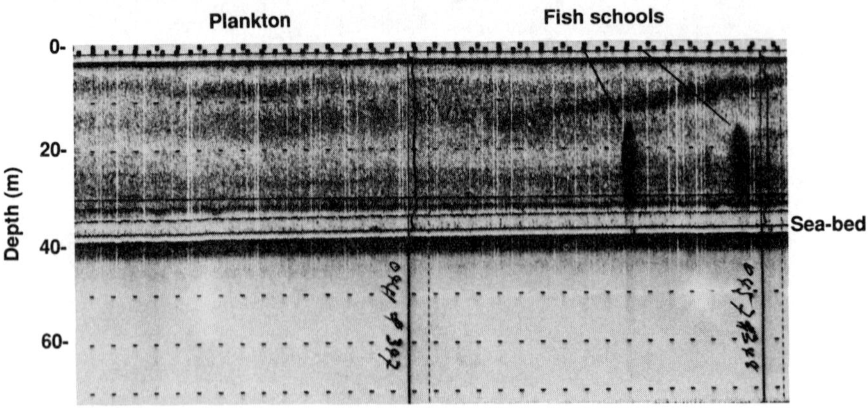

Fig. 4.15 Echogram showing diffuse background traces of plankton and two fish schools. Vertical white lines are caused by aeration below the transducer. Recorded one hour before dawn on 6 September 1986, South China Sea (5 °N, 112 ° E), Furuno FQ-50 echosounder.

This technique is of limited use to biologists, however, because it provides no information about the sizes or species composition of the individual animals.

Some independent evidence is necessary to reveal the species composition, but the size distribution of zooplankters may be determined acoustically by measuring S_v at several frequencies (Greenlaw, 1979; Holliday and Peiper, 1980; Kristensen and Dalen, 1986; Sameoto *et al.*, 1990). The method depends upon the frequency dependence of scattering, which is characteristic of the target size. In particular, Rayleigh scattering applies at low frequencies up to a limit which is inversely proportional to the target size (Chapter 2). The target strength is relatively small in the Rayleigh region. This means that the smaller targets are more strongly detected at the higher frequencies.

Theoretical models

The size of a zooplankter is often described by the equivalent spherical radius, *ESR*. This is the radius of a sphere having the same volume as the zooplankter. The problem is how to deduce the distribution of *ESR* from the S_v values observed at various frequencies. To do this, we need to know how the target strength changes with frequency and size. Some experimental work has been published (Greenlaw, 1977; Penrose and Kaye, 1979; Richter, 1985; Foote *et al.*, 1990) but simplifying assumptions are necessary to provide calculable functions which describe the target strength over a wide frequency range. The simplest model is to suppose that the zooplankter scatters sound like a fluid sphere of radius *ESR* (Anderson, 1950). The target strength depends upon g, the density of the sphere and h, the sound speed within it, expressed as proportions of the respective values of these variables in the surrounding water. Both g and h are a few percent more than unity for zooplankters. However, the fluid–sphere theory is a poor approximation at high frequencies where there are sharp resonances owing to the symmetry of the sphere. Zooplankters have irregular shape, and the resonance structure (if any) is likely to be less sharp. Various refinements have been proposed to overcome this problem, notably the high-pass model (Johnson, 1977) and the truncated Anderson model (Peiper and Holliday, 1984). In this context, truncation means that the high-order vibrations of the sphere are ignored, with the result that the target strength varies more smoothly with frequency (Fig. 4.16).

An important limitation of all the sphere models is that some zooplankters have an elongated shape, in which case the target strength will depend upon the orientation of the animal as it does for fish. Stanton (1988, 1990) has proposed a fluid–cylinder model which allows the orientation to be included as a parameter. Again, high–pass or truncated versions of the

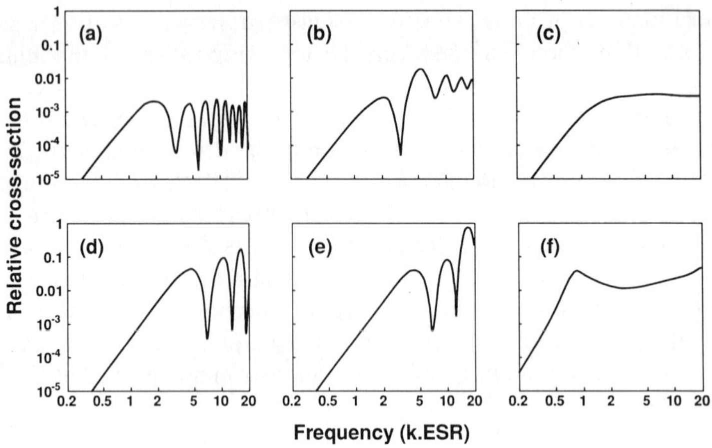

Fig. 4.16 Predicted backscattering strength of plankton modelled as fluid targets ($g = 1.041$; $h = 1.035$). Vertical axes, ratio of acoustic and geometric cross-sections; horizontal axes, dimensionless frequency (wavenumber times equivalent spherical radius). Sphere models: (a) complete Anderson; (b) truncated Anderson; (c) Johnson high-pass. Cylinder models, length 6.2 *ESR*, axis normal to the beam: (d) complete Stanton; (e) truncated Stanton; (f) KRIDA.

Stanton model may be adopted to improve the representation at high frequencies. The acoustic properties of the various models are compared in Fig. 4.16.

Peiper and Holliday (1984) have compared the three versions of the fluid–sphere theory mentioned above. They report that the truncated Anderson model gives the best correlation with biological samples. Sameoto *et al.* (1990) have described a successful application of the high-pass Stanton model in studies of copepods and euphausiids. However, Dalen and Kristensen (1990) maintain that a more empirical approach is necessary. They compare various theoretical models with measured target strengths of the krills *Euphausia pacifica*, *Meganyctiphanes norvegica* and *Thysanoessa* spp. This leads Dalen and Kristensen to suggest a hybrid approach, the KRIDA model. The basic concept is the fluid cylinder with truncation of high-rder modes, but the size and other properties of the cylinder are adjusted to match the observed target strengths. Furthermore, the average target strength predicted by KRIDA allows for a stochastic variation of the orientation of the krill with respect to the acoustic beam (Kristensen and Dalen, 1986); tilt angle is explained on p. 167.

There is no strong consensus as to which model is the most useful for interpreting the acoustic echoes from plankton. The best choice seems to depend upon the size of the scatterers. The truncated Anderson model is satisfactory for small organisms, e.g. copepods and calanoids, whose *ESR* is less than 1 mm. The krill are much larger than that, up to 40 mm in body length, and for them the sphere models are too crude and inaccurate. The semi-empirical KRIDA model is the best description of acoustic scattering by krill and its variation with frequency.

Experimental techniques

Acoustic surveys of plankton are conducted with sonars operating at up to 10 MHz. The effective range at 10 MHz is only a few metres, which means that lower-frequency sonars used concurrently are similarly restricted in range, in order to sample the same volume of water. One technique is to deploy several transducers in a probe which is lowered from a stationary ship. The transmissions are repeated as the probe descends, giving a depth profile of targets in the water column. The probe may be dropped at a series of stations along a transect, to reveal horizontal changes in the plankton distribution. Alternatively, the probe may be towed at a particular depth to map the horizontal distribution in more detail.

Holliday *et al.* (1989) have described a sophisticated instrument for plankton investigations, the multifrequency acoustic profiling system (MAPS). It employs 21 circular disc transducers operating at different

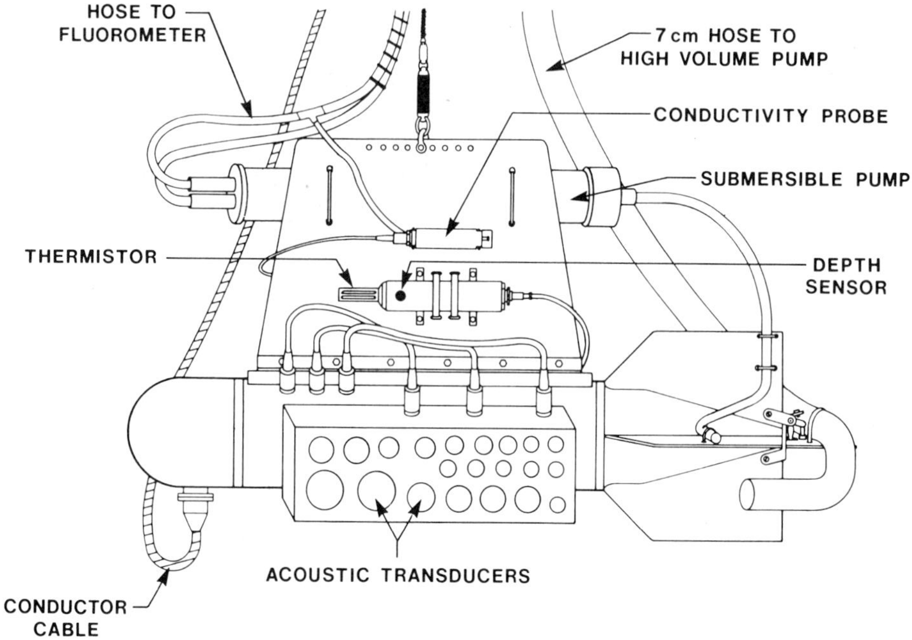

Fig. 4.17 The MAPS (multifrequency acoustic profiling system). The probe has 21 acoustic transducers operating at different frequencies, and sensors for temperature, conductivity and depth measurement. A hose is attached for simultaneous pump sampling. Redrawn from Peiper *et al.* (1990).

frequencies from 0.1 to 10 MHz (Fig. 4.17). The disc diameter is inversely proportional to the frequency, thus all the transducers have the same beam width. The 21 frequencies are chosen so that the ratio of any two adjacent frequencies is nearly constant. The transducers transmit 50-μs pulses in turn, and the cycle of sequential transmissions through the 100-fold frequency range repeats in 20–30 s. The sampling volume (as defined on p. 27) is about 0.1 m^3 at a distance of 1–2 m from the transducer faces. The acoustic measurements are converted to estimates of the scatterer density by size class, using the truncated Anderson model and an algorithm developed by Holliday (1980).

Costello *et al.* (1989) have compared the MAPS and pump-sampling techniques for zooplankton density estimation. Biological samples were collected by filtering the water from a pump whose inlet was attached to the MAPS probe so that the same populations would be sampled by both methods. The dominant species in the filtrate were the calanoids

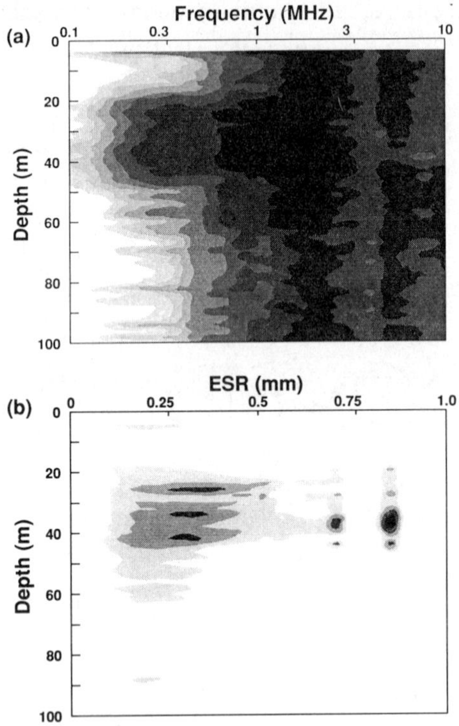

Fig. 4.18 Example of results obtained with the MAPS probe. The grey scales indicate relative values; black is high; white is low. (a) Volume scattering strength v. frequency and depth; (b) biovolume estimates by size class and depth. Redrawn from Peiper *et al.* (1990).

Clausocalanus spp. and *Oikopleura* spp. with modal sizes (*ESR*) in the range 0.1 to 0.25 mm. Close agreement was found between the abundances and size distributions indicated by the pump and acoustic measurements. Peiper *et al.* (1990) have described zooplankton distributions revealed by the MAPS. Figure 4.18 illustrates the results obtained by the MAPS, shown as grey-scale diagrams of the distribution by depth and size class. This kind of information can be displayed most effectively by colour coding (e.g. Plate 7).

Greene *et al.* (1989) have described an alternative approach to the problem of size estimation, using a dual-beam sonar operating at one frequency, 420 kHz. This technique depends upon the targets being sufficiently well separated for the individual echoes not to overlap. It is most likely to be successful in the study of large zooplankters, the krill and other crustaceans.

Chapter five

Observation and measurement of fish

5.1 INTRODUCTION

Active sonars and echosounders are instruments which transmit an acoustic pulse through a transducer into the water. The pulse travels away from the transducer, but the transmitted energy is concentrated in certain directions, within a beam whose width depends upon the size of the transducer. Any fish or other targets within the beam are said to be 'insonified' by the transmission, just as they would be 'illuminated' by a beam of light. The insonified targets scatter the transmitted energy, producing echoes which are subsequently detected by the receiving part of the instrument. The received signal will contain some information about the targets which generated the echoes, and the aim is to extract as much of this information as possible, by deciphering the signal as it were. This deciphering is known as the inverse scattering problem. It is relatively simple to calculate the signal generated by a known target (the scattering problem), but it is more difficult to work backwards, to say what targets might have caused the observed signal.

The information available to the observer depends upon the type of equipment in use. A simple echosounder with a single-beam transducer is capable of detecting the presence or absence of targets, but little else, whereas modern echo integrators, scanning sonars and the like can be much more revealing. Ideally, we would like to identify the detected targets, to say what species and sizes of animals are present in the water, and to measure quantities such as the biomass or the number of fish in the acoustic beam (Burczynski, 1979; Johannesson and Mitson, 1983). The more information that is required from the received signal, the more sophisticated the acoustic instrument has to be (Mitson, 1984a).

In this Chapter, we consider first what can be achieved using simple echosounders of the type commonly found on fishing vessels, and how to interpret the visual representation of targets as displayed on the echograms produced by such equipment. The technique of school counting is described, in which side-scan sonar is used to measure the dimensions of schools. Schools may also be counted using the vertical echosounder, and we show how the true size distribution of schools may be estimated from the observed sizes of marks on the echogram. We go on to discuss the measurement of fish abundance which relies upon more sophisticated techniques, intended primarily for applications in fisheries research and management (MacLennan and Forbes, 1987; MacLennan, 1990). The alternative techniques of echo counting and echo integration are compared. Their advantages and limitations are discussed with reference to practical applications. The theory of fish abundance estimation is presented, to the level necessary for the calculation of fish densities from acoustic measurements. In the case of echo integration, there is a simple relationship between the fish density and the backscattered energy, called the echo integrator equation. The theory depends upon a major assumption about the distribution of targets within the transducer beam, known as the linearity principle. We discuss the experimental evidence in support of linearity. We also consider the circumstances in which it might not apply, most importantly through the shadow effect which may occur in dense fish concentrations.

List of symbols

a_{rs}	Constant coefficients for school size calculation
$a_0(t)$	Exact TVG function at long range
A	Area of school horizontal cross-section
A_s	Area covered by the survey
A_1, A_2	Amplitude of a sine wave
$A(t)$	Actual TVG function of the echosounder
b	Signal due to a scatterer of unit strength
B	Width of the observation band
B_i	Coefficient derived from the beam-pattern
c	Speed of sound in water
C	Echo integrator calibration factor
d_i	Apparent area-density of fish in layer i
D	Distance travelled
e_i	Echo integral from one fish in isolation
E_i	Echo integral from one transmission
E	Echo energy; mean echo integral
$\langle E \rangle$	Ensemble average of E_i
$\bar{g}$	TVG correction factor

$g(R)$	TVG error at range R
$G(L)$	Frequency of schools of trace-length L or more
G_i	Observed number of schools in class i or larger
h	Thickness of a fish layer
H	School height
$H(R)$	Range-dependent echosounder gain-function
$H_e(R)$	Exact range-dependent function
k_i	Number of echoes in class i
L	Length of school parallel to the survey track
m	Size class of the largest observation
M	Number of layers in a depth interval
n_i	True number of schools or fish in class i
n, N	Number of transmissions, fish, etc.
N_i	Observed number of schools in size class i
$N(x)$	Expected frequency of trace length x
P	Proportion of surveyed area occupied by schools
Q	Quantity of fish in the surveyed area
R	Range from the transducer
R_1, R_2	Limits of the range interval
S	Total number of observed schools
t	Time
t_1, t_2	Limits of the time interval
T	Pulse duration of the transmitter output
TS	Target strength
u	Echo amplitude
$v(t)$	Echosounder output signal at time t
$v_1(t)$	Echosounder output with constant gain (no TVG)
$v(R,t)$	$v(t)$ due to one target at range R
V	Ship speed
$V(t)$	Transmitter output at time t
V_s	Total volume of schools
w	True width of school
w'	Width of trace on echogram
w_i	School size of class i
$w_E(u)$	Probability of echo amplitude u
$w_F(u)$	Probability function for backscattering by fish
$w_T(b)$	Probability of observing b from a unit scatterer
x, y	General variables
z	Depth, or a general variable
z_1, z_2	Limits of a depth interval
β	Acoustic absorption coefficient (in nepers m^{-1})
ε	Width of a size class
η	Phase difference
Ω	Solid angle

ϕ	Angular width of transducer beam
ψ	Equivalent beam angle of the transducer
ρ	Fish density (number per unit area)
ρ'	Fish density (number per unit volume)
$\langle\sigma\rangle$	Mean backscattering cross-section per fish
σ_e	Extinction cross-section of one fish
ω	Angular frequency

5.2 SIMPLE OBSERVATION METHODS

The single-beam echosounder is the simplest and cheapest of acoustic instruments. Nevertheless, it is capable of detecting any target that produces an echo above the background noise level. The detection of an echo indicates little more than the presence of some unidentified target, but one way to learn more is to view the signals received over a period of time or a number of transmissions. This is done by means of the echogram, which is a pictorial representation of the echosounder output. A typical example is shown in Fig. 5.1, which is a monochrome display, showing a pattern of dark marks on a light background.

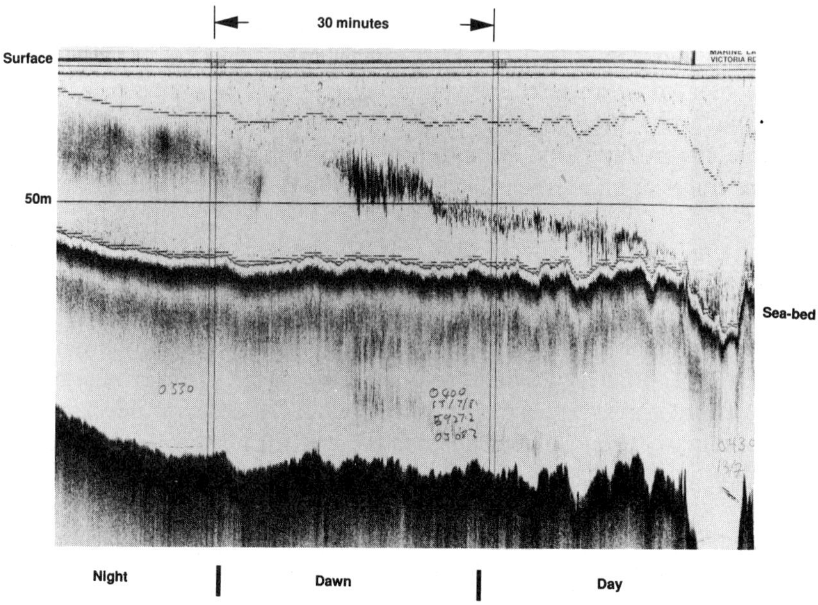

Fig. 5.1 Example of an echogram recorded on a ship which is moving at 10 knots in a straight line. The echogram represents a vertical cross-section of the water column below the ship and shows the changing behaviour of the fish layer between night and day.

Interpreting the echogram

Following each transmission, the receiver output is displayed as marks along a vertical line. The transmission signal appears at the top (Fig. 5.1). The lower marks indicate echoes; the heavier the mark, the stronger is the received signal. When the echo from a particular target arrives at the receiver, the corresponding mark appears at a distance below the transmission line which is proportional to the range of the target from the transducer. Of course, the echoes from many targets, fish in a school for example, may combine to produce strong marks extended in depth. The picture is built up by displaying the signals from each transmission along successive vertical lines which are separated horizontally by a small distance. If the transducer is stationary, the display represents a time series of observations of the same volume. In the case of Fig. 5.1, the signals are from a downward-looking echosounder on a ship which is travelling at constant speed in one direction. The display may then be considered as a vertical cross-section of the water column. The transducer is close to the sea surface which corresponds nearly enough to the line of transmission marks. The strong echo from the sea-bed is evident at the bottom of the figure. The marks in between are due to fish or other targets such as plankton which have produced detectable echoes. A sequence of contiguous marks from successive transmissions is called a trace. The important point is that by examining the traces displayed on the one echogram, we gain additional information about the targets and their distribution in space and time. Fish may be observed as scattered individuals, or clumped in schools or aggregations, which may themselves have characteristic features. In Fig. 5.1, the echogram covers the period through dawn. The fish are seen to be dispersed as diffuse traces at night, then they aggregate in denser concentrations with daylight.

Echograms may not provide sufficient information to identify the target species with certainty. Two species having similar gross anatomical features and occurring in similar school sizes could produce the same traces. In many cases, however, the size and shape of different traces on the echogram, and also their location in the water column, may suggest which species are present. To interpret the echogram in this way, the observer needs a good understanding of the local fish populations and their behaviour, especially as regards schooling. Even then, it is desirable to collect independent evidence (such as trawl samples) to be sure about the identification of targets.

Many echosounders, especially the older models, produce monochrome echograms which are seen as marks and background, perhaps with an intermediate shade evident in the case of weak echoes. The monochrome display has a limited range of contrasts as far as the human eye is concerned, and some of the information associated with the echo intensity will be lost. This limitation is overcome to a large extent when the echogram is displayed in colour (Plate 1). The signal intensity controls the colour. Red indicates very strong echoes, the yellow marks are less strong, and the blue ones are very weak. This is the idea behind the 'colour echosounder' in which the display is often shown on a cathode ray tube (CRT) without a permanent record. If a permanent record is required, a colour printer may be connected to the echosounder.

The underlying principles of colour and monochrome echograms are the same: they show how the acoustic reflectivity varies over a cross-section of the water column. The colour display provides a much better visual impression of the echo strength. However, for the particular task of comparing the size and shape of traces, when this is done by eye, it is easier to work with a monochrome display because the trace boundaries are more obvious.

Sonar-mapping

When the species being studied forms schools which are easily identified on the echogram, the technique of school counting can provide useful information about the fish distribution and density (Smith, 1970; Mais, 1977; Rusby, 1977; Bazigos, 1981). If the schools are close to the surface, good results can be obtained by using a side-scan sonar (p. 59), which is essentially a single-beam echosounder whose transducer is turned to point sideways, so that the beam is almost horizontal and perpendicular to the ship's track. Thus a wide area to the side of the ship can be surveyed. This method of school counting is sometimes referred to as sonar mapping (Hewitt *et al.*, 1976). It has been developed by Smith (1970, 1977) and others in the course of work on the northern anchovy, *Engraulis mordax*, in the California Current System.

Survey procedure

As the ship moves in a straight line, the signals received by the side-scan sonar are displayed on an echogram which represents a rectangular area to one side of the survey track. Schools may be seen anywhere in this area, but they are included in the count only when they appear entirely within the 'observation band' which is a range interval, R_1 to R_2 say (Fig. 5.2). This is done to avoid bias in the results. At long range the echoes can be distorted by inhomogeneities in the transmission path, or hidden in background noise, while at short range the beam may not be deep enough to detect all the schools (Fig. 5.3). The observation band should be chosen so that the number of detected schools per unit area is the same at all ranges, within the margin of statistical error.

Smith (1970) used a 30 kHz sonar with 10° vertical beam width. He observed schools most frequently in the range interval 250–500 m, while

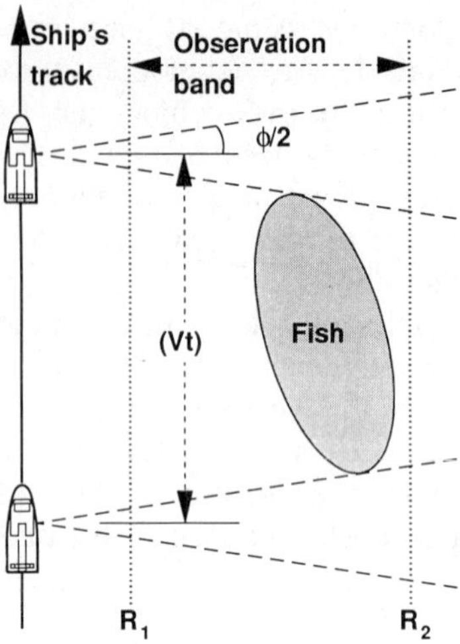

Fig. 5.2 Geometry of the side-scan sonar beam. Schools are observed within the observation band between ranges R_1 and R_2. (Vt) is the distance travelled while in contact with the school, and is more than the actual school length because of the finite beam width.

few were seen close to the ship and beyond 750 m. He therefore chose 200–450 m as the observation band within which the probability of school detection was assumed to be independent of the range.

School dimensions

The first step in the analysis is to decide which marks on the echogram are schools, then the dimensions of each mark are measured. The horizontal width (perpendicular to the ship's track) is determined directly from the echogram scale. The length (parallel to the track) is calculated from the distance travelled by the ship as it passes the school. Suppose V is the ship speed and t is the time between the first and last detections of the mark. (Vt) is the distance travelled, but the school is not quite so long. It is necessary

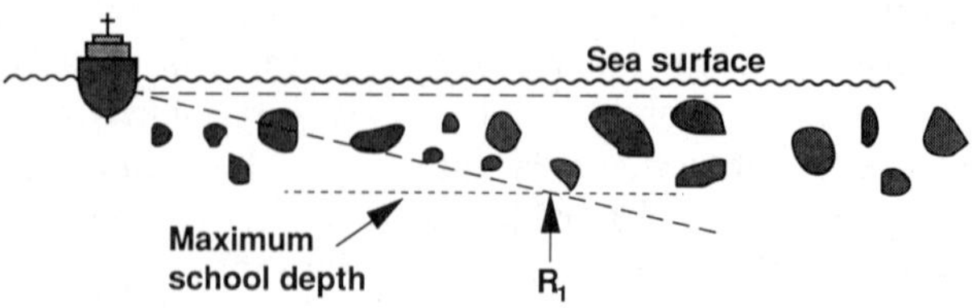

Fig. 5.3 Horizontal view in the direction of the survey track. The schools are in a near-surface layer whose depth determines the near edge of the observation band. At ranges less than R_1, some schools are outside the beam.

to take account of the finite size of the beam (Fig. 5.2). If R_a is the range to the far side of the school, and R_b is the range to the near side, the true length L is approximated by the formula

$$L = Vt - (R_a + R_b)\tan(\phi/2) \tag{5.1}$$

where ϕ is the horizontal beam width of the transducer. A real transducer does not have the well-defined shape of the ideal beam suggested by Fig. 5.2, but it is good enough for present purposes to assume that ϕ is the angle between the half-power points of the beam pattern, as quoted by the transducer manufacturer. For example, suppose that a school is observed for 20 s at ranges 250–350 m while the ship passes at 10 knots (5.13 m s^{-1}). If ϕ is 4°, the school length is $102.6 - 21.0 = 81.6$ m, about 20% less than the apparent length of the mark on the echogram.

The apparent width of the school should also be corrected to take account of T, the pulse duration. The correction is more important for small schools. If w' is the mark width as displayed on the echogram, and c is the sound speed, the true width is

$$w = w' - cT/2 \tag{5.2}$$

Size-distribution of schools

Observed schools may cover a wide range of sizes, and a particular school is likely to have an irregular shape. The width and length will be different. But we suppose that the long axis of the school lies in any direction with equal probability. This is a reasonable assumption unless the schools adopt a preferred orientation owing to migration behaviour or reactions to the survey vessel. If we assume that the orientation is random, the frequency of school numbers among size classes should be the same whether it is the widths or the lengths that are used to determine the distribution. If many schools were to be superimposed, the irregularities would tend to cancel one another, leaving a circular shape as the mean cross-section of the ensemble (this term is defined on p. 151). The expected area of a school of width w is $A = \pi(w/2)^2$. Thus w is the diameter of the average school. If the area of each school is required directly, then the formula for the area of an ellipse should be used, namely $A = \pi wL/4$.

The results of school-counting are normally presented as a histogram of the number by size class. Suppose N_i' schools are counted in size class i, and w_i is the width (or the diameter) at the class midpoint. It is necessary to correct the count for the edge effect, which reduces the apparent frequency of large schools. The problem arises because of the rule that schools are counted only when they are entirely within the observation band. If any part of a school is outside the band, the school is ignored. In Fig. 5.4, the area corresponding to the counted schools is within the dashed lines. The larger the school, the smaller is the effective observation

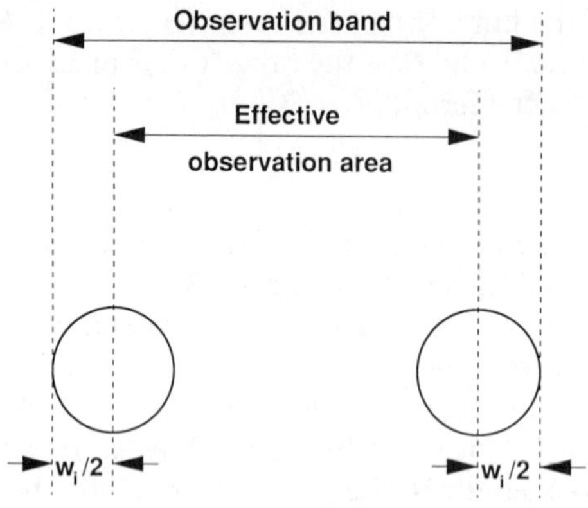

Fig. 5.4 Correction of the school-count for the edge effect. The effective observation area is narrower than the observation band by an amount equal to the school diameter, w_i.

area. If B is the width of the observation band, and N_i is the corrected count which refers to the same area irrespective of the size class, then

$$N_i = N_i' B / (B - w_i) \tag{5.3}$$

This formula depends upon the assumption that N_i' is large enough for the average school to be sensibly circular in shape. We suppose that sufficient schools have been measured to be reasonably certain of the mean and other statistics of w. Figure 5.5 shows the size distribution of schools of northern anchovy, as an example of the results which Smith (1970) obtained by school counting. The histogram is skewed to the left. Most of the schools are small, but a few are very large. The occasional large school which is the only one in a size class ($N_i' = 1$) should be examined

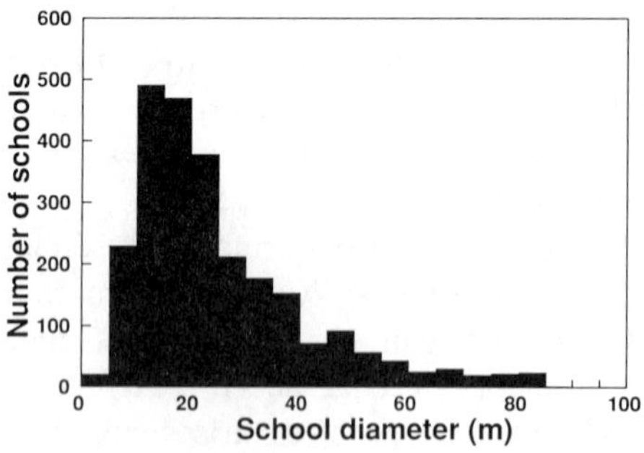

Fig. 5.5 The size distribution of schools of northern anchovy, *Engraulis mordax*, after Smith (1970), from side-scan sonar counts corrected for edge bias. The observations included 63 schools >85 m diameter (not shown) which occurred too infrequently for the edge bias to be corrected reliably.

individually to take account of the irregular shape in calculating the school volume or biomass.

Abundance estimation

The side-scan sonar records give a qualitative indication of the location of fish concentrations and the structure of schools. Beyond that, another objective might be to estimate the abundance of the target species. If D is the total distance travelled by the ship, the proportion of the surveyed area occupied by schools is

$$P = \sum_i N_i \pi w_i^2 / (4BD) \tag{5.4}$$

To estimate the abundance, it is necessary to know the quantity of fish in the typical school. There are several approaches to this problem: (1) echo integration of the side-scan signal; (2) commercial catch data, e.g. from a purse seine fishery in which complete schools are captured; and (3) knowledge of fish behaviour which may indicate the packing density, within the schools (Breder, 1976; Cushing, 1977; Love, 1981). Blaxter and Hunter (1982) have reviewed field observations of the packing density which Pitcher and Partridge (1979) suggest is typically one fish per cubic body length. The side-scan sonar gives no information on the school height, which is required to calculate the volume, but the vertical size may be inferred from the horizontal cross-section if the fish are known to form schools with a characteristic three-dimensional shape (Squire, 1978). The area density of fish in schools, ρ, may be deduced from one or more of these sources of information. If A_s is the area covered by the survey, and the school-count in the observation band is considered to be representative of the whole area, then the total quantity of fish is estimated as

$$Q = \rho A_s P \tag{5.5}$$

Shore-based sonar

As an alternative to the mobile sonar carried on a ship, the equipment may be installed at a fixed location near to a base on land, and fish may be observed as they move through the sonar beam. This technique has been applied successfully to fish monitoring in freshwater lakes and dams (Gaudet, 1990; Mesiar *et al.*, 1990). The fixed installation offers the possibility of continuous observation over a long period – months or even years. Furthermore, large and heavy transducers, which would be impractical on a ship but which are capable of high powers and low frequencies, can be deployed to detect targets at distances of many miles.

Weston and Andrews (1990a,b) have described experiments with such a sonar off the Cornish coast. The transmissions were directed horizontally to seaward, and the returning echoes were received by a hydrophone array 18 m long. The operating frequency was 1 kHz at first and later changed to

2 kHz, giving horizontal beam widths of 4° and 2° respectively. Fish traces could be distinguished from the background reverberation by the change of their range with time as the fish migrated or drifted back and forth with the tide. Useful records were obtained at ranges up to 40 km or so from the transducer, depending on the propagation conditions and disturbance by the wind. The observations continued (except for short intervals) over a period of 5 years, giving a remarkable long-term record of fish concentrations in the area. Weston and Andrews showed that schools could be counted, although the size of individual schools could not be estimated reliably. Nevertheless, the school count (assumed to be *Sardina pilchardus*) showed a strong seasonal dependence which correlated with the catch rate of the local fishery. The numerical school density was greatest in the summer, typically three schools per km^2. The installation was also used to study the schooling behaviour, particularly the dispersal and re-forming of the schools at dusk and dawn respectively. Various characteristic behaviours, noted at these changeover times, appeared to be motivated by the season as well as the ambient light conditions.

Echosounder-mapping

The side-scan technique is only applicable to fish which school in a well-defined layer close to the surface. In the case of fish which are more widely distributed in the water column, it is possible to count schools using the conventional downward-looking echosounder (S. Olsen, 1969; Forbes and Nakken, 1972). This can be done with very basic equipment – a cheap echosounder, a ship or boat to move over the area being surveyed, and some means of estimating the speed of travel. We call this technique echosounder mapping. It employs the same principles as sonar mapping, except that targets are viewed in a vertical instead of a horizontal plane. Usually the echogram is recorded on paper, so that the school marks can be measured at any time and previous measurements may be checked. The schools could be counted as they appear on the CRT display of a colour echosounder, but this practice is not recommended. It is much better to work with a permanent record on paper.

In echosounder mapping, the marks on the echogram represent vertical cross-sections of the schools. The height and length of each cross-section are derived in a manner similar to that described on p. 131 by correcting the observed mark dimensions for the effect of beam width and pulse duration.

The dimensions of schools can be measured directly using multi-beam sonar or other instruments which are sensitive to the direction as well as the range of targets (Misund and Beltestad, 1989; Chapter 3). Direct measurement of school size will be more accurate than the statistical treatment which is necessary in the case of measurements made with the simple echosounder.

Size-distribution of schools

In the case of large schools, most of the fish will be to the side of the survey track, beyond the transducer beam. Only a small proportion of the fish will have contributed to the observed traces, and the histogram of trace–length frequencies will not be an accurate indication of the true size distribution. The root of the problem is that the survey track crosses each school along a line which lies at an unknown distance from the centre of the school. The track is unlikely to cross the middle of the school. Consequently, the length of the trace will be less than the school diameter, supposing for the present that the schools are circular in the horizontal cross-section.

The true size distribution may be expressed in terms of the frequencies of school diameters by size class. Although the diameters have not been measured directly, it is possible to convert the observed trace–length histogram to one of true diameters, by the method described in Appendix C. An example of the results which can be obtained from this approach is shown in Fig. 5.6. However, the calculated histogram of school diameters

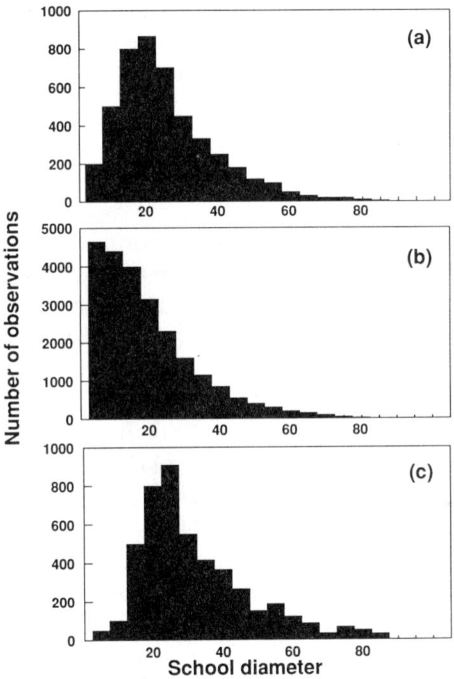

Fig. 5.6 Estimation of the true distribution of school sizes from measurements of echogram traces. Example with hypothetical data. (a) Observed number of traces by length class corrected for the beam width; (b) cumulative distribution of trace lengths; number greater than the indicated diameter; (c) estimated number of schools by diameter class, calculated by the method in Appendix C.

will generally be a crude representation of the true size distribution. The method assumes that the beam width is small compared to the school diameter, so it is not applicable to very small schools. The results for the largest schools may also be anomalous since they are derived from a small number of observations, making an accurate estimate of the size impossible. Nevertheless, the frequencies of schools in the middle size classes can be predicted accurately, provided many schools have been detected within each size class. When the occasional large school is encountered, the best plan is to map the true dimensions directly, by crossing the school several times from different directions. Furthermore, the largest schools are most likely to have an irregular shape, in which case the simple theory based on circular cross-sections is not applicable.

Abundance estimation

It may be assumed that the survey track is equally likely to cross any part of the school. If many schools are observed and the mean trace length is $\bar{L}$, it can be shown that on average the school diameter is $(4/\pi)\,\bar{L}$. Furthermore, the average school area is $(3\pi/8)\,\bar{L}^2$. These results are explained in Appendix C. If H_r is the height of the rth school as measured from the echogram, assumed to be representative of the entire school and not just the observed part, and L_r is the corresponding trace length, an unbiased estimate of the volume V_s of all the observed schools is

$$V_s = (3\pi/8)\sum_r L_r^2 H_r \qquad (5.6)$$

The quantity of fish is immediately obtained from V_s if the packing density of the schools is known, as described for sonar mapping. When the area to be surveyed is large, many schools remote from the cruise track will not be detected. To estimate the total abundance, we need to know the area which has been sampled. This area may be estimated as the mean school diameter times the length of the cruise track; suppose it is a proportion P of the whole area being surveyed. The total abundance is then estimated as V_s/P.

An unbiased measure of the fish abundance can also be obtained by noting the proportion of the cruise track over which schools are detected. For example, if the distance covered while schools are observed is 5% of the track, it may be assumed that regardless of the school shape or size, 5% of the surveyed area is occupied by similar schools. In the case of irregularly shaped schools, this technique should be more accurate than the alternative based on Equation 5.6 which depends upon the assumption that the cross-section of the average school is circular.

The mapping and school counting techniques work best when the stock consists of many distinct schools which are not too different in size and shape. The techniques determine the volume occupied by the schools, from which the abundance can be deduced if the packing density is known from other sources. However, much better results can be obtained by methods which measure the fish density directly by acoustic techniques, as described in the following sections.

5.3 ECHO COUNTING

In the case of fish which are sparsely distributed in the water, as opposed to being clumped in schools or layers, it may be possible to detect the echoes from individual fish as first observed by Trout *et al.* (1952). The count of these echoes might be used to determine the density of fish within the acoustic beam. This was first attempted by Midttun and Saetersdal (1957), although they did not determine the sampled volume, that within which targets are counted. The sampled volume depends upon the size of the target (p. 141). Thus the analysis of the echo-count to determine the fish density must take account of the size distribution of the surveyed population. Several theoretical approaches to this problem have been developed over the years (e.g. Craig and Forbes, 1969; Ehrenberg, 1972; Kieser and Mulligan, 1984; review, Kieser and Ehrenberg, 1990).

The amplitude of the echoes can be measured to provide information about the fish size distribution. In general, larger fish give bigger echoes, but the relationship between fish size and echo amplitude is not simple. Suppose that a number of echoes from the same fish are measured. The echo amplitude will depend upon the position of the fish within the acoustic beam and upon the beam pattern of the transducer. But even if the effect of the beam pattern is removed, which can be done directly using an echosounder of the split-beam or dual-beam type, or indirectly from the echo statistics (p. 142), the observed echo amplitudes will very likely cover a range of values. The reasons for this variation are discussed more fully in Chapter 6. For the present, the important point is that the echo amplitude from a given fish must be considered as a stochastic parameter, one that can have a range of values described by a probability distribution. The same is true of the acoustic cross-section and the target strength, since these parameters are defined in terms of the echo amplitude.

The observed echo amplitude distribution is further complicated when the echoes have come from a range of fish sizes, and more than one species may be present. Moreover, in the case of the single-beam echosounder, the target direction is unknown, so the beam pattern also contributes to the observed variation.

Single-target echoes

The echo-counter has to distinguish those echoes coming from isolated fish, which we call single-fish echoes, and those which result from two or more fish being detected concurrently. Before proceeding, it is necessary to define what constitutes an echo. Any part of the received signal whose amplitude is always above a pre-determined threshold is considered to be an echo from one or more targets (Fig. 5.7). Signals below the threshold are ignored. The reason for applying the threshold is to reject noise and signals from very small targets which are of no interest.

It is only the single-fish echoes which are useful in echo counting. Fortunately, it is usually possible to determine whether the echo has come from one or more fish from the shape of the echo envelope, which depends upon the number of targets contributing to the received signal. The original pulse generated by thc transmitter may have had an ideal rectangular envelope (Fig. 5.8(a)), but the echo envelope has a more rounded shape owing to the electronic filters in the receiver and the frequency dependence of the transducer sensitivity.

In the case of the echo from an isolated fish, the envelope might look like the example shown in Fig. 5.8(b). The precise shape will depend upon the characteristics of the echosounder, notably the bandwidth of the receiver. However, when two or more fish are contributing to the received signal, the echo may be distorted by two effects. Suppose T is the duration of the transmitter pulse. Firstly, if there are targets at different ranges within a pulse length, the echo duration may be much more than T (Fig. 5.8(c)). Secondly, in the case of targets at nearly the same range but in different directions, the individual target echoes may interfere, causing the envelope to fluctuate, as in Fig. 5.8(d).

One criterion is to accept only those echoes whose duration is within some interval around T. For example, it might be decided to reject echoes longer than $1.5T$. This would eliminate the echoes from diffuse targets or aggregations spread over one-quarter of the pulse length or more in range. Other criteria are necessary to eliminate echoes from multiple targets at nearly the same range. The decision may be based on measurement of the

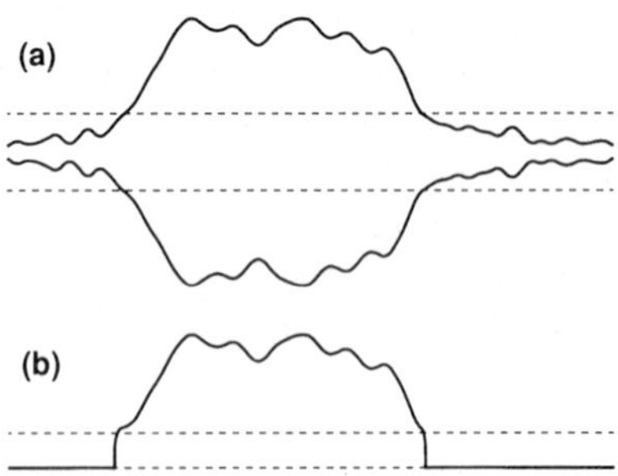

Fig. 5.7 Echo detection. (a) The received signal envelope; dashed lines indicate the threshold which is applied to reject noise, so (b) the detected echo is that part of the signal whose amplitude is above the threshold level.

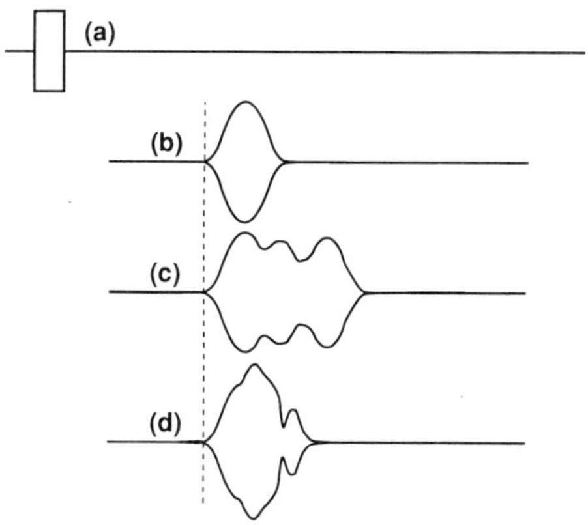

Fig. 5.8 Examples of signal envelopes. (a) The transmission pulse; (b) echo from an isolated fish; (c) overlapping echoes from targets at different ranges; (d) several targets at nearly the same range.

echo rise-time or the phase coherence within the echo, factors which will remain within known limits for single-fish echoes.

These simple techniques for detecting single-fish echoes are not perfect. Some of the detected echoes may not be from isolated fish, for example if two fish were close together but far from other neighbours. However, the main requirement is to ensure that the amplitude distribution of the detected echoes is representative of the echoes from isolated fish, and that the error rate is low enough not to distort the statistics of the distribution.

The best conditions for echo counting occur when the fish are randomly distributed at a low average density. The worst conditions occur when the fish are in dense schools or layers. It is possible to determine whether the fish distribution is suitable for echo counting, by examining certain features of the received signal such as the distribution of peak amplitudes (Spindel and McElroy, 1973). According to Stanton (1985a), there are fewer peaks at small amplitudes when the signal is predominantly multiple-fish echoes (Fig. 5.9). Practical applications of this effect have been discussed by Stanton (1985b) and Stanton and Clay (1986).

Range compensation

The echo amplitude decreases with the range of the target due to beam spreading and absorption as the sound waves propagate through the water. However, the receiver includes a time-varied-gain (TVG) amplifier whose purpose is to compensate the output signal for these propagation losses. The gain is increased in proportion to the TVG function $a(t)$, where t is the time from the start of the transmitter pulse (Fig. 5.10).

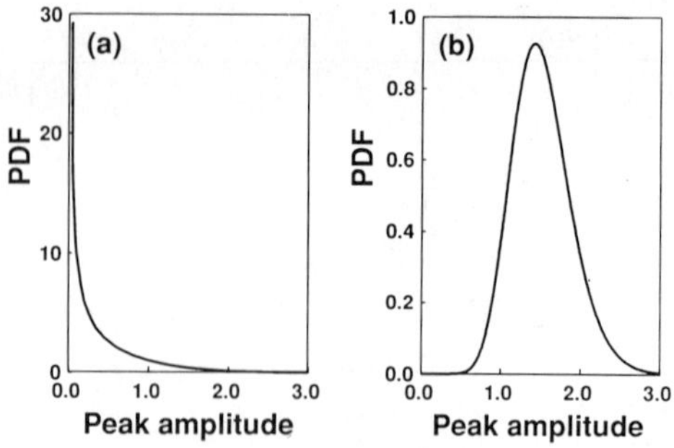

Fig. 5.9 Probability distributions of peak amplitudes in the signal received from (a) isolated targets; (b) multiple targets with overlapping echoes. PDF = probability density function. Redrawn from Stanton (1985a).

In the case of one isolated target whose range is much greater than the pulse length, the appropriate TVG function is

$$a_0(t) = (ct)^2 \exp(\beta ct/2) \tag{5.7}$$

where c is the sound speed and β is the absorption coefficient. This function is commonly known as '40 log R' TVG. The range dependence is obtained by substituting the range $R = ct/2$ as an approximation, and calculating $20 \log[a_0(t)]$, which is the TVG expressed in decibels.

The spreading losses increase with the range, as does the time delay until the echo is received. The echo amplitude is multiplied by the TVG function and the result should be an output signal which does not depend upon the target range. However, $a_0(t)$ is an approximation to the exact function $a(t)$, which has been derived by MacLennan (1987). The exact TVG function is

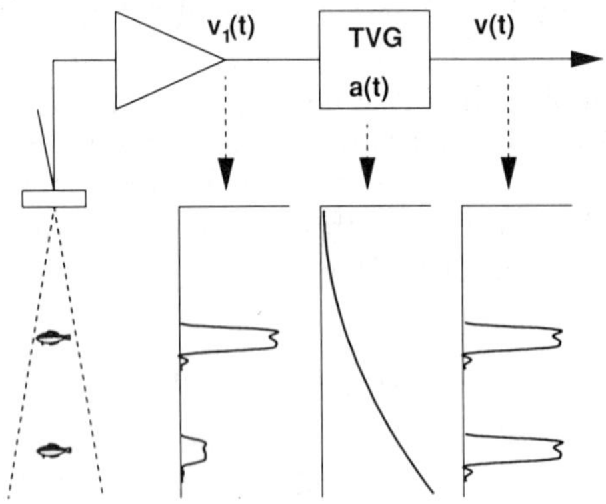

Fig. 5.10 Range compensation by TVG, single-target case. $v_1(t)$ is the uncompensated signal proportional to the echo amplitude. The receiver gain $a(t)$ increases with time so that the output $v(t) = a(t)\, v_1(t)$ is independent of range for similar targets. This is '40 log R' TVG.

only required for targets at ranges of less than a few pulse lengths. $a_0(t)$ is a good enough approximation for the ranges at which most fish targets will be observed, since the echoes are received at times $t \gg T$.

Single-beam echosounders

The count of single-fish echoes is not useful by itself, even when the range dependence has been removed, because the sensitivity of the single-beam transducer depends upon the direction of the target relative to the acoustic axis. The received signal provides no intrinsic information about the target direction. The same signal might result from a small fish on the acoustic axis or a large one near the edge of the main lobe. It is usual to accept only those signals which exceed some threshold (p. 138). The threshold is the same for targets in any direction, so the sampled volume (that within which all fish are counted) depends upon the target strength of the observed fish. The larger the fish, the greater is the portion of the beam that can 'see' it (Fig. 5.11). An important problem is how to determine the fish density from the count and other measurements of the received signal. The solution is not

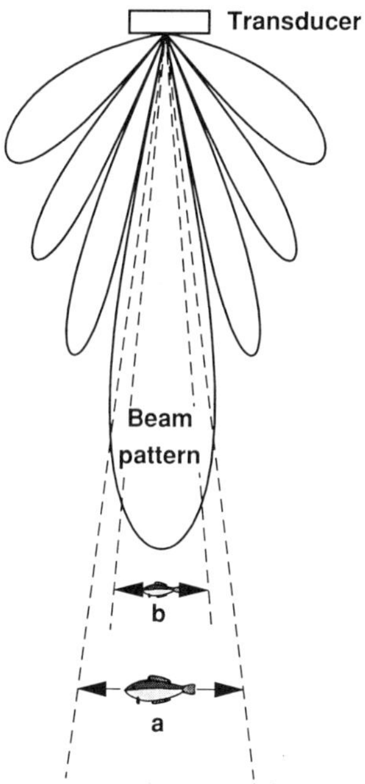

Fig. 5.11 Dependence of the sampled volume on the target size. The beam pattern is shown as a polar plot of the sensitivity against direction. Echoes above a receiver threshold are detected. Large fish (a) are detected over a greater volume than small fish (b).

immediately obvious because the echo-count relates to some unknown average of the sampled volumes.

The first step is to determine the distribution of target strengths among the observed fish. Since the sampled volume is a known function of σ , the fish density (and its dependence on σ) then follows immediately. The target strength is defined in terms of the energy in the echo produced by the target (Chapter 6). The energy is the integral of the square of the echo amplitude, and it is the appropriate measure of the size of echoes in the present context (section 2.2).

Indirect target strength estimation

It is important to recognize that the target strength of a fish depends upon many factors apart from the size. In particular, the echo energy is likely to fluctuate between wide limits as the orientation of the fish changes. We return to this problem in Chapter 6. For the present, the key point is that the target strength must be considered a stochastic parameter. To estimate the fish density from the acoustic measurements, one first has to determine the probability distribution of the target strength. In the case of single-beam echosounders, the echo energy also varies due to the random position of the target in the beam. This effect has to be removed to reveal the residual variation due to the distribution of target strengths among the observed fish. The target strength distribution can be derived from statistics of the observed echo energies, as first suggested by Raitt (1948). This is the so-called 'indirect' method of which there are several variants. Those of Craig and Forbes (1969), Ehrenberg (1972) and Clay (1983) are described in the following sections. However, the indirect methods all have the disadvantage that many echoes have to be measured to yield satisfactory results. The problem arises from the assumed random distribution of targets. The actual number of fish in a given volume is subject to statistical fluctuation, so the observed echo energy includes stochastic errors. The solution is sensitive to these errors. Anomalous results can occur if the data-set is not large enough, such as negative frequencies being predicted for certain target strengths.

Craig–Forbes algorithm

The method of Craig and Forbes (1969) is an early approach to the problem of target strength estimation, but it is still popular in practice. An echosounder is used to measure the energies of N echoes which are counted in m consecutive classes (e_i to e_{i+1}, for $i=1, 2 \ldots m$), and '40 log R' time-varied gain is applied to remove the range dependence. e_1 corresponds to the smallest detectable echo, so any smaller than e_1 are excluded from the count. No echoes are detected larger than e_{m+1}. Suppose that k_i echoes are counted in the ith class. The $N = \Sigma k_i$ echoes have come from fish of various target strengths located in different parts of the beam. The echo energy varies due to the change in sensitivity across the beam as well as the stochastic target strength. It is assumed that the target strength and the position of the fish are uncorrelated.

Next, we consider the cross-section of the beam to be divided into regions of different sensitivity (Fig. 5.12). The central region Z_1 is the most sensitive, Z_2 is the second most sensitive and so on. Suppose there are n_j fish per unit volume with target strengths in the interval TS_j to TS_{j+1}. TS_j is defined as the target strength of a fish which produces an echo energy between e_j and e_{j+1} when it is in Z_1. The k_m echoes in the largest energy class have been produced exclusively by the strongest targets ($TS_m \leqslant TS < TS_{m+1}$) in Z_1. However, the k_{m-1} echoes in the second largest energy class comprise the strongest targets in Z_2 and the second strongest class of target in Z_1. It follows that none of the k_i echoes in the ith energy class can be associated with target strengths below TS_i.

It is assumed that for all j, the n_j fish are randomly distributed over the beam cross-section. The transducer sensitivity and its dependence on the

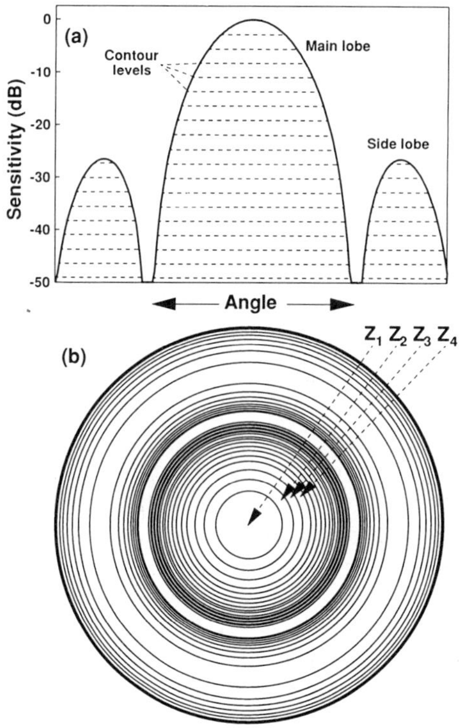

Fig. 5.12 Cross-section of an acoustic beam divided into regions by contours of equal sensitivity; (a) sensitivity v. angle from the acoustic axis; (b) the central region Z_1 is the most sensitive. In this hypothetical example, the beam is symmetrical and the contours are circles.

target direction (i.e. the beam pattern) must be known. The sensitivity factors B_r are calculated from the beam pattern. If R_1 and R_2 are the range limits between which echoes are counted, and Ω_r is the solid angle of directions in region Z_r, then

$$B_r = \Omega_r (R_1^3 - R_2^3) / 3 \tag{5.8}$$

The k_i may now be expressed as a linear combination of the n_j, ignoring those targets which are too weak to produce the required echo energy.

$$k_i = B_i n_i + B_{i+1} n_{i+1} + \cdots + B_m n_m \tag{5.9}$$

There are m such equations, obtained by substituting $i = 1$ to m, and they are easily solved for the n_j to determine the target-strength distribution.

Ehrenberg's method

Ehrenberg (1972) has derived an integral equation which describes the echo-energy distribution in terms of probability functions for the target strength and the location of targets. Unfortunately, numerical computation of the target strength distribution is likely to be unsatisfactory because of ill-conditioned simultaneous equations. The indirect methods all suffer from this problem. Various techniques have been suggested to overcome the difficulty, such as the polynomial-fitting technique of Robinson (1982) which smoothes the observed echo energy distribution, but with limited success.

Deconvolution technique

Clay (1983) has proposed a different approach (see also Stanton and Clay, 1986; Jacobson *et al.*, 1990). If $w_E(u)$ is the probability of observing the echo-amplitude u, then

$$w_E(u) = \int_0^1 w_T(b)\, w_F(u/b)\, (1/b)\, db \tag{5.10}$$

where $w_T(b)$ is the probability of observing b, the signal from a scatterer of unit strength; $w_T(b)$ may be deduced from the beam pattern of the transducer, which is assumed to be known. Equation 5.10 now has to be solved for w_F, the probability function which describes the scattering properties of the fish. $w_F(u)$ is equivalent to the target strength distribution. Clay applied the substitutions $b \to \exp(-x)$ and $u \to u_0 \exp(-y)$, leading to the simpler expression

$$w_E(y) = \int_0^\infty w_T(x)\, w_F(y - x)\, dx \tag{5.11}$$

which in mathematical language is called a convolution integral. It may be solved (or deconvolved) numerically to determine w_F. Algorithms for this purpose will be found in Clay (1983). According to Rudstam *et al.* (1988), the deconvolution technique for removing the beam-pattern effect has certain theoretical advantages over the traditional Craig–Forbes method, although the latter offers greater computational simplicity.

Direction-sensing echosounders

The precision of the target strength measurement can be improved by using instruments which determine the direction as well as the range of the target. This allows the signal to be compensated for the beam pattern. One target strength value is obtained directly from each echo that is identified as coming from a single target, hence the term 'direct method'. However, the problem of how to identify the single-fish echo is unchanged, and the target strength must still be considered as a stochastic parameter to be described by a distribution of values. The same rules for rejecting echoes from multiple targets may be applied in both the direct and indirect methods.

Two kinds of direction-sensitive equipment have been developed for echo counting, the dual-beam and split-beam echosounders. The working principles of these instruments have been described in Chapter 3. In addition to the direct measurement of target strength, the instruments allow the sampled volume to be fixed at a constant value. An 'acceptance cone' may be defined, and echoes from directions outside this cone are rejected, so the volume from which echoes are recorded does not depend upon the target size.

Ehrenberg (1983) has compared the theoretical performance of split-beam and dual-beam echosounders. While the split-beam technique is more difficult to implement, it is superior in the presence of noise. In practical terms there is probably little to choose between the two, and both techniques have given satisfactory results in experiments at sea (e.g. Traynor and Williamson, 1983; Foote, 1987).

Threshold problem

The use of a signal threshold to reject noise is an additional complication. Traditional echo counting methods suffer from a threshold-related bias (Weimer and Ehrenberg, 1975). The bias increases with the spread of the target strength distribution. Kieser and Ehrenberg (1990) have shown that satisfactory results can be achieved with traditional methods, although it is necessary to know the target strength distribution and to use a calibrated echosounder to compensate for the bias. Another approach is the duration-in-beam technique (Crittenden *et al.*, 1988; Thorne, 1988). If the fish density is sufficiently low, successive echoes known to be from the same target may be observed, and the transit time required for each fish to traverse the beam can be measured. Weaker targets will have shorter transit times since they are only detected in the central part of the beam. If the fish speed relative to the transducer is known, the sampled volume may be estimated from the distribution of transit times. The duration-in-beam technique does not require prior knowledge of the target strength.

Although the threshold problem can be overcome in theory, echo counting works best with large fish, which are strong scatterers, so that few echoes are close to the threshold level.

Applications

Echo counting as originally conceived was a method for estimating the abundance of fish. For this purpose it is necessary to know firstly the number of insonified fish, and secondly the sampled volume of water which contains these fish. The ratio gives the fish density, and the total abundance is estimated from the mean density multiplied by the volume of water in the area of interest.

However, the echoes being counted are only those from fish well enough separated from neighbours to avoid overlapping echoes, and this is an important limitation of echo counting. The technique is useful in the case of fish which are distributed randomly and whose mean density is low. If the statistics of the fish distribution are known (a Poisson model will be appropriate for a strictly random distribution), then it is possible to estimate the proportion of fish which are sufficiently isolated to be counted. The recorded count may then be corrected, to include those fish which are not isolated. The correction will be small when a large proportion of the fish are separated (a) vertically by more than half the pulse length and (b) horizontally by more than the width of the main lobe of the beam pattern. The width of the lobe cross-section increases with distance from the transducer. Hence the critical fish density, above which echo-counting will not give satisfactory results, is lower for targets further from the transducer. This effect limits the depth which can be surveyed quantitatively by echo counting. For example, consider an echosounder with 10° beam width and 1 ms pulse duration. Assuming the sound speed is $1500\,\mathrm{m\,s^{-1}}$, it is possible to resolve echoes from fish which are separated by 0.75 m in range or, at 10 m from the transducer, by 1.7 m across the beam. The critical fish density would then be of the order of $0.5\,\mathrm{m^{-3}}$, decreasing to $0.02\,\mathrm{m^{-3}}$ at 50 m from the transducer.

Notwithstanding these problems, many successful applications of echo counting have been reported, mainly from work on freshwater fish and migratory salmonids, in lakes and rivers (Lindem, 1981; Thorne, 1979; Thorne and Thomas, 1984). In the case of rivers, the normal technique of surveying an area with mobile equipment is less useful. An alternative approach is to install one or more transducers in fixed positions so that the passage of fish may be monitored as they migrate up or downstream

(Gaudet, 1990; Mesiar *et al.*, 1990). The problem of single-fish detection is somewhat different in this case, since the main source of unwanted echoes is the passage of inanimate objects, such as logs or lumps of ice. Johnston and Hopelain (1990) have described a method for discriminating fish moving upstream from other targets, based on the Doppler effect, which causes the frequency of the echo to change with the velocity of the target relative to the transducer (p. 65).

Satisfactory conditions for abundance estimation by echo counting are found more often in fresh water than in the sea. However, the echo counting technique can also be used to measure the target strength of fish in the wild, and for this purpose it is applicable to schooling marine species. The measurements are limited to individual fish detected on the edges of dense concentrations. It does not matter that the fish contributing to the observed target strength distribution are a small sample from the population, provided that the sample is unbiased.

5.4 ECHO INTEGRATION

In the case of fish which aggregate in schools or layers, the density is usually too high for echo counting to give reliable estimates of fish abundance. The alternative technique of echo integration, first proposed by Dragesund and Olsen (1965), has proved to be more generally applicable as a means of estimating the quantity of fish or other scatterers in the acoustic beam, whether or not the received signal contains overlapping echoes. The theoretical ideas behind echo integration developed over some 20 years from the mid 1960s, notably through the work of Thorne (1971), Moose and Ehrenberg (1971), Ehrenberg (1973, 1974a), Bodholt (1977), Foote (1981a, 1983) and Aksland (1986).

The echo integrator is simply an echosounder whose output is connected to a device which accumulates the energy in the received signal (Fig. 5.13). If $v(t)$ is the voltage produced by the echosounder at time t after the transmit pulse, the energy is the integral of the squared amplitude of $v(t)$ with respect to time. Thus the echo integrator output due to one transmission is

$$E_i = \int_{t_1}^{t_2} |v(t)|^2 \,\mathrm{d}t \tag{5.12}$$

and E_i is called the echo integral. In practice, the echo integrator accumulates the echo integrals from many transmissions. The fish density is calculated from the mean echo integral, for which we use the symbol $\langle E \rangle$ to denote the ensemble average (p. 151). If there are N transmissions, then

$$\langle E \rangle = \sum_{i=1}^{N} E_i/N \tag{5.13}$$

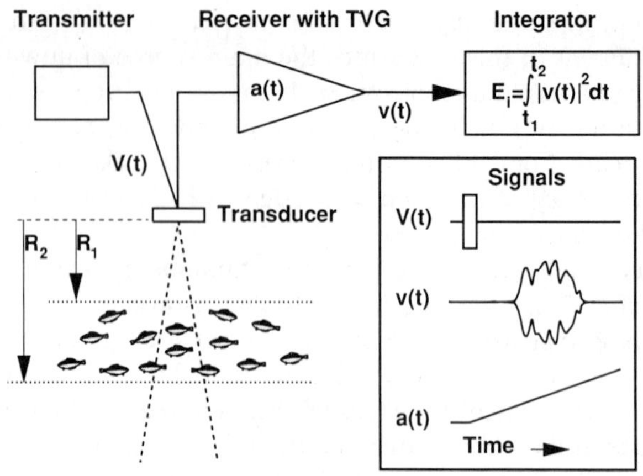

Fig. 5.13 Principles of echo integration. Signals from targets in the preset range-channel R_1 to R_2 contribute to the integrator output E_i. The integration is performed between the times $t_1 = 2R_1/c$ and $t_2 = 2R_2/c$ after the beginning of $V(t)$, the transmitter pulse. $a(t)$ is the TVG function and $v(t)$ is the receiver output.

In the following discussion, $\langle E \rangle$ is abbreviated to E. The time gate t_1 to t_2 is chosen to correspond to the depth channel which is to be sampled. If $(t_2 - t_1)$ is much longer than the pulse duration, almost all the targets contributing to E will be in the channel bounded by the depths $ct_1/2$ and $ct_2/2$. We shall now consider the relationship between E and the quantity of insonified targets in the depth-channel. A simple version of the theory is presented here; for a more rigorous treatment of the statistical theory of echo integration, the reader is referred to the classic dissertation on the subject by Ehrenberg (1973).

Range compensation

The TVG function appropriate to a single target, 40 log R, has already been described. In the case of many targets which are distributed randomly over the beam, whose mean density is to be derived by echo integration, a different function is required. Suppose the targets are fish in a thin layer extending sideways well beyond the beam (Fig. 5.14). As the range of the layer increases, so does the number of fish in the acoustic beam. The larger number of targets compensates the transmission losses to some extent. We want the signal from the layer to be independent of the range, assuming the fish density is constant. To achieve this in the case of randomly distributed targets, if the range is much greater than the pulse length, the approximate TVG function is

$$a_0(t) = (ct)\exp(\beta ct/2) \tag{5.14}$$

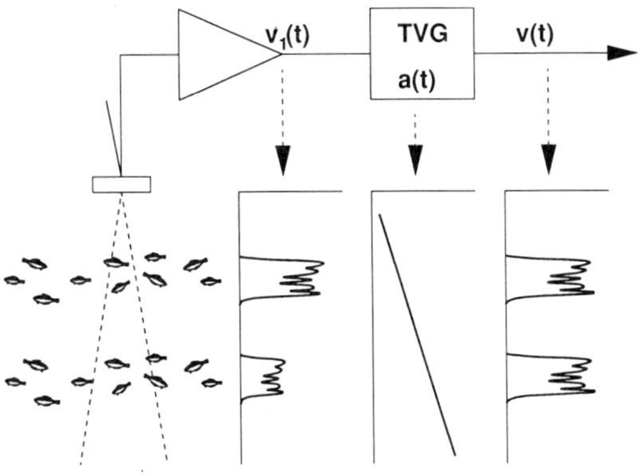

Fig. 5.14 Range compensation by TVG, in the case of targets distributed in layers. $v_1(t)$ is the uncompensated signal proportional to the echo amplitude. The receiver gain, $a(t)$, increases with time so that the output $v(t)=a(t)\ v_1(t)$ is independent of range for similar layers. This is '20 log R' TVG.

which is the same as Equation 5.7 but without the square in the (ct) term. This function is known as '20 log R' TVG. The range dependence is obtained by substituting $R=ct/2$ and expressing the result in decibels, as before. Again, $a_0(t)$ is a good enough approximation at the ranges of most fish targets. The exact functions may be required for targets close to the transducer, in particular the standard target which is suspended below the transducer when calibrating the equipment (Chapter 3).

When echo-counting, 40 log R is the correct type of TVG; when echo-integrating, 20 log R is always required and the controls on the echo-sounder should be set accordingly. It does not matter that the population being integrated may be so sparse that only one fish at a time is in the beam. The 20 log R function is still appropriate. The echo integrator accumulates the integrals from many transmissions, and therefore from many fish which are detected at random positions throughout the beam.

Echo integrator equation

The integration is performed between times t_1 and t_2, corresponding to the depth channel R_1 to R_2 as in Fig. 5.13, and E is the average integrator output over many transmissions. While the integration is in progress, the transducer may be on a moving ship and the fish are also mobile. Thus many different fish will contribute to E. If the number of detected fish is very large, and if they are randomly distributed over the beam cross-section, then E is proportional to the density ρ , the number of fish per unit area of the depth channel. This is the primary assumption of echo integration theory.

The relationship may be written in the form

$$\rho = [C\bar{g}/(\psi\langle\sigma\rangle)]E \qquad (5.15)$$

This is known as the echo integrator equation (MacLennan, 1990). E is the measurement produced by the echo integrator. To calculate ρ, we need to measure or to estimate the four terms within the square brackets. $\langle\sigma\rangle$ is the expected value of the acoustic cross-section which describes the scattering properties of the insonified fish. This is a major topic in its own right, and we shall postpone further discussion of $\langle\sigma\rangle$ until Chapter 6. C is a calibration factor which depends on the sensitivity of the transducer and the electronic components of the echo integrator. ψ is the equivalent beam angle, a measure of the beam width (section 2.3), and $\bar{g}$ is the TVG correction factor. The measurements of C, $\bar{g}$ and ψ come from the equipment calibration (Chapter 3).

The equivalent beam angle has been defined in section 2.3. It may be visualized as the solid angle at the apex of the ideal conical beam which would produce the same echo integral as the real beam from a large number of randomly distributed targets (Fig. 5.15). In this context, an ideal beam is one whose sensitivity is constant inside a cone, and zero outside. $\bar{g}$ is the average TVG error over the depth channel. It arises because the actual TVG applied to the signal, as generated by the receiver, is likely to differ from the function required for exact range compensation. $\bar{g}$ is determined from electrical measurements of the actual TVG function, and the waveform of

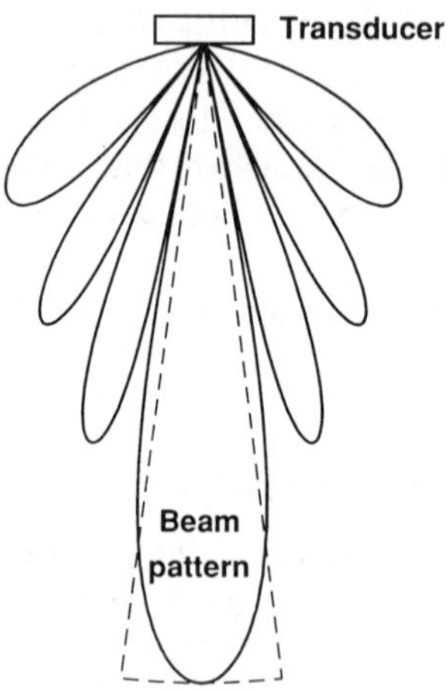

Fig. 5.15 Section through a transducer beam pattern (solid curve) and the equivalent ideal conical beam (broken curve) which has constant sensitivity within the cone and zero sensitivity outside. The equivalent beam angle ψ is the solid angle enclosed by the apex of the cone. The beam pattern has a log scale on the radial coordinate which emphasizes the side lobes.

the signal at the echosounder output. The formal definition of $\bar{g}$ is given in Appendix D.

When the range compensation is exact, $\bar{g}=1$ and there is no TVG error. However, some error is likely in practice, for two reasons. Firstly, the actual TVG may deviate from the intended function due to inaccuracy in the circuits which control the gain. Secondly, particular values of the sound speed and the absorption coefficient have to be assumed in implementing the intended function; both these parameters vary with the hydrographic conditions, notably the water temperature, and any deviation from the assumed values will result in TVG error.

Linearity principle

Much of the foregoing theory has been founded on the assumption that echo integration is a linear process. In other words, it is supposed that the output of the echo integrator is proportional to the quantity of targets having similar acoustical properties.

Definitions and assumptions

More formally, the linearity principle may be stated as follows. Consider an aggregation of targets which are insonified over a period of time by N transmissions (pings) generated by an echosounder. At any instant, the received signal consists of the overlapping echoes from targets within the sampling volume, sometimes called the pulse shell (Burczynski, 1979). This volume has been defined in section 2.4. It extends to half the pulse length in range, and across to include the equivalent beam angle. Suppose that n targets ($n>1$) are present in the sampling volume. Linearity requires that the mean sampled value of the echo energy from the n targets will be the same as the sum of all the echo energies that would have been observed if each target were insonified separately. The two main assumptions required for linearity are that (1) the targets are randomly distributed in the insonified space, ensuring that the phases of the echoes from each target are randomly distributed relative to each other, and (2) the targets move so that a new random set of phases is generated from each transmission. The echo energy from a single sample from the *i*th ping is E_i. All the E_i will be different because of the changing echo phases and target locations within the beam and, at least for live targets free to behave naturally, the variability of the echoes from each target arising from the stochastic nature of the target strength.

The E_i form an 'ensemble', which is a statistical term defined strictly as 'the infinite set of realisations of a stochastic process' (Kendall and Buckland, 1971). In the case of fish targets, the stochastic process is the random movement of the fish. The concept of the ensemble is frequently extended to apply to a sample set of realizations, as we do here in considering a finite number of pings. Between successive pings, the fish must move randomly relative to one another if a new independent member of the ensemble is to be obtained. An alternative applicable to a survey would be successive independent samples from different sampling volumes on the same ping, or for the transducer to move and observe a new set of n randomly distributed targets. The echo amplitudes from such an ensemble are described by the Rayleigh distribution (Skudrzyk, 1971), while the echo energies follow the exponential distribution. The arithmetic average of the E_i is written as $\langle E \rangle$, the angle brackets being the usual way of denoting the 'ensemble average'. $\langle E \rangle$ is an estimator of the mean of the distribution. In the limit of a large number of observations, $\langle E \rangle$ tends to the true mean, which is the expected value of the echo integral. Note that the ensemble average is derived from observations of a number of different random realizations, not just a number of observations of one realization (even if the targets are randomly positioned) or a single observation of a large number of targets.

Now suppose that all the targets except the jth one are removed, but it continues to behave as before. e_j is the expected echo integral from this target in isolation, that is to say the mean value taking account of random positioning and other stochastic effects which, except for the echo phase, apply equally to one or many targets. This 'thought experiment' is repeated for each target in turn, giving the average result $\bar{e} = \Sigma_j \, e_j/n$. The linearity principle states that the expected value of $\langle E \rangle$ is equal to the sum of the e_j. This implies the mathematical condition

$$\lim_{N \to \infty} [\langle E \rangle] = n\bar{e} \tag{5.16}$$

which holds for any value of n. The key feature of this equation is that $\bar{e}$ is a property of the individual targets and does not depend upon their density (number per unit volume). Thus if N is very large, $\langle E \rangle$ is proportional to the quantity of targets (Fig. 5.16(a)), and this is an essential requirement for the echo integration technique to be valid. Typically, N should be 20 or more to achieve a reasonably accurate asymptotic value of $\langle E \rangle$.

The linearity principle follows from the statistical properties of the Rayleigh distribution. The central hypothesis is that the randomness of the ensemble causes the individual target-echoes to be incoherent. Two echoes are said to be incoherent when their phases are completely unrelated. Any value of the phase difference is equally probable. Conversely, the echoes are coherent when the phase difference is constant over a series of pings. When the echoes are incoherent, the expected energy returned from the ensemble is the sum of the echo energy from the individual targets. The energy is proportional to the square of the echo amplitude, and to the echo integral, so Equation 5.16 follows immediately. As a simple illustration of this result, consider two echoes of angular frequency ω and phase difference η which are received individually as the signals $A_1 \sin(\omega t)$ and $A_2 \sin(\omega t + \eta)$, where

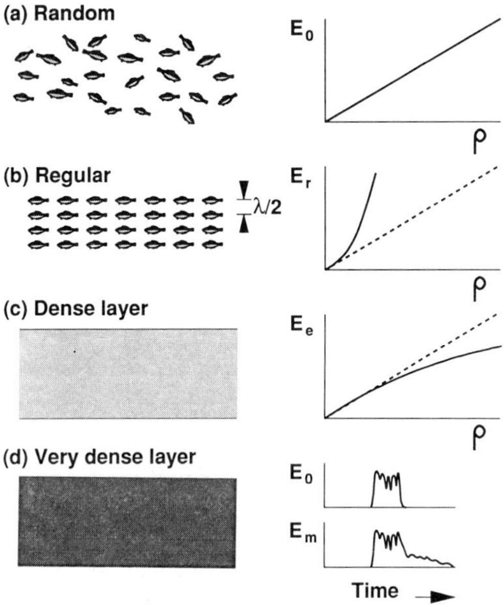

Fig. 5.16 Types of fish distribution. (a) Random, low density; the echo energy $E = E_0$ is proportional to the target density ρ, and linearity applies. (b) Regular, half-wavelength spacing; $E = E_r$ (solid curve) increases more rapidly than ρ. (c) Shadow effect, $E = E_e$ increases less rapidly than ρ. (d) Multiple scattering, $E = E_m$ is extended in time compared with E_0.

A_1 and A_2 are the amplitudes and t is the time. When the echoes are received together, the signal is

$$v(t) = A_1 \sin(\omega t) + A_2 \sin(\omega t + \eta) \tag{5.17}$$

which is a sine wave whose squared amplitude is

$$A_T^2 = A_1^2 + A_2^2 + 2A_1 A_2 \cos \eta \tag{5.18}$$

Suppose now that many pairs of echoes are observed which have random phases. Any value of η is equally likely, so on average $\cos \eta$ is zero. When Equation 5.18 is summed over many observations, the $\cos \eta$ terms cancel out. Thus the mean squared amplitude from the ensemble is

$$\overline{A_T^2} = \overline{A_1^2} + \overline{A_2^2} \tag{5.19}$$

The truth of the linearity principle is not self-evident in the case of fish targets, although Stanton (1985a) has published echo data which match the Rayleigh distribution. It is possible, for example, that the alignment of fish in a school might be sufficiently regular for coherent scattering to occur. If the echoes were all in phase, as in the special case of targets spaced at half-wavelength intervals in range, the amplitude of the school echo would be the sum of the amplitudes from the individual targets. The echo energy returned from the school would then increase as the square of the number of targets (Fig. 5.16(b)).

A further assumption required for linearity is that acoustic extinction and multiple scattering are negligible. These phenomena are discussed below (pp. 155–7). Linearity supposes that second-order echoes, which occur when the primary echo from one target is scattered by another target, contribute little to the echo integral of the aggregation. This will certainly be true at low densities, but it has to be asked whether the highest packing densities observed in fish schools are above the level at which multiple scattering becomes important (Lytle and Maxwell, 1983; Stanton, 1983).

Experimental evidence

The experimental verification of linearity is an essential test of the echo integration method and the validity of Equation 5.15. The experiment is difficult to perform because of the need to measure the ensemble and the individual targets under identical conditions, and for long enough to obtain precise means of the echo integrals. Hamilton *et al.* (1977) have described an early attempt to verify the linearity principle, known as the MP-1 experiment. They measured the echoes from sets of 10 cm polystyrene floats moored at randomly selected positions in a tank of water. The results did not support the linearity assumption, but it is now believed that the design of the MP-1 experiment was flawed. In particular, the spatial distribution of the stationary floats was insufficiently random to be representative of moving fish, and the calibration methods available at the time were incapable of the precision required (Swingler and Hampton, 1981).

Subsequently, Foote (1983) conducted a definitive test of linearity with live fish. He measured the echo integral from caged, free-swimming aggregations of herring, *Clupea harengus*, and pollack, *Pollachius pollachius*, at densities up to 57 fish m^{-3}. At the same time, he observed the behaviour of the fish (notably the tilt angle distribution) by photography. Finally, he anaesthetized the fish and measured the target strength of individuals as a function of the tilt angle, using the method of Nakken and Olsen (1977). $\langle \sigma \rangle$ is estimated as the mean cross-section over the tilt angle distribution. An acoustic estimate of the fish density now follows from Equation 5.15 and may be compared with the true density, which is the number of fish divided by the cage volume. The estimated and true densities agreed within the bounds of experimental error; Fig. 5.17 shows the results obtained at four frequencies between 38 and 120 kHz.

The Foote experiment demonstrates that the linearity principle is applicable to fisheries acoustics. The conditions which have been tested cover the range of frequencies, transmitted power levels and the normal fish densities likely to be encountered during acoustic surveys.

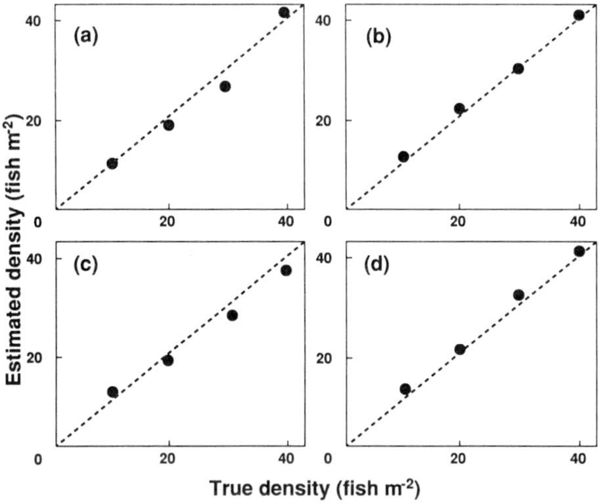

Fig. 5.17 Comparison of the estimated and true densities of caged fish, redrawn from Foote (1983). Linearity requires that within experimental error, the observations should lie on the 1:1 line. Results at four frequencies, (a) 38, (b) 50, (c) 70 and (d) 120 kHz.

Non-linear effects

There remains the question of whether linearity fails at very high fish densities. In experiments with caged fish, Rottingen (1976) found that the echo integral is proportional to the fish density up to a certain limit, after which the relationship becomes non-linear. As the density is further increased, E continues to rise but at a slower rate (Fig. 5.16(c)).

Shadowing

Foote (1978a,b) suggests that changes in fish behaviour with packing density might alter the scattering properties of the aggregation, causing some non-linearity. However, we believe the main cause of non-linearity is the so-called shadow effect. The fish nearest the transducer attenuate the acoustic energy so that the more-distant fish contribute less to the received signal. The amount of the signal reduction is determined by the area-density of the intervening fish and σ_e, the extinction cross-section, which is a measure of the energy removed from the beam by each fish. Lytle and Maxwell (1983) have considered the estimation of fish density in the presence of significant shadowing. They present a number of approximate solutions based on different theoretical models. The best solution depends upon the 'optical thickness', which is the distance over which the transmitted intensity is reduced by the factor $\exp(1) = 2.718$ due to shadowing, over and above the losses arising from spreading and water absorption.

Foote (1983) maintains that first-order scattering theory, with the addition of an attenuation term to allow for shadowing, will be adequate for the fish densities found in the wild. In this model, for the case of constant fish density within an aggregation, the strict linearity condition (Equation 5.16) is modified to become

$$\lim_{N\to\infty} [\langle E\rangle] = n\bar{e}\,[1-\exp(-2h\rho'\sigma_e)]/(2h\rho'\sigma_e) \tag{5.20}$$

where ρ' is the number of fish per unit volume, averaged over the thickness h of the aggregation. If the factor $(h\rho'\sigma_e)$ is very small, shadowing is not important and conditions 5.16 and 5.20 are nearly identical.

An approximate method of estimating the fish density in the presence of shadowing, assuming that σ_e and $\langle\sigma\rangle$ are both known, is to calculate the density in steps of increasing range (Foote, 1990a; MacLennan *et al.*, 1990a). Suppose the fish are observed in the depth interval z_1 to z_2. This interval is divided into M consecutive layers of thickness $(z_2-z_1)/M$. The layers are numbered 1, 2 ... M sequentially downwards. E_i is the echo integral from the ith layer, and Equation 5.15 is used to calculate d_i, the apparent area-density of fish in that layer. The true area-density ρ_i will be greater than d_i if shadowing is important. However, in the case of the first layer, which is at the top of the aggregation, there is no shadowing and $\rho_1 = d_1$. The apparent density of the second layer is corrected to take account of shadowing by the first layer, and so on down through the aggregation. The formula for the ith layer is

$$\rho_i = d_i\left(2\sigma_e \sum_{j=1}^{i-1} \rho_j\right) \Bigg/ \left[1-\exp\left(-2\sigma_e \sum_{j=1}^{i-1} \rho_j\right)\right] \tag{5.21}$$

This approach has been applied to the estimation of fish held in sea-pens (Furusawa *et al.*, 1984; Burczynski *et al.*, 1990). Very high densities are common in the enclosures used by fish farmers. The enclosures are large and there is need for accurate and unobtrusive methods of determining the quantity of the captive fish. Echo integration is one such method, but it is necessary to take account of the high-order scattering problem at the densities found on fish farms.

The accuracy of Equation 5.21 depends upon the layers being thin enough for shadowing within one layer to be neglected. For satisfactory results, the factor $(2\sigma_e\rho_i)$ should be much less than unity. It is possible to derive more complicated formulae which are theoretically correct when $(2\sigma_e\rho_i)$ is large, but they are not much use in practice. When shadowing is very strong, the echoes returned from the bottom of the aggregation are so weak that accurate density estimation is impracticable.

The shadow effect depends upon the area-density ρ rather than the volume-density ρ'. Rottingen (1976) reports that for caged sprat of mean length 12.1 cm, shadowing becomes significant at $\rho' = 1900\,\mathrm{m}^{-3}$, corresponding to $\rho = 4560\,m^{-2}$ over the 2.4 m deep cage. This volume-density is much greater than that observed in wild schools, typically one fish per cubic body-length (Pitcher and Partridge, 1979). For 12.1 cm sprat, that would be equivalent to $\rho' = 560\,m^{-3}$, and a school 8.1 m thick would have an area-density of $4560\,m^{-2}$. The implication is that if a school of sprat or similar fish were thicker than this limit, the apparent density should be corrected for the shadow effect.

Multiple scattering

The acoustic field at a given target is produced by the incident wave and the scattering fields of all the other targets. When the scattering fields dominate the total field, high-order multiple scattering becomes important. If this occurred in a school of fish, acoustic energy from the incident wave would be retained inside the school by repeated scattering between the fish. The energy would gradually escape, and the echo would continue for longer than predicted by first-order scattering theory (Fig. 5.16(d)). In effect, the school would ring like a bell.

The presence of high-order multiple scattering may be indicated by the appearance of the echogram. The lengthened echo shows as a diffuse ragged tail below the more solid mark of the school (Fig. 5.18). The tail (arrowed in Fig. 5.18a) might be due to a few fish scattered below the school, but this explanation is most unlikely, especially in Fig. 5.18(b) where a tail is seen on the sea-bed echo as well. Both the downward transmission pulse and the returning sea-bed echo are subjected to multiple scattering while passing through the school.

Stanton (1983, 1984) has considered higher-order scattering in aggregations of randomly distributed targets. He shows that second-order scattering partially offsets the shadow effect, and he predicts an upper bound for the echo integral of a dense school.

Applications

Echo integration has become well established as a practical technique for the measurement of fish abundance. It can provide quick results and up-to-date information about the pelagic fish distribution in the area covered by the survey. The technique is more widely applicable than echo-counting because it does not depend upon having to identify single-fish echoes.

Echo integrators are in widespread use by fishery research institutions throughout the world. Many applications have been reported from work in tropical, temperate and Arctic waters (Thorne, 1983b; Jakupsstovu, 1984). Echo integrators have been used to study many kinds of fish, notably clupeoids (Aksland, 1983; Thorne *et al.*, 1983; Bailey and Simmonds, 1990), gadoids when they are found in mid-water (Dalen and Smedstad, 1983), occasionally salmonids (Mulligan and Kieser, 1986) although echo-

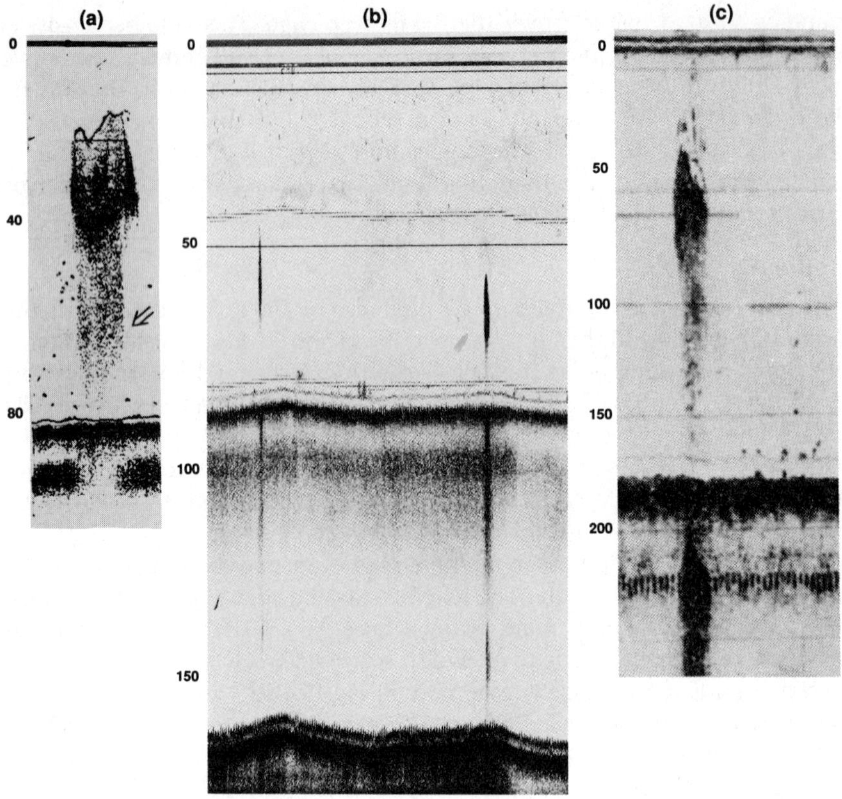

Fig. 5.18 Examples of multiple scattering in fish schools. (a) Herring, Alaska, 105 kHz, reproduced with permission from Stanton (1984); the multiple-scatter tail is arrowed; (b) herring, North Sea, 38 kHz; (c) capelin, NE Arctic, 38 kHz, reproduced with permission from Rottingen (1976). Vertical scales indicate depth below the surface in m.

counting is more usual in this case, and other aquatic species such as squid, *Loligo opalescens* (Jefferts *et al.*, 1987) and the Antarctic krill, *Euphausia superba* (Everson, 1982).

The technique has a number of important limitations, however. It is not possible to detect fish close to the sea-bed because of the obscuring effect of the sea-bed echo, and fish close to the sea surface are similarly inaccessible to integration. When the fish occur in dense concentrations, non-linear effects (notably shadowing) may degrade accuracy. Another problem is that the echo integrator by itself provides no information to identify the species or the size composition of targets. When the aim is to measure the abundance of one species and it is known that others are present in the survey area, the species composition has to be determined by other means such as trawl samples. The collection of unbiased samples is difficult. Also, the accuracy of echo integration depends critically on having well-calibrated instruments with a stable performance, good data on the scattering properties of fish and knowledge of acoustic propagation losses in the water. We shall return to these and other practical problems in later Chapters.

APPENDIX C The true size-distribution of schools

We assume that the schools are circular in horizontal cross-section, but the size (as measured by the school diameter) is variable. S is the number of schools detected by the echosounder, and $n(z)\Delta z$ is the number whose diameter is in the interval $(z \pm \Delta z/2)$. If S is very large and Δz is very small, $n(z)$ is the true size distribution of the schools.

When a school is observed on the echogram, L is the trace length corrected for the effect of the beam shape. L is less than the diameter because the survey track projected onto the horizontal cross-section crosses the school on a chord which is off-centre (Fig. C5.1). $N(x)\Delta x$ is the expected frequency of trace lengths in the interval $(x \pm \Delta x/2)$. The cumulative function $G(L)$ is the frequency of schools with trace-lengths of L or more. Thus

$$G(L) = \int_L^\infty N(x)\,dx \tag{C5.1}$$

The true size distribution is to be estimated from the observed frequencies of

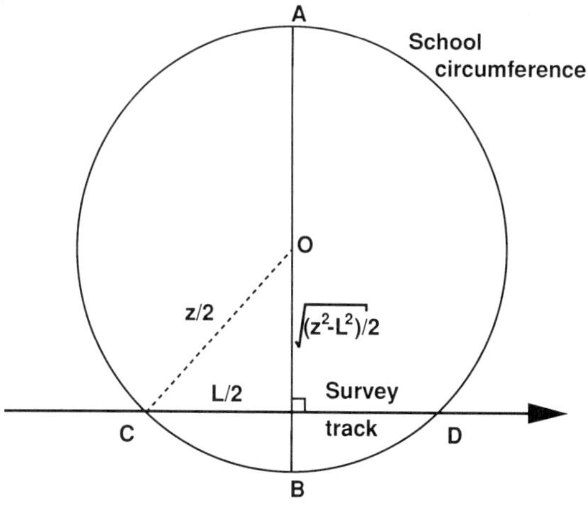

Fig. C5.1 Geometry of the survey track projected onto the horizontal, circular cross-section of a school. z is the school diameter AOB and L is the length of the chord CD.

traces in consecutive size classes. ε is the class interval, which is not necessarily small. N_r is the observed frequency of the rth class, which includes trace lengths in the range $(r-1)\varepsilon$ to $r\varepsilon$. The largest observed trace is in class m. The observed cumulative frequencies are

$$G_r = N_r + N_{r+1} + N_{r+2} + \cdots + N_m \tag{C5.2}$$

$G_1 = S$ is the total number of observations, and $G_{m+1} = 0$.

The school diameters are considered within the same size classes. If n_r is the frequency of schools with diameters in the rth class, then

$$n_r = \int_{(r-1)\varepsilon}^{r\varepsilon} n(z)\,dz \tag{C5.3}$$

The school centres are assumed to be randomly located with respect to the survey track. The chord CD (Fig. C5.1) may cross any point on the perpendicular diameter AOB with equal probability. This implies that of all the schools with diameters close to z, the proportion $\sqrt{[1-(L/z)^2]}$ will produce trace lengths of L or more. Thus $G(L)$ and $n(z)$ are related through the integral equation

$$G(L) = \int_L^{\infty} n(z)\sqrt{[1-(L/z)^2]}\,dz \tag{C5.4}$$

Substituting $z = x\varepsilon$ and $L = (r-1)\varepsilon$, this equation can be expressed as a sum over the class intervals,

$$G_r = \sum_{s=r}^{m} \int_{s-1}^{s} n(x\varepsilon)\sqrt{\{1-[(r-1)/x]^2\}}\,dx \tag{C5.5}$$

An approximate numerical solution is obtained by neglecting the variation of $n(z)$ within each size class. $n(x\varepsilon)$ is replaced by the constant n_s as the true frequency by diameter class, leading to the following set of linear equations

$$\begin{aligned} G_m &= a_{mm} n_m \\ G_{m-1} &= a_{m-1\,m-1} n_{m-1} + a_{m-1\,m} n_m \\ G_r &= a_{rr} n_r + a_{rr+1} n_{r+1} + \cdots + a_{rm} n_m \\ G_1 &= a_{11} n_1 + a_{12} n_2 + \cdots + a_{1m} n_m \end{aligned} \tag{C5.6}$$

Values of the coefficients a_{rs} up to $r = 10$ are given in Table C5.1. They are obtained from the formulas

$$\begin{aligned} a_{rs} &= \int_{s-1}^{s} \sqrt{[1-(1/x)^2]}\,dx \\ &= \sqrt{[s^2-(r-1)^2]} - \sqrt{[(s-1)^2-(r-1)^2]} \\ &\quad + (r-1)\{\arccos[(r-1)/(s-1)] - \arccos[(r-1)/s]\} \end{aligned} \tag{C5.7}$$

Table C5.1 Values of the coefficients a_{rs} required to estimate the true distribution of school diameters from the observed cumulative frequencies of trace-lengths

r	$s=1$	$s=2$	$s=3$	$s=4$	$s=5$	$s=6$	$s=7$	$s=8$	$s=9$	$s=10$
10										0.300
9									0.316	0.536
8								0.335	0.564	0.674
7							0.359	0.596	0.707	0.774
6						0.388	0.635	0.744	0.808	0.850
5					0.426	0.682	0.786	0.845	0.882	0.907
4				0.478	0.741	0.836	0.886	0.916	0.935	0.949
3			0.554	0.816	0.894	0.931	0.951	0.964	0.972	0.978
2		0.685	0.913	0.957	0.975	0.983	0.988	0.991	0.993	0.994
1	1.000	1.000	1.000	1.000	1.000	1.000	1.000	1.000	1.000	1.000

To solve Equation C5.6, first calculate $n_m = G_m/a_{mm}$. This is substituted in the next equation which is solved for n_{m-1}, then by repeated substitutions through the set, the true frequencies $(n_1, n_2 \ldots n_m)$ are obtained explicitly.

Statistics of the distribution

Suppose there are many schools of diameter z. The trace lengths associated with these schools could have any value between 0 and z. Chord CD crosses the school diameter at a distance y from the centre, and $L=\surd(z^2-4y^2)$. Any value of y between 0 and $z/2$ is equally probable, so the mean length is

$$\bar{L}(z)=(2/z)\int_0^{z/2}\surd(z^2-4y^2)\,dy$$
$$=(\pi/4)z \qquad \text{(C5.8)}$$

The overall mean of the traces due to all schools of any size is obtained by integrating Equation C5.8 over the size distribution. It is

$$\bar{L}_0=(\pi/4)\int_0^{\infty} z[n(z)/S]\,dz \qquad \text{(C5.9)}$$

The integral in Equation C5.9 is simply the mean school diameter, $\bar{z}$. Thus

$$\bar{z}=(4/\pi)\bar{L}_0 \qquad \text{(C5.10)}$$

The cross-sectional area of each school is $A=\pi(z/2)^2$ and the mean can be derived by a similar analysis. In this case it is the squared trace lengths which have to be considered. For schools of diameter z,

$$\overline{L^2}(z)=(2/z)\int_0^{z/2}(z^2-4y^2)\,dy=(2/3)z^2 \qquad \text{(C5.11)}$$

and for all the schools, irrespective of size,

$$\overline{L_0^2} = (2/3) \int_0^{\infty} z^2 [n(z) / S] \, dz \qquad \text{(C5.12)}$$

Again, the integral in Equation C5.12 is the mean of the squared diameters, or $(4/\pi)$ times the mean area which is estimated as

$$\overline{A} = (3\pi/8)\overline{L_0^2} \qquad \text{(C5.13)}$$

APPENDIX D Calculation of the TVG error

$A(t)$ is the actual TVG function of the echosounder. It may differ from the intended '20 log R' function. $\bar{g}$ is the error in the echo integral due to the imprecise TVG. The error depends upon the depth channel being sampled, say R_1 to R_2.

Suppose that $v(R, t)$ is the signal at the echosounder output due to an isolated target at range R, at time t after the transmission pulse. We define the range-dependent gain-function to be

$$H(R) = \int_{t_1}^{t_2} |v(R, t)|^2 \, dt \Big/ \int_{t_1}^{t_2} |v(R, t) / A(t)|^2 \, dt \qquad \text{(D5.1)}$$

where the integrals are taken over the duration of the echo, t_1 to t_2. When the range is much greater than the transmitted pulse length, so that $A(t)$ is nearly constant over the echo duration, the integrals of $v(R, t)$ cancel. $(2R/c)$ is substituted for t, giving the approximate form which does not include the time explicitly.

$$H(R) \approx A(2R/c) \qquad \text{(D5.2)}$$

The more complicated relationship, Equation D5.1, is important only at short ranges. For exact range-compensation, $H(R)$ should equal the exact function, which is

$$H_e(R) = R^2 \exp(2\beta R) \qquad \text{(D5.3)}$$

Now because $g(R)$, the TVG error at range R, is the ratio of the exact and actual range-dependent gain-functions,

$$g(R) = [R^2 \exp(2\beta R)]/[H(R)] \qquad \text{(D5.4)}$$

Finally, the correction factor for the echo integrator equation is obtained from the average of the reciprocal of $g(R)$.

$$\bar{g} = (R_2 - R_1) \Big/ \int_{R_1}^{R_2} [1 / g(R)] \, dR \qquad \text{(D5.5)}$$

At long range, when Approximation D5.2 is a good approximation, this becomes

$$\bar{g} = (R_2 - R_1) \Big/ \int_{R_1}^{R_2} \{A(2R/c) / [R^2 \exp(2\beta R)]\} \, dR \qquad \text{(D5.6)}$$

Chapter six

Target strength

6.1 INTRODUCTION

When a fish is insonified by a sonar which transmits acoustic energy into the water, some of the transmitted energy is reflected as an echo which is subsequently detected by the sonar receiver. The target strength of the fish is a number which indicates the size of the echo. The greater the target strength, the stronger is the echo relative to the transmission. When acoustic instruments are used to observe fish, the objective is usually to learn as much as possible about the fish or other targets whose echoes have contributed to the signal at the receiver output. It is relatively easy to measure electrical features of this signal, such as the amplitude of echoes when they are clearly separated from one another, or the energy content of the signal when it consists of superimposed and indistinguishable echoes. The problem then is, what do these measurements tell us about the insonified targets? In order to deduce quantitative information about fish targets, such as the number per unit volume, an important requirement is to know the value of target strength appropriate to those fish that have contributed to the received signal.

In this Chapter, we first describe three techniques for the experimental investigation of the target strength of fish or other aquatic animals. The techniques differ in the state of the animals being examined, according to whether they are (1) immobilized and unconscious, (2) active but confined within a cage, or (3) wild and free to behave normally in their natural environment. Much research has been conducted using all three techniques, and we discuss what can be learned and the limitations of each one. Examples of the results obtained from target strength experiments are presented. Collected results show that target strength depends upon fish size, but there are differences between species for the same size of fish. The importance of fish behaviour and physiology in explaining the variability of target strength is discussed. Fish are classified in groups of species having similar acoustic properties, according to the type of swimbladder possessed. Finally, published experimental results are presented in a form suitable for easy reference.

List of symbols

- a Constant in Love's formula for target strength
- b Constant in formulas relating target strength to fish length
- f Operating frequency of an echosounder or sonar
- L Body length of a fish (normally the total length)
- m Constant multiplying 'log L' in formulas for target strength
- N Number of independent measurements
- TS The target strength of one fish
- λ Wavelength of sound
- σ The acoustic cross-section of one fish
- $\bar{\sigma}$ Mean acoustic cross-section

6.2 TARGET STRENGTH MEASUREMENT TECHNIQUES

Sound waves propagating through water are scattered by any object whose density is different from that of the surrounding medium. The scattered waves travel away from the object in all directions (Fig. 6.1), and their intensity decreases with the distance from the object owing to spherical

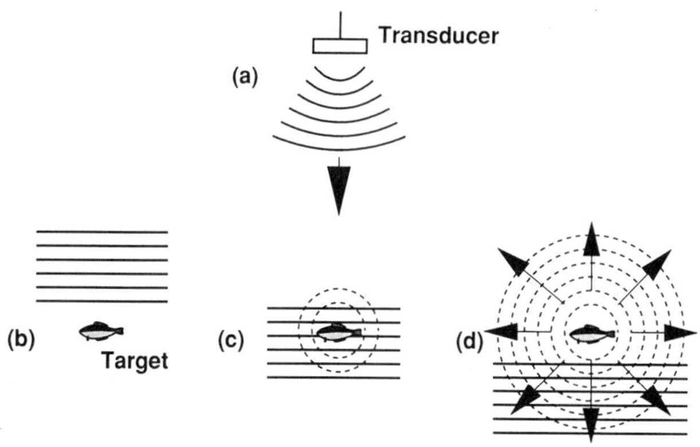

Fig. 6.1 Scattering of acoustic waves. (a) The transducer transmits a sinusoidal pulse which propagates in the direction arrowed; solid curves indicate wave-fronts; (b) at long range the wave-fronts are almost plane; (c) the incident pulse interacts with the target and scattering generates spherical wave fronts (broken curves) centred on the target; (d) after the incident pulse has passed, the scattered waves continue to propagate in all directions.

spreading and the absorption of energy by the water. Most echosounders and sonars have one transducer, which is used both to transmit acoustic pulses into the water and to detect echoes. In that case it is only the backscattered waves that are important, namely those travelling in exactly the opposite direction to the incident waves generated by the transmitter.

The target strength is defined by the intensities of the incident and the backscattered waves, as is the acoustic cross-section; these equivalent quantities are defined formally in section 2.5.

In principle it is possible to compute the target strength of a particular fish from acoustical theory and knowledge of the parts of the fish body which contribute to the echo, notably the swimbladder (Cooper and Temple, 1983; Clay and Heist, 1984; Foote, 1985; Furusawa, 1988; Do and Surti, 1990). According to Foote (1980b), the swimbladder (in species that possess one) reflects 90% or more of the backscattered energy. However, there is considerable variation between individual fish, even those of the same size and species. The problem is that the echo depends on the internal physiology – the shape of the swimbladder, for example, which can be very different between fish which are similar in external appearance. Furthermore, target strength depends also upon fish behaviour, especially the orientation of the body with respect to the transmitted beam (MacLennan *et al.*, 1990b). The scattering of sound by fish is too complicated for useful target strength values to be derived from theoretical considerations alone. For practical purposes, it is necessary to measure the target strength by experiment.

The essential requirements for target strength experiments are (1) a calibrated echosounder, and (2) knowledge of the species, quantity and size distribution of the insonified fish. The target strength is highly variable. Even for the same fish, the target strength is unlikely to be constant, owing to changes in the orientation of the body, the physiological state of the swimbladder and possibly other factors, too. It is best to think of target strength as a stochastic parameter, a quantity having a distribution of values and a mean value which is the average of a large number of measurements, while individual measurements will be widely scattered around this mean. During an experiment, a large number of echoes must be measured to establish the statistics of the target strength distribution, and in particular, the mean value for the observed fish. Results from many experiments with different fish will be required to determine the target strength properties of the population as a whole, the size dependence for example.

It is important to note that in this context, the 'mean target strength' must be calculated from the average of the acoustic cross-sections, and not from the average of the individual target strength values. Suppose that we have N independent measurements of the cross-section. The nth measurement is σ_n and the subscript n takes the values $1, 2 \ldots N$. The average cross-section is

$$\bar{\sigma} = (\sigma_1 + \sigma_2 + \cdots + \sigma_N) / N \qquad (6.1)$$

and the appropriate mean target strength is $10 \log(\bar{\sigma}/4\pi)$. This is not the same result as would be obtained by first calculating the individual target strength values, namely $10 \log(\sigma_n/4\pi)$, and then their arithmetic average.

Three experimental techniques have been used extensively for the measurement of target strength. They differ mainly in the state of the targets which are being studied. In the case of fish, they may be (1) immobilized and unconscious, (2) active but confined by a cage, or (3) wild and free to behave normally in their natural environment (Midttun, 1984). The first two techniques have also been applied in studies of plankton and crustaceans.

Immobile fish

The earliest experiments were performed on dead or stunned fish, which are held by thin wires so that the body does not move, except when the wire lengths are adjusted. Many investigations have been reported from pioneering work in several countries, notably Midttun and Hoff (1962), Haslett (1970), Shibata (1971) and Love (1971). We shall describe the technique by reference to the comprehensive experiments conducted by Nakken and Olsen (1977), whose apparatus is illustrated in Fig. 6.2. Only one fish is observed at a time. It is positioned upside-down, and near to the water surface for ease of access. Two transducers are positioned below, angled so that the fish is in the centre of both beams. The transducers are connected to separate echosounders operating at 38 kHz and 120 kHz respectively, so target strength data at both frequencies may be collected during the one experiment.

The lengths of wire can be adjusted to observe the change of target strength with body orientation, particularly the tilt, which is the angle between the head-to-tail line of the fish and the plane at right angles to the acoustic axis of the transducer. Figure 6.3 shows the kind of variations observed by Nakken and Olsen (1977). The tilt dependences at the two frequencies are different, the 120 kHz curve showing more variability, although in both cases the maximum target strength occurs when the fish is a few degrees head-down.

Live fish in cages

The variation of target strength with the tilt angle is very great. Since free-swimming fish will change their orientation from time to time in their natural behaviour, the results of experiments with moribund fish are useful

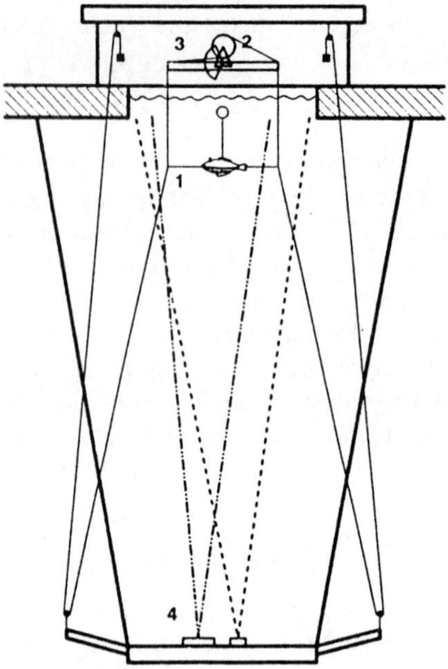

Fig. 6.2 Apparatus for target strength measurements on immobilized fish, redrawn from Nakken and Olsen (1977). (1) Fish suspension wires; (2) hoisting system; (3) tilting system; (4) two transducers, 38 and 120 kHz. A small float holds the fish upside-down.

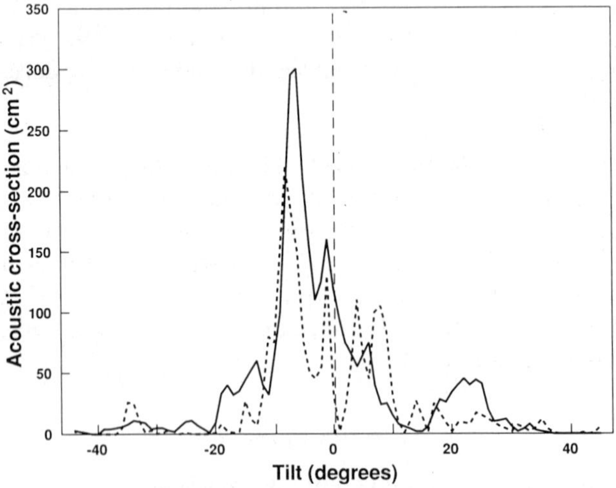

Fig. 6.3 Acoustic cross-section of a cod, *Gadus morhua*, in dorsal aspect at two frequencies; solid curve 38 kHz; broken curve, 120 kHz; fish length 45 cm. Positive tilt angles are head-up. Redrawn from Nakken and Olsen (1977).

only with knowledge of the tilt angles adopted by fish in the wild (Foote, 1980a). Experiments with active, live animals are more likely to produce results relevant to acoustic surveys, assuming that the treatment of the observed fish (by confining them in a cage, for example) has had little effect upon their natural behaviour.

The earliest observations on live animals were made using a small cage containing only one fish (McCartney and Stubbs, 1971; Foote, 1986; Goddard and Welsby, 1986). However, the need to study aggregations of fish was soon recognized. Johanesson and Losse (1977) first proposed the modern technique in which many fish were confined by a cage large enough to allow free-swimming behaviour. This technique was subsequently developed in a series of experiments reported by Edwards and Armstrong (1983, 1984).

Materials and methods

The apparatus used by Edwards and Armstrong is illustrated in Figs 6.4 to 6.6. The cage was suspended below a raft moored in Loch Duich, a sheltered sea loch on the west coast of Scotland. Again, two transducers were used, operating concurrently at different frequencies and mounted on a gimbal table which could be rotated to align the acoustic beams and the fish cage. A reference target, such as a tungsten carbide sphere of known target strength, was suspended on monofilament nylon twines below the gimbal table to calibrate the signal from the fish.

The raft was connected by cables (1000 m long) to a shore station which housed the ancillary electronic equipment – the electronic parts of the echosounder, recording devices and so forth. Both the transmitter and the receiver were located in the shore station. The loss of signal amplitude in transmission over the long cable was not small, but the accuracy of the results was unaffected because the signals from the reference target and the fish were attenuated to exactly the same extent. Television and still cameras were located below the fish cage, firstly to observe the horizontal distribution of the fish, which might be random or structured in some way, and secondly to measure the orientation of individual fish by means of stereo photography.

Cameras positioned below the fish cage, as in Fig. 6.4, do not interfere with the fish echoes. Alternatively, Foote (1983) has described how the work can be done with the cameras to the side of the cage, in a position corresponding to a minimum of the transducer beam pattern, so that the cameras produce very small echoes. This arrangement improves the precision of the tilt angle measurement, but it requires very careful positioning of the cameras, which must not be allowed to move into the main lobe of the beam pattern.

The sensitivity of the echosounder to an individual fish will depend upon the horizontal position of the fish within the cage, since the fish direction as seen from the transducer will change with the distance from the centre of

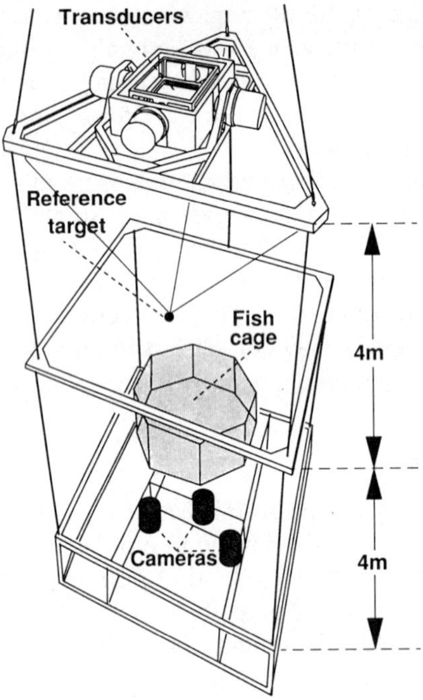

Fig. 6.4 Apparatus for target strength measurements on live caged fish. The reference target provides continuous calibration of the fish echo. The cage is typically 2 m diameter, 1 m deep and 5 m below the transducer.

the cage. In calculating the mean target strength of the caged fish, it is usually assumed that the fish are randomly distributed over the horizontal cross-section of the cage. If this is not true, for example if the fish prefer to swim around the cage circumference, the results will be biased. Visual observation of the horizontal fish distribution will indicate whether a correction is necessary to take account of non-random behaviour such as schooling.

The horizontal sensitivity can be made more uniform by increasing the distance between the cage and the transducer, so that the acoustic beam is wider at the cage. However, if the beam is too wide, the received signal may include echoes from wild fish and other unwanted targets outside the cage.

The received signal will include echoes from the cage as well as from the enclosed fish, and the cage contribution must be subtracted from the measured data. This may be done by recording the echo integral from the empty cage before the fish are introduced, and preferably again once the fish have been removed at the end of the experiment. The cage should not be made from heavier material than necessary, to minimize the acoustic

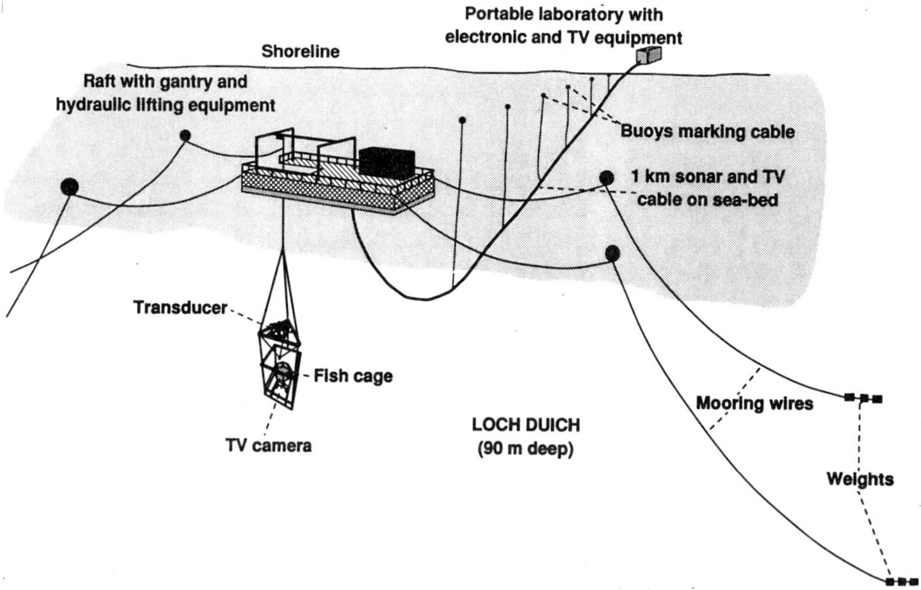

Fig. 6.5 Sketch of the Loch Duich site as used for target strength experiments. The fish cage is suspended from a moored raft which is connected by signal cables to an onshore laboratory.

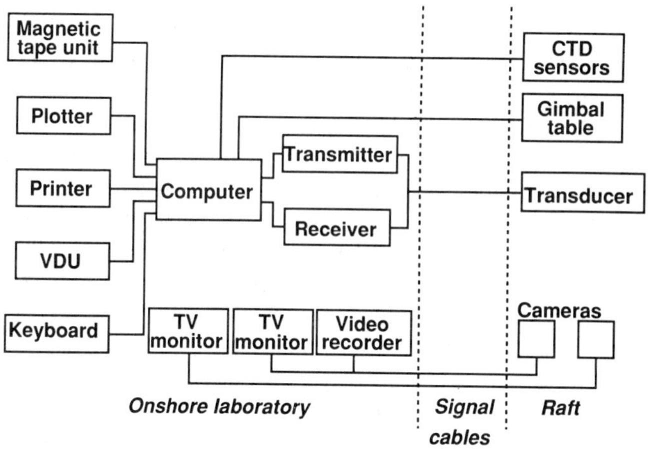

Fig. 6.6 Block diagram of the electronic equipment used to measure target strength. The data are collected under computer control and stored on magnetic tape. The sensors may include instruments to monitor the water temperature, salinity and the ambient light level. When operating at two frequencies, there is a second transducer, transmitter and receiver (not shown).

reflectivity. The cage contribution is not usually important in experiments with fish that have a relatively high target strength, those with a swimbladder, but otherwise it can be a problem.

Treatment of the fish

The capture and handling of live fish is an important part of the experimental technique. The fish must be kept healthy and subjected to the minimum of damage and stress from the moment of capture. The difficulty is that often it is necessary to catch the fish some distance from the experimental site. The fish then have to be transported from the place of capture. The whole operation should be planned in advance, with due attention to the fragility of the species in question and the choice of equipment and techniques. There is an extensive literature on fish transport which should be consulted at the planning stage (e.g. Solomon and Hawkins, 1981; Berka, 1986).

Sometimes fish may be transported quickly by road, when simple containers might be adequate, such as plastic bags cooled with ice. On other occasions, slow transport by sea may be necessary, calling for more sophisticated equipment to safeguard the fish. Figure 6.7 shows a barge made from an 8 m lifeboat hull, designed by Edwards and Armstrong (1984) for the long-distance transport of clupeoid fishes. The barge contains a 10 m^3 tank which is continuously flushed with seawater by electric pumps.

Experimental procedure

At the start of each experiment, the fish are transferred into the cage at the surface, then the transducer and cage assembly (Fig. 6.4) is lowered to the required depth, around 20 m in the case of the Loch Duich raft. The cage depth must be sufficient to prevent disturbance by surface waves or the tidal flow, and to avoid the low-salinity near-surface layer which is sometimes found in enclosed sea lochs. The transfer of the fish and the descent to the working depth is likely to be stressful, and anomalous results may be observed during the first day or two. Furthermore, systematic changes in target strength are often observed according to the time of day, and are associated with changes in the light level. This phenomenon is discussed later, but the important point relevant to the design of caged-fish experiments is that to obtain a satisfactory average value of target strength, the measurements must continue for several days. Short experiments which run for hours rather than days will produce little useful information. The need for a long series of measurements again emphasizes the importance of careful handling to ensure the survival and consistent condition of the fish.

At the end of each experiment, the cage is raised and the fish are recovered, at which stage their lengths and weights are measured. It is not practical to collect these data beforehand.

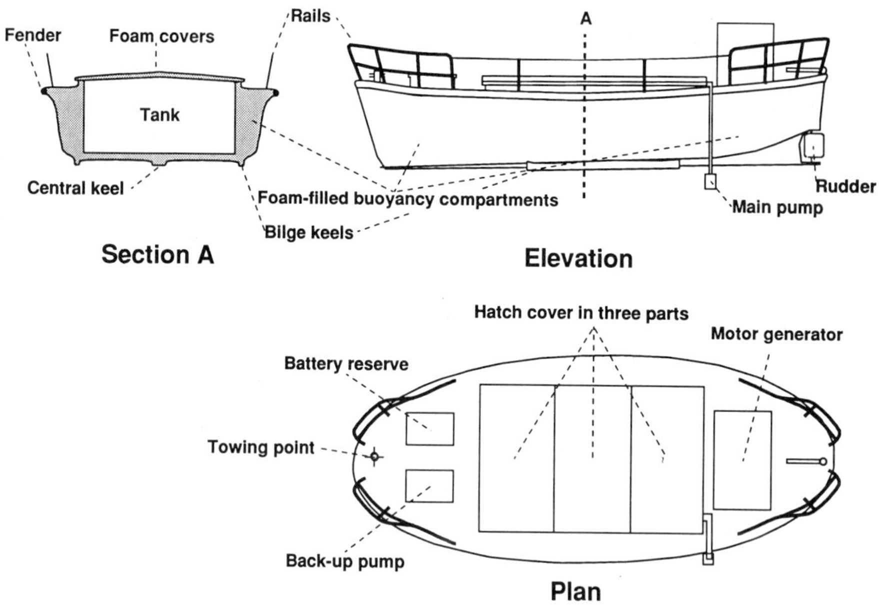

Fig. 6.7 Barge for the transport of live fish.

Wild fish

Experiments with caged fish have provided much insight into the nature of the target strength and how it depends on physiological and environmental factors. However, it is possible that the stress of capture or confinement in the cage has some indirect effect on the target strength. For example, the captive fish might not school in the same way as wild fish, so the distribution of tilt angles might also be different. The measurements made on caged fish might not be representative of the target strength of wild fish in open water.

Ideally, target strength data should be obtained from measurements on fish that are typical of the population to be surveyed, both as regards the physiological state of the animals and their environment. The observations should be made unobtrusively, without disturbing the fish or their natural behaviour. This kind of measurement is said to be performed *in situ*, meaning that target strength is determined while the fish remain in place and, all being well, unaware of what is going on. Several techniques have been developed for the *in situ* measurement of target strength.

The comparison method

The echo integrator provides an acoustic measurement from which the fish density may be determined if the target strength is known. Conversely, if the fish density can be measured independently, the target strength may be estimated from the echo integrator data. We call this the comparison method. Misund and Øvredal (1988) have described how it can be applied to herring. The fish must lie in a well-defined school which is easily captured by purse seining. First, the purse seiner crosses the school on several transects while the echo integrator data are collected. At the same time, the dimensions of the school are mapped to determine the total volume. This can be done using an ordinary echosounder to record the vertical extent of the school along the transects, while the horizontal dimensions are determined from the distances covered by the boat while crossing the school. However, it is helpful to have a scanning or multi-beam sonar operating at a shallow angle so that the horizontal extent of the school can be observed directly. The sonar display will also show whether the school is disturbed by these activities. After the acoustic measurements are complete, the purse seine is deployed to capture the whole school. The size of the catch divided by the school volume gives the density, which is required to calculate the mean target strength.

When the fish are known to be confined to a small area with limited access, it may be easier to obtain an independent estimate of the population. Mulligan and Kieser (1986) have applied the comparison method in their work on the sockeye salmon, *Oncorhynchus nerka*, in Long Lake, British Columbia. The fish migrate into the lake over a short period in the summer months, and they are counted as they pass through a weir at the single entrance. The weir count is an accurate estimate of the total population which can be compared with echo integration or echo counting data, provided most of the population is accessible to acoustic survey, not too close to the surface, the bottom or the shore.

Another technique is to compare the echo integrator data against the catch rate of a trawl or other sampling device which is towed through the depth stratum sampled by the echo integrator. Pieper (1979) and Sameoto (1980) have applied this technique to studies of euphausiids. It is necessary to assume that the trawl acts like a sieve, filtering a known volume of water, and that all the animals within this volume are captured. If the target species exhibits an avoidance reaction, so that some proportion escapes capture, the results will not be valid.

Direct and indirect target strength measurement

It is not often that an accurate independent measure of the population density is conveniently available, and the more usual kind of *in situ* experiment is conducted to determine the target strength of individual fish, without prior knowledge of the population size. To do this, it is necessary to know the position of the fish in the acoustic beam as well as the echo energy, to compensate for the effect of the beam pattern of the transducer. This can be done with special instruments, such as the split-beam or dual-beam echosounders described in Chapter 3, by the so-called direct method. It is also possible to estimate the distribution of target strength values when many targets are observed by a conventional single-beam echosounder. This is known as the indirect method, in which it is assumed that the detected targets are at randomly located positions within the beam. If the beam pattern is known, the target strength distribution may be deduced from the observed echo energy distribution by the methods discussed in section 5.3.

There are two important limitations of the *in situ* technique as applied to individual fish. Firstly, it measures only those fish that are far enough from other fish to avoid overlapping echoes. It is essential to select for recording and analysis only those echoes that come from isolated fish. The superimposed echoes from two fish near to each other will have a combined energy very different from that coming from one fish; they must be rejected by the signal processor, using algorithms based on measured features of the received signal such as the echo duration. The same problem arises in the technique of echo counting (section 5.3).

The selection of isolated targets means that in the case of species which form schools, the recorded measurements will be made on solitary fish which are on the periphery of the main concentrations. These fish may not be typical of the average school member; they may adopt a different tilt angle for example, because the visual cues from nearest neighbours will be less strong.

The second difficulty with the *in situ* method is that it is necessary to catch a representative sample of the fish which have been observed acoustically, to determine the length and species composition. Only then can the observed target strength measurements be related to specific fish targets. The sample is often taken by fishing with a trawl. Unfortunately, it is very difficult to ensure that the trawl catches fish from the same schools that have been observed acoustically. The usual procedure is to operate the acoustic equipment first, and then to deploy the trawl to collect the sample. The two operations are not concurrent. The ship must retrace the track covered during the acoustic measurements and may or may not encounter the same schools. Moreover, the trawl is not a random sampler; it may catch some sizes of fish more efficiently than others (Wardle, 1983). Other kinds of fishing gear such as the purse seine may be used, but there is still doubt as to whether the catch will be representative of the selected fish whose target strengths have been measured. However, the capture of representative samples is not difficult in favourable circumstances, when there are large monospecific aggregations of fish which are nearly the same size, as is found for example in surveys of the walleye pollock, *Theragra chalcogramma*, in the northern Pacific (Foote and Traynor, 1988).

Notwithstanding the practical problems with the *in situ* method, measurements on fish in their wild state should in principle provide the most accurate target strength data for the purpose of estimating fish abundance.

6.3 EXPERIMENTAL RESULTS

Many experimental investigations have been reported using all of the techniques described above. Each technique has advantages and disadvantages, and we shall now consider how the different approaches have contributed to our knowledge of the acoustic scattering properties of fish and other aquatic species.

Immobile fish

Among the early investigations, the comprehensive work of Love (1971, 1977) is notable. He collected data on many species from 14 different families, using echosounders operating at a range of frequencies from 15 to 1000 kHz. The body of the fish was exposed to the acoustic beam in different aspects, from dorsal through side-on to ventral. Love supposed that the target strength would be systematically dependent on both the body length L and the frequency f. He proposed the following formula for the target strength.

$$TS = \mathrm{m} \log L + \mathrm{a} \log f + \mathrm{b} \tag{6.2}$$

In this formula, m, a and b are constants for a given aspect, to be determined by the best fit to the experimental data. However, Equation 6.2 is now mainly of historical interest. It became apparent from later work that the determinants of the target strength are too complicated to be described by one simple relationship, applicable to all species.

Nakken and Olsen (1977) conducted experiments with many individual fish of 17 different species, at two frequencies, 38 and 120 kHz. In each case, target strength was measured over a range of tilt and roll angles to observe the change with the orientation of the fish body in the acoustic beam. The results could then be presented as graphs of target strength against angle, as in Fig. 6.3, or as tables (Foote and Nakken, 1978). It was found that the tilt angle had a much larger effect than the roll. Further, target strength would sometimes change rapidly over a small range of tilt angles, and these changes were greater than could be explained by simple geometrical ideas of the aspect of the fish body exposed to the acoustic beam. However, there was normally a well-defined maximum in the target strength–tilt function, and this value was referred to as the 'maximum dorsal-aspect target strength'. Different species and sizes of fish could be compared in terms of the maximum target strength. In general, larger fish of the same species had larger target strengths, but there was considerable scatter in the results.

The early investigators did recognize that the maximum target strength measurement would be of limited practical use, because active free-swimming fish would adopt a range of tilt angles. For wild fish, target

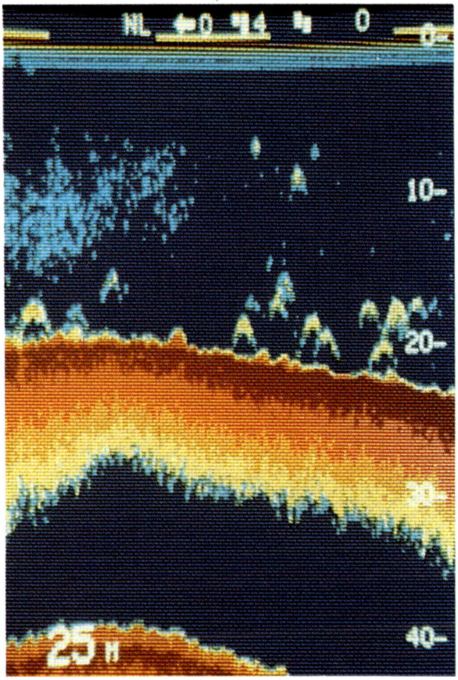

Plate 1 Colour echogram showing fish traces above the sea-bed in shallow water (20 m deep). The colour indicates the echo strength: red is strong and blue is weak. Furuno FCV-500 echosounder. Reproduced with permission from Furuno Electric Co., Japan.

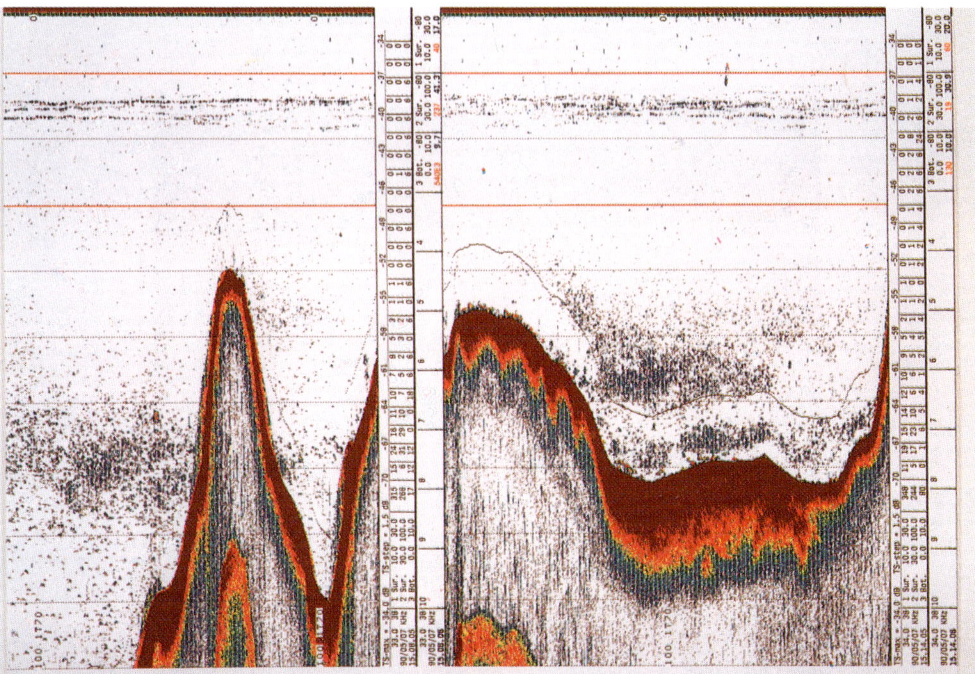

Plate 2 Echogram recorded on a colour printer showing low-density aggregations of small fish in the Oslo fjord, Norway. F/K *Simrad*, 15.14 hrs on 7 May 1990. Split-beam echosounder (Simrad EK500), 38 kHz and 40 log R time-varied gain. Vertical scale, 0–100 m depth. Numerical data printed on the echogram show measured target strengths of individual scatterers in three depth layers. Reproduced with permission from Simrad A/S, Norway.

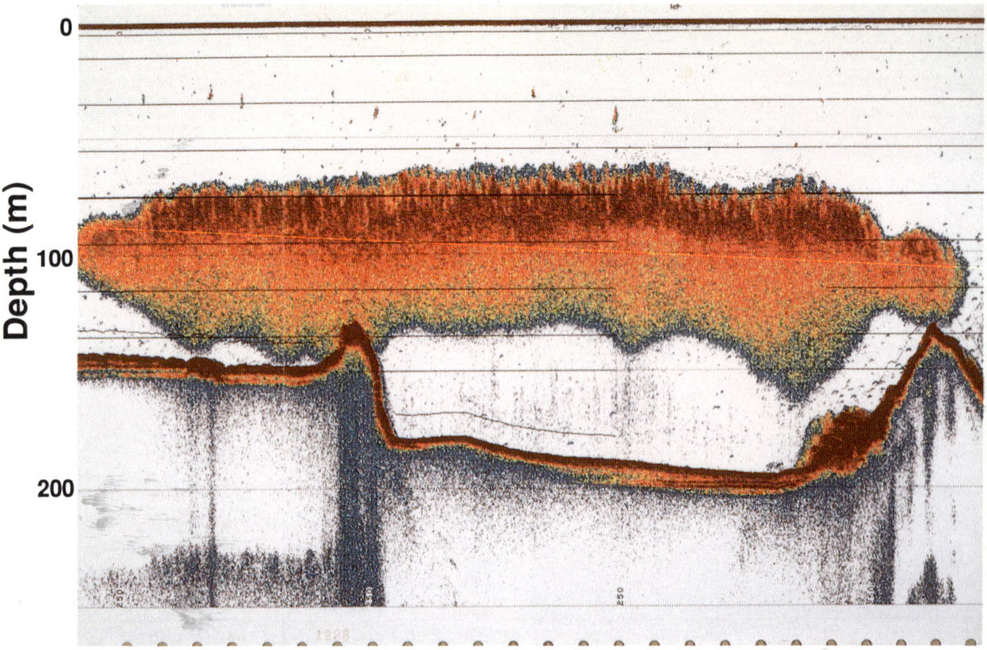

Plate 3 Echogram of a large herring school in Gratangan fjord, Norway, recorded on RV *Michael Sars* using a Simrad EK500 split-beam echosounder operating at 38 kHz. The school is about 5.5 km long and 60 m deep on average.

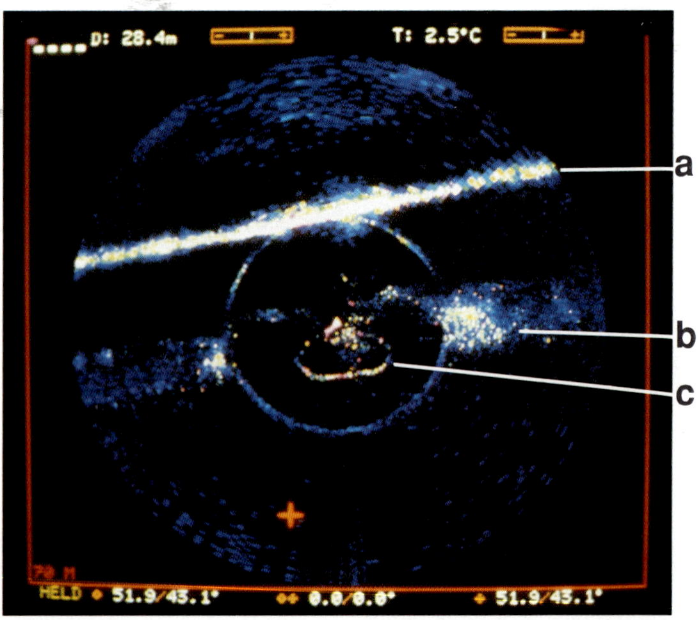

Plate 4 Scanning netsonde. Plan view of the netmouth of a pelagic trawl. The transducer is on the headline of the net and the display shows (a) the sea surface, (b) a fish layer and (c) the footrope of the trawl. Reproduced with permission from Dr E. Ona, Marine Research Institute, Bergen, Norway.

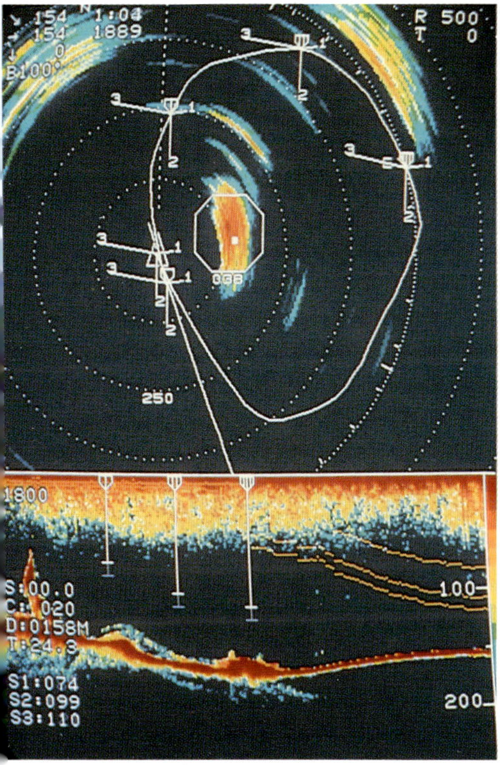

Plate 5 Omni-sonar display. The upper picture shows the track of a fishing vessel (white curve) as it shoots a purse seine around a sardine school (indicated by the white octagon). The lower picture is a conventional echogram showing the water and sea-bed below the vessel. The yellow curves on the right indicate the descent of the purse net. Reproduced with permission from the Furuno Electric Co., Japan.

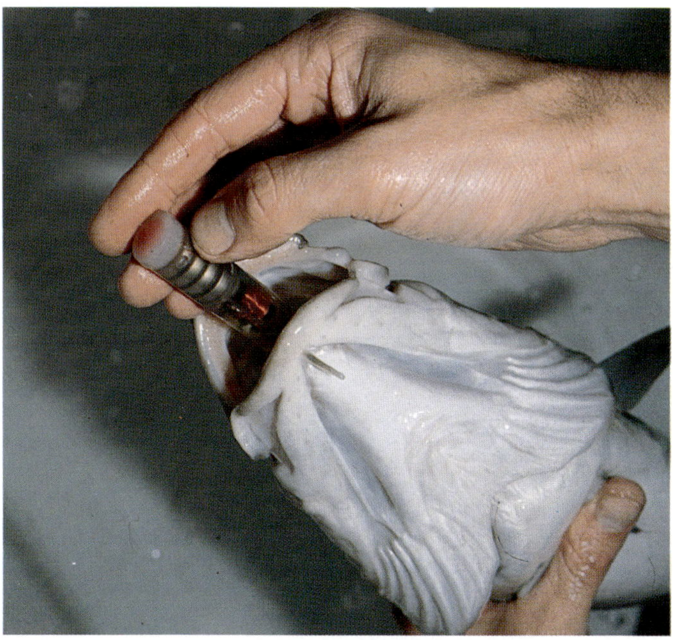

Plate 6 Acoustic tracking of fish. An acoustic pinger is shown being inserted into an anaesthetized cod, *Gadus morhua*. The pinger weighs 10 g in air and 2 g in water. When the fish is released to the sea, the pinger transmits regular pulses at 76 kHz modulation frequency. The fish is located by measuring the arrival time of the pulses at a number of hydrophones in the vicinity.

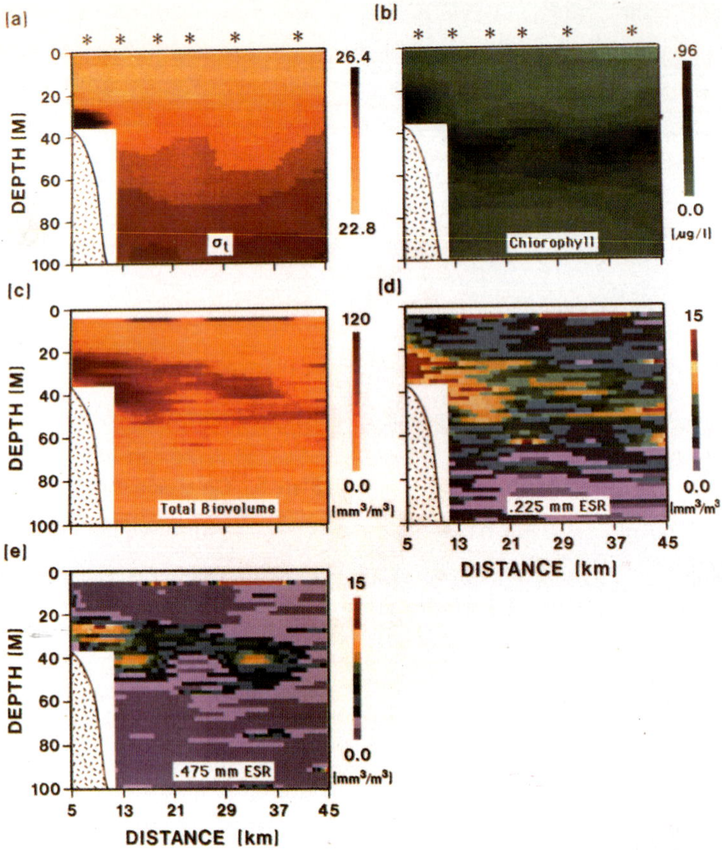

Plate 7 Data displayed by the multifrequency acoustic profiling system (MAPS) from a transect of Santa Monica Bay, California. The white insets show the sea-bed near the shore. Asterisks indicate station positions. Colour codes are shown to the right of each picture. (a) σ_t, a measure of water density; (b) chlorophyll concentration; (c–e) zooplankton volumes estimated from acoustic measurements; *ESR*, equivalent spherical radius. Reproduced with permission from Peiper *et al.* (1990).

Plate 8 Portable dual-beam echosounder (Biosonics, 420 kHz) deployed in a small boat, with towed body (paravane) containing the transducer. Reproduced with permission from Biosonics Inc., Seattle, USA.

strength would vary between individuals, and the average value would be less than the maximum recorded in dorsal aspect. Love (1971) suggested that the target strength averaged over the tilt angle range $\pm 45°$ would be representative of wild fish. By reference to tilt angle distributions derived from underwater photographs of live fish, Nakken and Olsen (1977) considered that the average target strength appropriate to wild fish would be 6 dB less than the maximum dorsal-aspect value. However, this conclusion was based on very limited data.

Foote (1985) has conducted experiments on immobile pollack, *Pollachius pollachius*, in his investigation of acoustic scattering by the swimbladder. Having measured the target strength-tilt function over a range of angles, he froze the fish quickly in a bath of liquid nitrogen, then extracted the swimbladder and made a cast of identical shape. From geometrical measurement of this cast, he constructed a digital map of the swimbladder surface, and went on to calculate the theoretical scattered sound field of the real swimbladder, as a function of the tilt angle. There was good agreement between the experimental and theoretical target strength-tilt functions (Fig. 6.8). The implication is that the rapid variation of target strength with tilt is caused mainly by the irregular shape of the swimbladder, since the theoretical calculations did not take account of scattering by other organs and body tissues. Foote (1983) has also made use of the Nakken-Olsen technique in his demonstration of the linearity principle (section 5.4).

Other studies using immobilized targets cover a wide range of species and frequencies. They include observations on conscious specimens of the common shiner, *Notropis cornutus*, and the mummechog, *Fundulus heteroclitus*, at 220 kHz (Huang and Clay, 1979), on dead euphausiids and the shrimp *Sergestes similis* at several frequencies in the range 220–1100 kHz (Greenlaw, 1977), on the yellowtail, *Seriola quinqueradiata*, the Japanese horse mackerel, *Trachurus japonicus*, and the sea bream, *Pagrus major*, at 50 and 100 kHz (Miyanohana *et al.*, 1986), on the cod, *Gadus morhua*, at 30 kHz (Fedotova and Shatoba, 1983), on the squids *Todarodes pacificus* and *Ommastrephes bartrami* at 50 and 200 kHz (Arnaya *et al.*, 1989a) and on several freshwater species at 70 kHz (Borisenko *et al.*, 1989).

Live fish in cages

A considerable number of target strength experiments have been conducted at the Loch Duich field station of the Aberdeen Marine Laboratory. The equipment used now is that illustrated in Figs 6.4 to 6.6, although it has evolved through technical development during the course of the work. The aim has been to collect representative target strength data for species which might be encountered on acoustic surveys of fish populations, and to

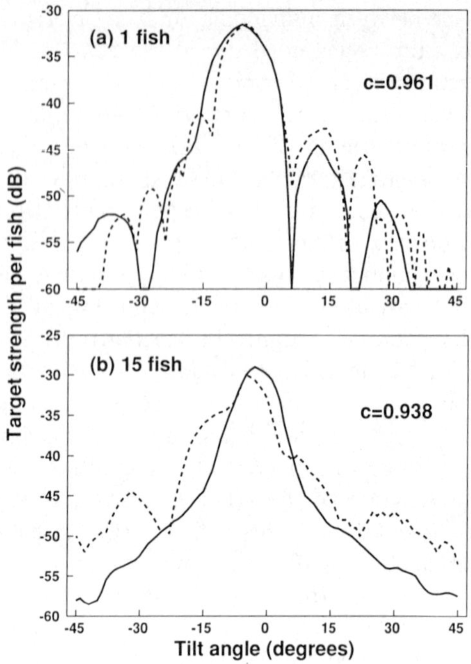

Fig. 6.8 Target strength functions of pollack, *Pollachius pollachius*, in dorsal aspect at two frequencies. Positive tilt angles are head-up. Solid curves, computed functions; broken curves, measured functions; (a) single fish, $L = 31.5$ cm; (b) composite results for 15 fish, mean $L = 37.0$ cm. c is the correlation coefficient between computed and measured functions. Redrawn from Foote (1985).

investigate the variation of target strength in relation to environmental and biological factors. The experiments have covered several species of the Gadidae, the Clupeidae, the Atlantic mackerel, *Scomber scombrus*, and others, mostly at 38 kHz but including some measurements at 120 kHz (Edwards *et al.*, 1984).

A similar procedure has been followed in most cases. At the start of each experiment, the fish are placed in the cage at the surface, then the cage is lowered to about 20 m depth where it remains during the measurements, which continue for several days. A few examples of the results obtained at 38 kHz are shown in Fig. 6.9 as graphs of target strength against time. The curves represent running averages of the measurements made during the previous hour, or some longer period as noted in the legend. The echosounder transmits three pulses per second, but the target strength measurement obtained from one transmission is extremely variable. The results are presented as averages over many transmissions, to reduce the random component and to show the systematic time dependence which is evident in several of the examples shown.

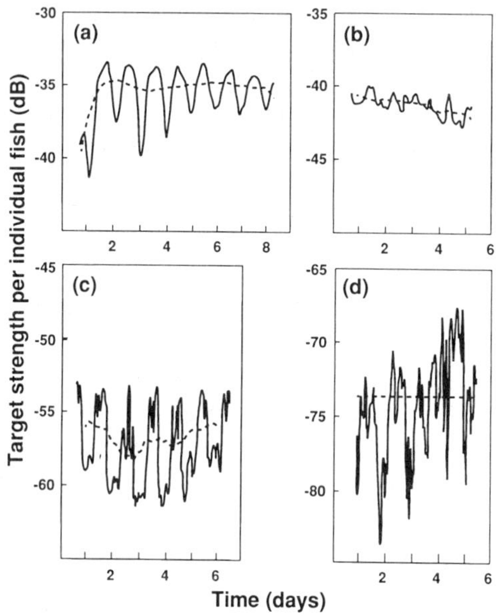

Fig. 6.9 Target strength of caged fish at 38 kHz (solid curves). Examples of experimental results for (a) cod, $\bar{L}=28.5$ cm, N=41; (b) herring $\bar{L}=24.3$ cm, N=78; (c) mackerel, $\bar{L}=32.5$ cm, N=100; (d) sandeel, $\bar{L}=12.3$ cm, N=3750. The marks on the time axes are at midnight. The broken curves are 24-hour moving averages in (a) to (c) and the overall mean target strength in (d).

Variation with time

In the case of the cod, *Gadus morhua* (Fig. 6.9(a)), the target strength is low at the start, then it gradually increases over the first day or two. This happens because the fish are initially acclimatized to atmospheric pressure, having been kept in a holding pen at the sea surface. When the fish are lowered to 20 m depth, the swimbladder is compressed to less than half its initial size. The bladder volume is gradually restored by gas secretion, and as it inflates, so the target strength increases. Superimposed on this trend, however, is a pronounced diurnal cycle. The target strength is higher during the day than at night. This is probably caused by some change in fish behaviour associated with the light level. The ability of the fish to see each other could affect their orientation or schooling behaviour. However, the 24-hour running average of target strength is steady once the initial acclimatization period is over.

This diurnal cycle in the target strength is often observed with fish in cages, but there are important differences between species. In the case of the herring, *Clupea harengus*, there is no evidence of an acclimatization period following the change in depth at the start of the experiment (Fig. 6.9(b)). However, there is a slow decline in the 24-hour average of the target strength. This decline is caused by the diffusion of gas out of the swimbladder, whose volume slowly decreases with time. The effect is more pronounced in smaller fish. This is to be expected, since in approximate terms, the gas diffusion rate is proportional to the swimbladder surface area, and hence to the square of the fish length, whereas the volume of the swimbladder depends upon the cube of length. Over the same time, a smaller fish will lose a greater proportion of the gas in the swimbladder through diffusion.

The Atlantic mackerel, *Scomber scombrus*, has no swimbladder, and the target strength of this fish is much lower than that of clupeoids or gadoids of similar size. The problem of gas pressure adjustment does not arise. Consequently, there is no initial acclimatization period, nor is there any long-term trend in the daily average target strength (Fig. 6.9(c)). On the other hand, the diurnal cycle is particularly noticeable in this example. Another species with no swimbladder is the sandeel, *Ammodytes* sp. Figure 6.9(d) shows some evidence of diurnal changes of the sandeel target strength, although other effects are clearly contributing much random variation. Furthermore, there is a small rise in the measured target strength during the experiment. This is not believed to represent a real change of reflectivity, and is probably due to the difficulty of measuring such low target strengths in a cage experiment. The sandeel is a much smaller fish than the mackerel and so the individual-fish target strength values are not directly comparable. However, the measurements may be compared by adjusting target strength to represent the same weight of fish. To do this, the quantity $10 \log(M/S)$ is added to the sandeel data, where M is the mean weight of the mackerel (287 g), and S is that of sandeel (6.75 g). The added term is $10 \log(287/6.75) = 16.3$ dB. For the same weight of fish, the average target strength of sandeel is a few dB higher than that of mackerel, but the scatter about this average overlaps to a large extent. Although the two species are very different in many physiological respects, it appears that their acoustical properties for the same fish weight are not too dissimilar.

Variation with depth

In the case of fish with a swimbladder whose volume changes with the ambient pressure, the target strength will depend upon the depth of the fish, the amount of time it has remained there, and whether the fish arrived at that depth by swimming downwards or upwards. These effects have been studied by Edwards and Armstrong (1984) who repeatedly lowered and raised fish in a cage while monitoring the target strength continuously. Figure 6.10 compares the results from three experiments with different species. In each case, the fish have been acclimatized to near-surface pressures before the start of the experiment by being held in a pen at a depth of 2–3 m for several weeks before transfer to the experimental cage.

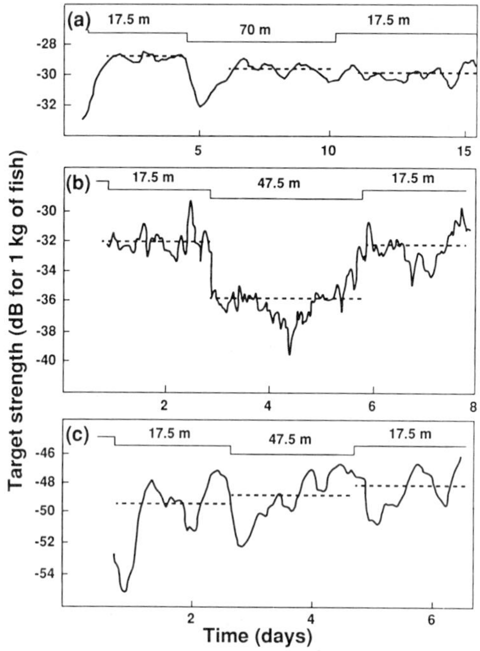

Fig. 6.10 Depth dependence of the target strength. The fish are in a cage which is moved down and up. (a) Haddock; (b) herring; (c) mackerel. The step-changes of depth are shown above each target strength record. Broken lines indicate the mean target strength over the acclimatized period at each depth. Redrawn from Edwards and Armstrong (1984).

Each experiment begins when the cage is dropped to 17.5 m depth. At intervals of several days, the cage is dropped further, then raised back to 17.5 m.

The target strength of the haddock, *Melanogrammus aeglefinus*, falls when the fish are lowered from 17.5 m to 46.5 m, then recovers to its original value over 36 hours (Fig. 6.10(a)). A similar recovery can be seen at the start of the experiment following the transfer of the fish from the keep cage at the surface to the experimental cage at 17.5 m. When the fish are raised from 47.5 to 17.5 m, however, the effect on target strength is small and transient. The haddock and the cod show similar characteristics in this respect, but the response of the herring, *Clupea harengus*, to the same test is quite different (Fig. 6.10(b)): the target strength changes systematically with depth and does not recover with time. The Atlantic mackerel, *Scomber scombrus*, does not show a consistent response to the changes in depth, as might be expected for a fish with no swimbladder (Fig. 6.10(c)). There are, however, large transient changes in target strength, particularly in the few hours following the trauma of the transfer to the experimental cage.

Crustacea and mollusca

Few experiments have been conducted on invertebrates, but Foote *et al.* (1990) have measured the target strength of the Antarctic krill, *Euphausia superba*, at 38 and 120 kHz. They report target strength values which are very small indeed, about −86 dB at 38 kHz for krill 31–38 mm long. Because of the low target strength, the echo from the krill is weaker than that from the cage, and it is difficult to determine the cage contribution accurately. This can be done in two ways, firstly by measuring the empty cage and subtracting the mean cage integral from each cage-plus-krill measurement, and secondly, by regressing the cage-plus-krill data on the number of krill and estimating the target strength from the slope of the regression line. The two methods gave target strength values which differed by several dB. The alternative approach of modelling the acoustic scattering properties of krill and other zooplankton has been discussed in section 4.5. The use of a scattering model together with experimental data at a few frequencies allows the target strength to be predicted over a wide frequency range.

Arnaya *et al.* (1989b) have studied the surume ika, *Todarodes pacificus*, a small squid. They observed various numbers up to 110 squid in a cage 1.0 m deep and 0.9 m diameter. The dorsal mantle length ranged from 10 to 20 cm with a mean of 16 cm. The echo energy was found to be proportional to the density of squid in the cage, confirming the linearity of acoustic scattering by this species. Arnaya *et al.* measured the target strength at two frequencies. They estimated the mean target strength (for $\bar{L} = 16$ cm) to be −51.3 dB at 28.5 kHz and −54.6 dB at 96.2 kHz.

Wild fish

The target strength of individual wild fish has been measured directly using both the dual-beam and split-beam methods (Dickie *et al.*, 1983; Williamson and Traynor, 1984; Burczynski and Johnson, 1986; Foote *et al.*, 1986; Foote and Traynor, 1988), and indirectly from observations made with the conventional single-beam echosounder (Ehrenberg *et al.*, 1981; Haldorsson, 1983). These *in situ* measurements are unobtrusive inasmuch as the transducer will normally be more distant than the visibility range from the target, thus avoiding visual cues. The *in situ* technique requires some method of rejecting those signals due to the superimposed echoes of neighbouring fish. When an echo is detected as coming from an isolated fish, with no interference from other targets, one measurement of the target strength of that fish is obtained. In normal circumstances, when the fish occur in moderately dense concentrations, the signal processor will record the target strengths of only a small proportion of the insonified fish, those which are sufficiently isolated to produce single-fish echoes.

Interpretation of in situ *data*

In practice, *in situ* experiments are performed by recording many target strength measurements over a period of time while a particular layer or school of fish is under observation. The measurements usually cover a wide span of target strength values. It is convenient to display this sort of information as a histogram of target strength frequencies, as in Fig. 6.11, which shows measurements collected from a layer of fish to the west of the British Isles in the springtime, using a dual-beam echosounder (Forbes, 1985). The target strength distribution is bimodal, and we suppose that each peak is associated with one species and size range of fish. There may of course be more than one species present, or more than one year class of the same species. The important question to be answered is this: which species and/or size range is associated with which peak of the target strength histogram?

In the case of Fig. 6.11, it has been discovered by trawling that the layer contains the blue whiting, *Micromesistius poutassou*, in the size range 25–30 cm, and also the myctophid, *Maurolicus muelleri*, which is a much smaller fish. Probably the blue whiting have been feeding on the myctophids. Because of the size difference between the two species, it is relatively easy to identify the modes in the target strength histogram; the right-hand peak is associated with the blue whiting, and the left-hand one is the myctophids.

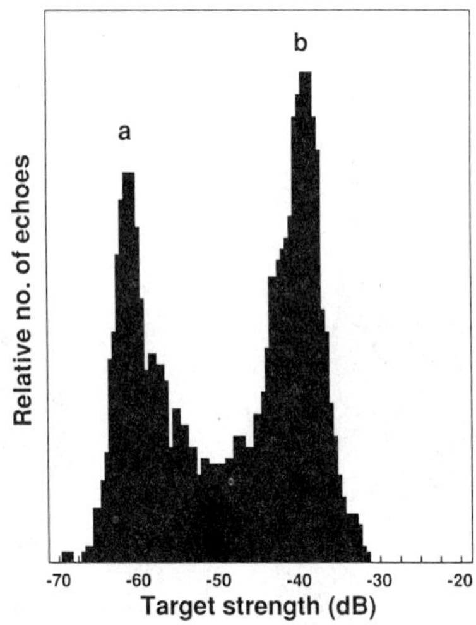

Fig. 6.11 Target strength histogram from measurements on a fish layer, redrawn from MacLennan and Forbes (1987). The data were collected using a dual-beam echosounder. The two peaks correspond to fish of different sizes, identified by trawling as (a) myctophids and (b) blue whiting, *Micromesistius poutassou*.

The essential problem in the interpretation of *in situ* target strength data is to relate the size modes of fish from the trawl catch to the modes observed in the target strength histogram. This is not difficult when there are only a few well-separated modes. However, Fig. 6.12 shows an example which requires more careful interpretation. Again, blue whiting are known to be present in the observed layer. The length distribution of the fish is determined by trawling (Fig. 6.12(b)). Three length modes are evident in the trawl samples, corresponding to successive year classes of fish. In the case of the target strength histogram (Fig. 6.12(a)), three modes are also indicated at target strengths appropriate to blue whiting, and may correspond to the length modes of the trawl samples, but the quality of the data is not good enough to be certain about this conclusion.

The best conditions for *in situ* target strength measurement occur when the observed fish are homogeneous, consisting of one year class of one species, and when the size distribution can be confirmed by fishing. Some examples from the literature are presented in Fig. 6.13. It is notable that the observed target strengths cover a large range, spanning 20 dB or more, even when the fish are nearly the same size. The implication is that target strength exhibits a great deal of variation, although we might expect that the average of many target strength measurements would have a systematic dependence on the size and species of the fish targets.

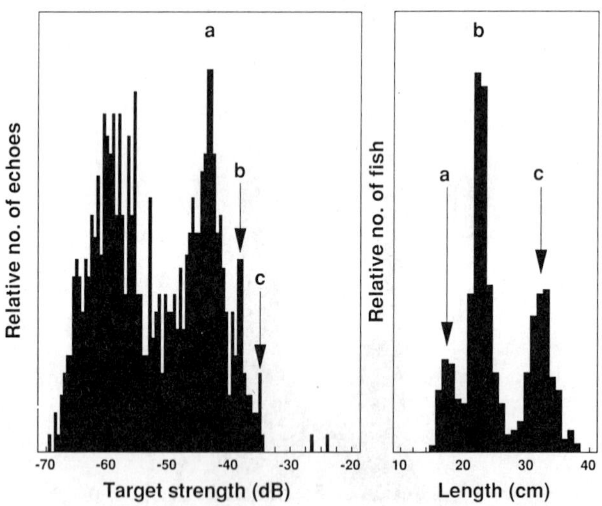

Fig. 6.12 Comparison of target strength measurements and trawl samples of blue whiting, redrawn from Forbes (1985). The target strength histogram (left) shows evidence of three modes (a, b and c); so do the trawl samples (right), which may allow the target strength modes to be related to fish from successive year classes (a, b and c) although the correspondence is uncertain in this example. Additional modes below −50 dB are associated with smaller targets, e.g. myctophids.

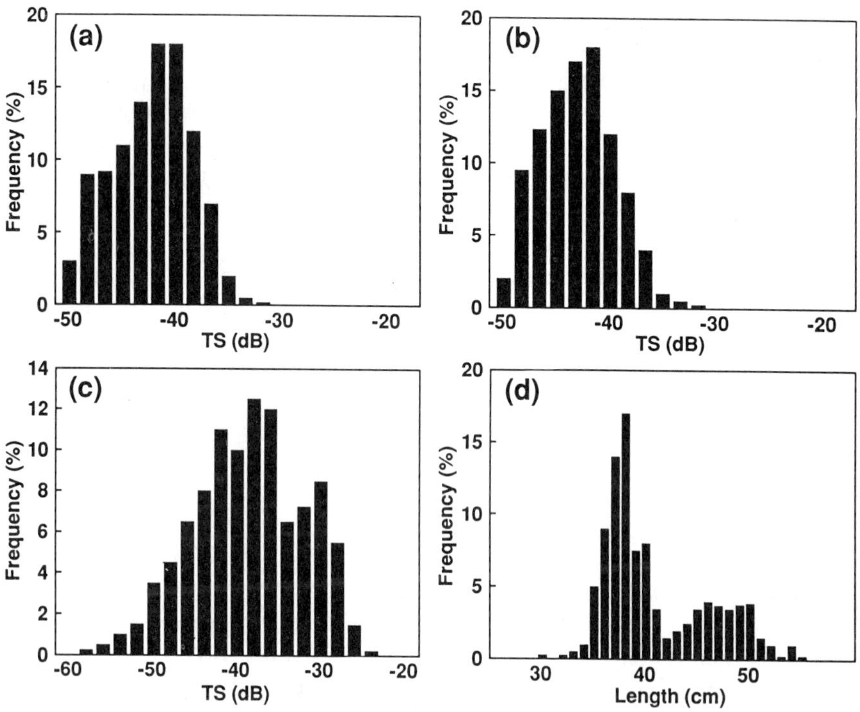

Fig. 6.13 Target strength of wild fish at 38 kHz. Examples of experimental results with split-beam echosounders. (a) Herring, *Clupea harengus*, at 65–95 m depth, $\overline{TS} = -43.4$ dB, $\bar{L}$ (from trawl samples) = 28.5 cm, N = 6545; (b) herring at 15–45 m, $\overline{TS} = -42.6$ dB, $\bar{L} = 28.5$ cm, N = 2687; (c) walleye pollock, *Theragra chalcogramma*, $\overline{TS} = -33.6$ dB; (d) trawl samples related to (c). Herring data from Foote *et al.* (1986); others from Foote and Traynor (1988).

Size-dependence of target strength

The results of target strength experiments are often expressed in terms of the target strength dependence on the body length L using the equation

$$TS = \mathrm{m} \log L + \mathrm{b} \tag{6.3}$$

where m and b are constants for a given species. This is essentially the same as Love's model (Equation 6.2) without the frequency term. Equation 6.3 is now generally accepted as a reasonable if not necessarily precise description of the target strength length dependence (Foote, 1979a,b). From observations made on a particular group of fish, we get one estimate of the mean target strength which is associated with the mean length of the fish. When data are available from several experiments of this kind, made on different groups of fish whose mean lengths cover a range of values, the slope m and the intercept b can be estimated by linear regression of target strength on log *L*. Note that Equation 6.3 does not imply a deterministic relationship which can be applied to the individual fish. *L* is the mean length of the observed group, and *TS* is the expected value of the target strength.

We use *L* to denote the total length of the fish, from the front of the head to the tip of the tail. Some writers have described their results in relation to the fork length, measured from the head to the notch in the tail, in which case the target-strength–length relationship would not be directly comparable with those presented here.

It is convenient to summarize the results from a series of experiments by quoting the regression parameters m and b. Tabulated results are presented at the end of this Chapter: m is generally between 18 and 30, and often close to 20; b is more variable, in the range −60 to −90 dB, depending on the species.

6.4 DISCUSSION

An important conclusion to be drawn from the experimental work is that the target strength is highly variable. MacLennan *et al.* (1989, 1990b) have shown that the target strength of similar fish may change with time or between individuals, due to behavioural or physiological reasons which are not well understood. It is necessary to think of target strength as a stochastic parameter which is described by a probability distribution. This means that the result of any one target strength measurement is unpredictable. However, as far as the practice of acoustic surveying is concerned, the inherent variability of target strength is not too important. It is the expected value (of the backscattering cross-section) that is required to calculate the fish density. The expected value is the mean of the probability distribution, and we suppose that it is determined by observable characteristics such as the size and species composition of targets.

Comparison of target strength measurement techniques

Experiments with immobile fish do not reliably estimate the target strength of active fish. Uncertain physiological changes are likely to occur after the animal is rendered unconscious, and the effect of this treatment on the acoustic properties of the body is unclear. More accurate results are obtained from studies with live fish. However, the immobile-fish technique has the advantage of allowing the orientation of the target to be measured precisely, so that the functional dependence of target strength can be studied in detail, especially the variation with the tilt angle.

Experiments with live fish in cages have provided much insight and practical information as to how the target strength depends upon the physiology of the animals, their behaviour, and environmental factors such as the ambient pressure and the light level. The technique has the advantage that the observed fish are available for examination. During the experiments, the behaviour of the fish may be observed visually, and the distribution of tilt angles or other behavioural parameters may be determined by photography. Afterwards, it is a simple matter to record precise length and weight distributions from post-mortem examination. The acoustic measurements may be related to observed features of the fish aggregation, to search for determinants of the target strength. One limitation is that the behaviour and condition of captive fish may not be representative of those in the wild, owing to the stress of capture and the restraints of captivity. However, if the behaviour of wild fish in their natural free state could be observed as precisely, it should be possible to deduce their distribution of target strengths by reference to the results of experiments on captive fish of similar size and species. This would require a good understanding of the relationship between the target strength and behavioural factors such as the tilt angle distribution. Of course, the unobtrusive observation of behaviour in the wild is a difficult task, given that fish are easily disturbed by the sight of cameras or flashlights.

In principle, the *in situ* methods should provide the most accurate estimates of the target strength of fish in their natural state, which would be directly applicable to acoustic surveys of wild populations. The data can be collected by remote instruments so that the fish need not be disturbed in any way. However, there are practical difficulties. Apart from the comparison method which is applicable in a few cases, the *in situ* measurements are collected on isolated fish which may not be representative of the bulk of fish inside dense concentrations. Further, it is essential to collect samples by fishing to determine the size and species composition, but it is not easy to sample the same fish that have been observed by the acoustic instruments. In comparing the acoustic and fishing data, it is necessary to make assumptions about the randomness of the sampling which are often difficult to justify.

Of all the target strength measurement techniques which we have considered, the *in situ* approach is the best in theory but the most difficult to apply in practice. The indirect methods can be applied with relatively simple equipment, but the direct methods are more accurate and statistically more robust (Ehrenberg, 1983). The split-beam technique performs better than the dual-beam alternative in the presence of noise (Ehrenberg, 1979), but in practice, we find that satisfactory results of comparable quality have been obtained with both dual-beam and split-beam echosounders.

Classification of fish targets

The importance of the swimbladder in determining the target strength has already been mentioned. The bones, the liver and fatty tissues also reflect sound, but much less strongly than any gas-filled organs that may be present, because the echo energy depends primarily on the difference in density between the reflecting organ and the water surrounding the fish. It does not matter whether the density of the reflecting organ is higher or lower than that of water; the greater the density difference, the stronger is the echo. The gas in the swimbladder has a very low density, compared to the body tissues. That is why the target strength of fish that do not have a swimbladder, such as the mackerel, *Scomber scombrus*, is much smaller than that of the clupeoids and gadoids, which do have gas-filled organs.

Two kinds of swimbladder are found in fish. First, there are the physoclistous fish, those with closed bladders. The cod, *Gadus morhua*, and other gadoids belong to this group. They have glands through which gas extracted from the water may be secreted into the bladder. If the fish descends, it may compensate for the pressure increase with depth by 'pumping up' the bladder to restore its volume and to maintain neutral buoyancy. In the cod, this adjustment is accomplished slowly, over one or two days following a descent from the surface to 20 m depth, and in the meantime the fish will be heavier than water. If the fish ascends rapidly, the gas is retained by the bladder above the ambient pressure in the surrounding water, until the excess gas has been removed by glandular action, or if the pressure difference is very large, by rupture of the bladder wall.

Secondly, there are the physostomatous fish, which have a pneumatic duct connecting the swimbladder to the alimentary canal and thence to the surrounding water (Whitehead and Blaxter, 1989). The bladder cannot sustain any excess pressure because, if the fish ascends for example, the excess gas is vented through the pneumatic duct. Many schooling species such as the clupeoids are physostomes. These fish are unable to secrete gas into the bladder while underwater, or if they do have such a mechanism, it acts very slowly indeed. In the case of the herring, *Clupea harengus*, Blaxter and Batty (1984) suggest that the fish replenishes the bladder gas by swallowing air during occasional visits to the sea surface. While the fish is underwater, gas is lost continuously by diffusion through the bladder wall. The effect of ambient pressure on bladder shape means that the target strength is likely to change if the fish moves up or down in the water column.

There is a third group to be considered, namely the fish with no swimbladder. The target strength of these fish will be relatively small, but it is not expected to be depth dependent for physiological reasons. However, it is possible that changes in behaviour with depth, due to changes in the light level, might result in an indirect depth dependence of the target strength.

The experimental evidence suggests that fish within each of the three categories – physoclists, physostomes and those with no bladder – have broadly similar acoustic properties. On the other hand, there are substantial

Table 6.1 Typical values of the target strength (*TS*) of species representative of fish classified according to the type of swimbladder. σ is the equivalent acoustic cross-section. Values relate to fish 30 cm long

Species	*TS* (*dB*)	σ (*cm*^{2})
Cod, *Gadus morhua* (physoclist)	−33.4	50.0
Herring, *Clupea harengus* (physostome)	−41.8	8.3
Mackerel, *Scomber scombrus* (no swimbladder)	−57.8	0.2

differences between the categories, particularly in the magnitude of target strength for the same size of fish. Typical target strength values for species representative of the three categories are given in Table 6.1. The cod has the highest target strength, more than 8 dB (6 times) higher than that of herring, and 24 dB (250 times) higher than that of mackerel. The comparison is more striking when it is made in terms of the acoustic cross-section.

Variation with fish size

A simple geometrical model of the size dependence of target strength would be to assume that the backscattered energy is proportional to the horizontal cross-sectional area of the organs contributing to the echo. This area would be proportional to the square of the fish length L, assuming for the sake of argument that all parts of the body grow at the same rate. This implies that σ is proportional to L^2, and that target strength is equal to 20 log L plus a constant term. According to this model, for example, the target strength of a 30 cm fish would be 6 dB higher than that of a similar fish 15 cm long.

An alternative theory suggests that the echo strength depends upon the volume of the scattering material, rather than the cross-section. This implies that σ is proportional to L^3 so that target strength would be expressed as 30 log L plus a constant term. However, the wavelength of the sonar transmissions commonly used in fisheries acoustics is commensurate with the size of discrete scatterers such as a whole fish or a swimbladder. Because of this, it is not possible to decide from *a priori* reasoning whether the area-scattering or the volume-scattering model is correct. Empirical evidence must be considered if we are to decide between the alternative theories.

Love (1971, 1977) extended the geometrical model to include the wavelength of the sonar transmission. Physical similarity might suggest that if the ratio (L/λ) is the same between two experiments, e.g. if a second fish of half the size is insonified at twice the frequency, then the ratio (σ/λ^2) should also be the same for both fish. However, experimental observations do not support this simple similarity approach. Other factors apart from the geometry complicate the frequency dependence of the target strength.

Consequently, it is important to measure target strength at the same frequency as that employed in the acoustic survey or other application. Target strength values that are extrapolated from measurements at a different frequency may not be reliable. It is now believed that the length dependence is best described by the simple Equation 6.3. The parameters of this equation, m and b, may change with frequency in an unpredictable way.

Physostomes

Figure 6.14 shows the collected results of experiments on two clupeoid species at 38 kHz, the herring, *Clupea harengus*, and the sprat, *Sprattus sprattus*, reported by Edwards *et al.* (1984). Although the data are scattered, the least-squares regression of target strength on log L is found to have a slope which is not significantly different from 20, the dependence predicted by the area-scattering model.

Foote (1987) has suggested the following relationship for clupeoid target strength at 38 kHz as the best estimate which is well supported by the published evidence (whether experimental or theoretical in origin, and including the results of both caged-fish and *in situ* experiments).

$$TS = 20 \log L - 71.9 \tag{6.4}$$

The line represented by this equation is shown in Fig. 6.14. It is slightly below the regression line of the caged-fish data, but the difference is not significant (95% level). In the case of the clupeoids, the results of caged-fish and *in situ* experiments are consistent, and in particular they have shown the same length dependence of the target strength.

There are many reasons why the individual results of target strength experiments might differ from the average expressed by Equation 6.4. The

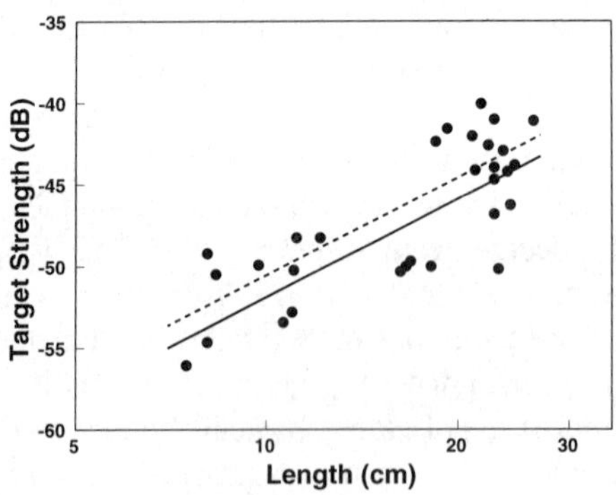

Fig. 6.14 Variation of target strength with fish length. Results from measurements on herring and sprat at 38 kHz. ●, average result from one caged-fish experiment; broken curve, linear regression line (Edwards *et al.*, 1984); solid curve, best-fit equation $TS = 20 \log L - 71.9$ as recommended by Foote (1987).

schooling behaviour might depend on environmental conditions or the season, especially around spawning time. Furthermore, there could be morphological differences between the swimbladders of similarly sized fish according to their condition and state of maturity. Also the experiments have been conducted at many different places and seasons of the year.

Physoclists

The results reported for physoclists have been more divergent. The target strength appears to increase more rapidly with length, according to Midttun (1984), at a rate of about 25 log *L*. Figure 6.15 shows the collected results of experiments on caged cod reviewed by MacLennan (1981a). The slope of the regression line is clearly greater than 20. However, *in situ* experiments suggest that the length dependence for gadoids is close to 20 log *L*. The normal behaviour pattern of wild fish unrestrained by a cage may reduce the target strength slope. Foote (1987) believes that more weight should be placed on the *in situ* results. He proposes the following equation for the target strength of wild gadoids at 38 kHz.

$$TS = 20 \log L - 67.4 \qquad (6.5)$$

As will be seen from Fig. 6.15, the target strengths predicted by Equation 6.5 are much less than those of the caged fish over the whole length range. The difference is greater for the larger fish, which we might expect to be more affected by confinement, because the same size of cage has been used in all the experiments.

There is little direct evidence to indicate the length dependence of target strength for other species. However, the area-scattering model has been accepted as a reasonable description of the well-studied gadoids and

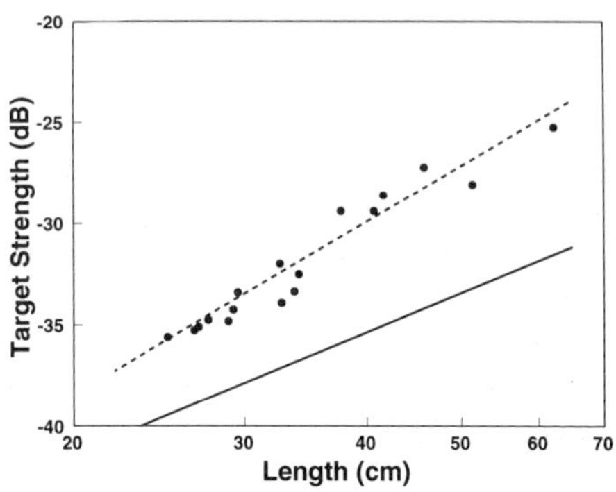

Fig. 6.15 Variation of target strength with fish length. Results from measurements on cod at 38 kHz. ● and broken curve, caged fish data and regression line from MacLennan (1981a); solid curve, best-fit equation $TS = 20 \log L - 67.4$ as recommended by Foote (1987).

clupeoids. In the absence of evidence to the contrary, we may suppose that the 20 log L dependence is a fair assumption for other species too.

Behaviour and physiology

Hawkins (1981) has described how the echo from a fish depends upon the size, density and relative position of the various organs and tissues within the body. In addition to these physiological factors, the echo depends upon the orientation of the fish body in the acoustic beam, and so behavioural factors have to be considered also.

Substantial differences in the target strength of individual fish of the same length may be explained by the natural variation of the maturity state and the condition factor within the population (Ona, 1990). It is less easy to explain changes with time, e.g. between night and day, in the target strength of a particular animal or school being observed. These changes are unlikely to be caused by physiological factors, and probably some change in the behaviour of the fish is responsible, affecting the spatial distribution of the school or the orientation of the individual fish bodies relative to the acoustic axis of the transmitting transducer.

Tilt angle

Experiments with caged fish have shown that the diurnal change in target strength correlates well with the tilt angle of the body, while the correlation with other behavioural factors, such as the nearest-neighbour distance, is unimportant (MacLennan *et al.*, 1990b). Figure 6.16 suggests an explanation as to why the tilt angle has such a pronounced effect on the target strength. When an acoustic pulse is reflected by a long, thin target such as a swimbladder, the energy in the reflected wave will change as the target tilts. There are two reasons for this effect. Firstly, the apparent size of the target, as seen from the transmitting transducer, decreases as the tilt increases. Secondly, while the apparent size varies as the cosine of the tilt angle, the echo energy may change even more rapidly because of interference between wavelets reflected from different parts of the target. When the target is parallel to the incident wavefront, as in Fig. 6.16(a), all the reflected wavelets are in phase and they reinforce each other. When the target tilts, the wavelets originating at opposite ends of the target become progressively out of phase and the summed amplitude is reduced, eventually to zero in the case of two wavelets that are exactly out of phase. This interference effect is only important when the acoustic wavelength is comparable to, or smaller than, the length of the target. In the case of very long wavelengths, the path difference is never great enough to produce a large phase difference.

As we have seen earlier, target strength changes much more rapidly with tilt angle than would be expected from geometrical considerations alone (Fig. 6.3). Interference effects are undoubtedly significant, and more so at

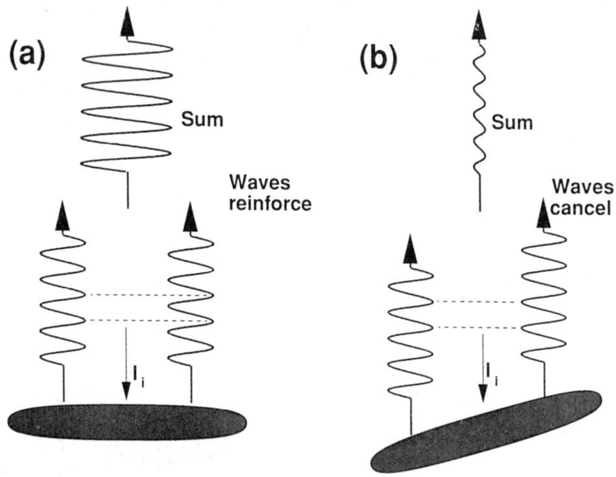

Fig. 6.16 Effect of tilt angle on the echo reflected from an oblong target. Destructive interference reduces the echo strength when the path difference to opposite ends of the target is a significant fraction of a wavelength.

higher frequencies. Moreover, experiments have shown great differences in the target strength–tilt relationship between apparently similar fish, probably owing to unseen differences in the shape of the swimbladder. Fortunately, it is the average over all the detected fish which is important for the interpretation of acoustic surveys. Thus the large changes observed for individual fish will tend to cancel when the target strength–tilt functions of all the detected fish are combined (Fedotova and Shatoba, 1983; MacLennan *et al.*, 1989).

Other factors

Changes in the horizontal distribution of the fish might explain some of the observed target strength variation in cage experiments. Figure 6.17 shows photographs of herring taken from below the cage. The fish are randomly distributed at night, but they school during the day as soon as the light level is sufficient for visual contact between nearest neighbours. However, although the night and day photographs are quite different in appearance, the statistical fish distribution is rather similar. Figure 6.18 shows a series of landscape presentations in which the height indicates the frequency of fish presence at each point in the cage. The night and day landscapes look similar to the eye, and any small statistical difference between them would be insufficient to explain the substantial diurnal change in the observed target strength.

While it has been shown that the echo strength is correlated with the mean tilt angle of fish in a school, MacLennan *et al.* (1989) found that the tilt does not explain all the observed target strength variation. It is not clear

(a)

(b)

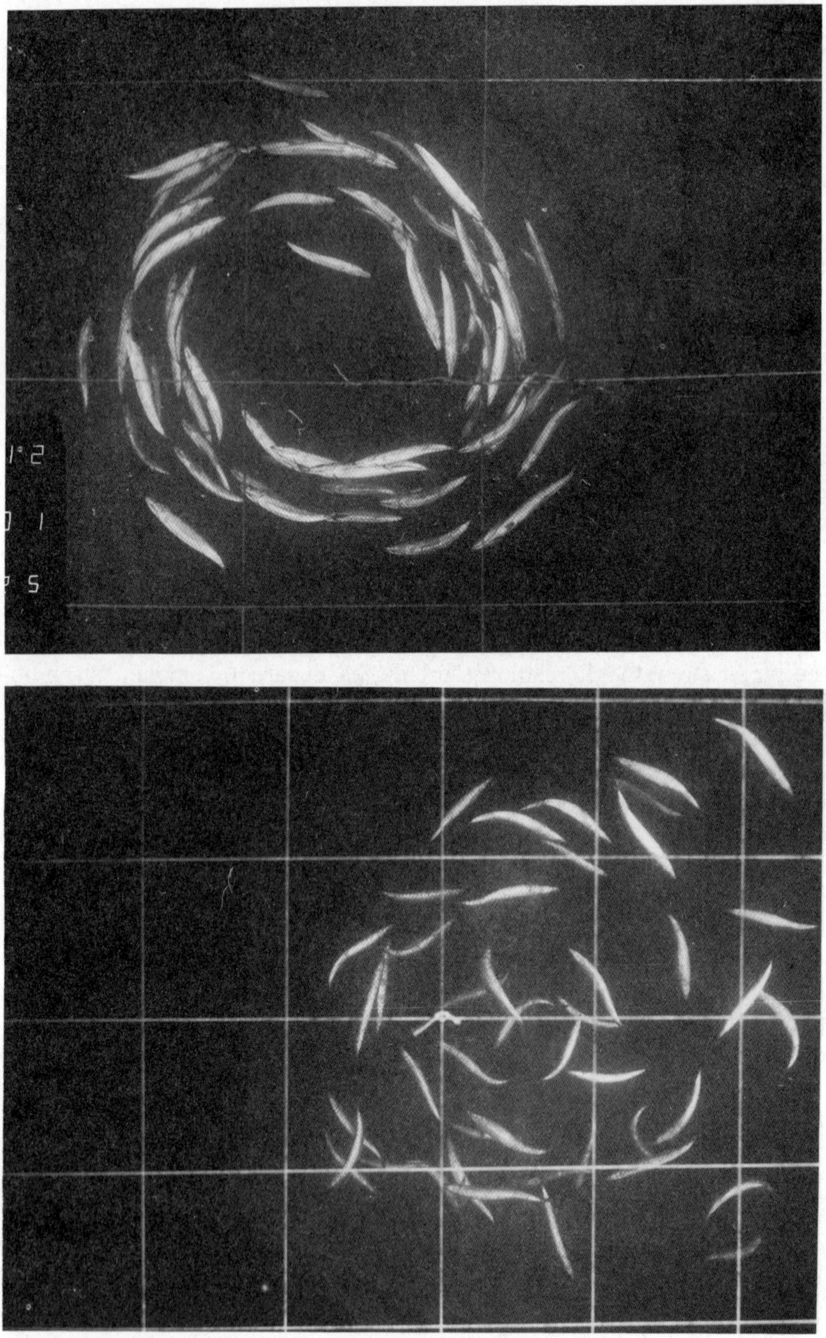

Fig. 6.17 Herring in a cage, photographed from below. (a) During the day, the fish swim in a circular school; (b) at night, the school has broken up.

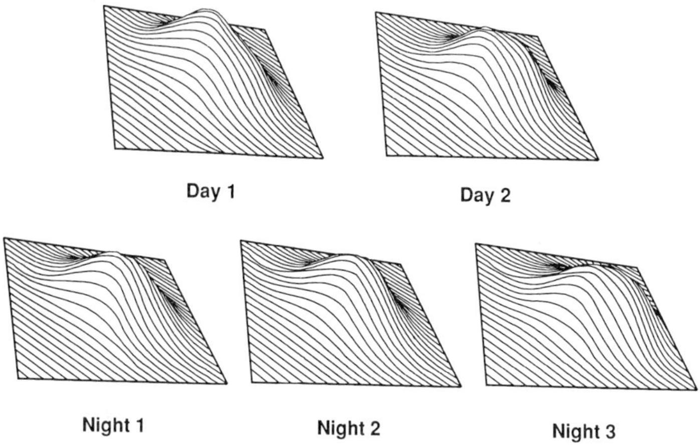

Fig. 6.18 Landscape plots showing the horizontal distribution of herring in a cage. The height of the landscape is proportional to the frequency of fish occurrence at that point in the cage. Each landscape represents accumulated observations from different 4-hour periods, around midday or midnight.

what other behavioural or physiological effects contribute to the variation, but the problem does emphasize the stochastic nature of target strength, and the need to measure many fish whose behaviour and physiology can be considered representative of those observed during the survey or other applications for which the target strength data are required.

6.5 COLLECTED TARGET STRENGTH DATA FOR SURVEY APPLICATIONS

We have seen that even for the same size and species of fish, the observed target strength can cover a wide range of values. Nevertheless, many experimental investigations have been reported in the literature. For some species at least, consideration of the mass of published data leads to a reliable indication of the expected value of target strength and how it depends upon the fish size (Foote, 1987). It is the expected value which is required for the interpretation of acoustic surveys.

It is generally accepted that target strength depends upon the fish size according to the equation $TS = \mathrm{m} \log L + \mathrm{b}$, where m and b are constants for a given species and frequency. When L is the mean length of a large number of fish, this equation gives target strength as the expected value, or the average of many observations (p. 151).

For wild fish m appears to be close to 20, the value predicted by the area-scattering model. Some writers (but not all) assume $\mathrm{m} = 20$, and then

Table 6.2 Collected results from target strength measurements of live gadoids and other physoclists. Assumed length dependence is $TS = (20 \log L + b_{20})$ unless otherwise stated

Species	*Location*	*Length*		*Time of day**	*F* (kHz)	*Mean TS* (dB)	b_{20}† (dB)	*Method*‡	*Ref* §
		Range (cm)	*Mean* (cm)						
Cod *Gadus morhua*	Norwegian Sea	50–105	81.6	N	38	−30.7	−68.9	SB	1
Saithe *Pollachius virens*	Norwegian Sea	45–91	57.2	N	38	−30.6	−65.8	SB	1
Redfish *Sebastes marinus*	Norwegian Sea	9–43	19.7	N	38	−40.6	−67.1	SB	1
Norway pout *Trisopterus esmarkii*	Norwegian Sea	12–19 10–21	14.8 17.6	N N	38 38	−44.9 −42.2	−68.3 −67.1	SB SB	1 1
Great silver smelt *Argentina silus*	Norwegian Sea	25–50	37.2	N	38	−36.6	−68.0	SB	1
Haddock *Melanogrammus aeglefinus*	Norwegian fjords	43.8 ± 4.2		N	38	−35.1	−67.9	SB	2
Walleye pollock *Theragra chalcogramma*	Bering Sea	35–52	41.7	D+N	38	−33.6	−66.0	SB, DB	3
Pacific whiting *Merluccius productus*	Pacific	47–60	51.8	N	38	−36.4	−70.7	DB	4

Blue whiting	NE Atlantic	21–37	31.1	N	29	−42.1	−71.9	Ind	5
Micromesistius poutassou	NE Atlantic	–	17.0	D	38	−42.0	−66.6	DB	6
	NE Atlantic	–	22.5	D	38	−37.5	−64.5	DB	6
	NE Atlantic	–	31.0	D	38	−35.5	−65.3	DB	6
Capelin	Gulf of St Lawrence	–	16.5	D	120	−41.0	−65.3	DB	7
Mallotus villosus	Iceland	–	11.5	N	38	−56.9	−78.1	Ind	8
	Iceland	–	14.5	N	38	−55.6	−78.8	Ind	8
Cisco	Canadian lakes	10–35	19.5	N	70	−38.9**	−64.7	Ind	9
Coregonus artedii									
Sockeye salmon	Canadian lakes	–	51.0	N	38	−27.7	−61.9	Comp	10
Oncorhynchus nerka	Canadian lakes	–	4.5	N	420	−52.3	−65.4	DB	11
	Canadian lakes	–	7.2	N	420	−49.8	−66.9	DB	11
Gadoids	Various	9–105	–	D+N	38	–	−67.4	All	12

*D, day only; N, night only; D+N, day and night.

† $b_{20} = (TS - 20 \log \bar{L})$.

‡ Comp, comparison with independent data; DB, *in situ* observations with dual-beam echosounders; SB, *in situ* observations with split-beam echosounders; Ind, indirect *in situ* observations.

§ References: 1, Foote *et al.* (1986); 2, Ona and Hansen (1986); 3, Foote and Traynor (1988); 4, Williamson and Traynor (1984); 5, Robinson (1982); 6, Forbes (1985); 7, Rose and Leggett (1988); 8, Haldorsson and Reynisson (1983); 9, Rudstam *et al.* (1987); 10, Mulligan and Kieser (1986); 11, Burczynski and Johnston (1986); 12, Foote (1987).

** Fitted equation is $TS = 21.9 \log L - 67.2$.

Table 6.3 Collected results from target strength measurements of live clupeoids. Assumed length dependence is $TS = (20 \log L + b_{20})$ unless otherwise stated

Species	Location	Length		Time of day*	F (kHz)	Mean TS (dB)	b_{20}† (dB)	Method‡	References §
		Range (cm)	Mean (cm)						
Herring	Northern North Sea	24–34	28.5	N	38	−43.0	−72.1	SB	1
Clupea harengus	Iceland	9–33	21.5	N	38	−46.6	−73.2	Ind	2
	N Baltic, coastal	6–24	13.0	N	70	−47.6**	−69.9	Ind	3
	Norwegian fjord	22–23	28.0	D	70	−43.4	−72.3	Comp	4
Sprat	North Sea	9–15	12.6	N	30	−47.1	−69.1	Ind	5
Sprattus sprattus	North Sea	–	7.15	N	30	−53.6	−70.7	Ind	5
Mixed herring/sprat	NE Atlantic	7–27	18.8	D+N	38	−45.8	−71.3	CF	6
	NE Atlantic	12–21	16.6	D+N	38	−49.0	−73.4	CF	6
	NE Atlantic	12–21	16.6	D+N	120	−51.6	−76.0	CF	6
	Kattegat/Skagerak	19–26	21.0	D+N	38	−46.2	−72.6	Ind	7
	SE Baltic	7–19	13.8	N	120	−50.3	−73.1	Ind	7
Clupeoids	Various	6–34	–	D+N	38	–	−71.9	All	1

*D, day only; N, night only; D+N, day and night.

† $b_{20} = (TS - 20 \log \bar{L})$.

‡CF, caged-fish experiment; Comp, comparison with independent data; Ind, indirect *in situ* observations; SB, *in situ* observations with split-beam echosounder.

§References: 1, Foote (1987); 2, Haldorsson and Reynisson (1983); 3, Rudstam *et al.* (1988); 4, Misund and Øvredal (1988); 5, Robinson (1983); 6, Edwards *et al.* (1984); 7, Degnbol *et al.* (1985).

** Fitted equation is $TS = 21.7 \log L - 75.5$.

Table 6.4 Collected results from target strength measurements of live crustaceans, molluscs and fish with no swimbladder. Length dependence is assumed to be $TS = (20 \log L + b_{20})$

Species	*Location*	*Length*		*Time of day**	*F* (kHz)	*Mean TS* (db)	b_{20} (db)†	*Mean weight* (g)	*TS re* 1 kg	*Method*‡	*Reference*
		Range (cm)	*Mean* (cm)								
Mackerel	NE Atlantic	31–35	32.7	D+N	38	−54.6	−84.9	275	−79.3	CF	1
Scomber scombrus	NE Atlantic	31–32	31.4	D+N	38	−59.0	−88.9	244	−82.8	CF	1
	NE Atlantic	31–32	31.4	D+N	120	−57.0	−86.9	244	−80.7	CF	1
Sandeel	NE Atlantic	11–14	12–25	D+N	38	−71.9	−93.7	6.46	−50.0	CF	2
Ammodytes spp.			–								
Krill	Antarctic	2.8–4.0	3.27	D+N	38	−85.1	−95.4	–	–	CF	3
Euphausia superba		2.8–4.0	3.27	D+N	120	−76.1	−86.4	–	–	CF	3
Euphausids	E. Pacific	1.4–1.9	1.65	D	102	−69.6	−74.0	5.52	−47.0	Comp	4
(mostly *E. pacifica*)	(California)	1.1–1.9	1.39	D	102	−71.8	−75.7	5.22	−49.0	Comp	4
Squid	E. Pacific	8–23	11.6**	D	120	−58.6	−79.9	23.7	−42.3	DB	5
Loligo opalescens	(Oregon)										
Squid	Japan	10–20	16	D	28.5	−51.3	−75.4	95.1	−41.1	CF	6
Todarodes pacificus		10–20	16	D	96.2	−54.6	−78.7	95.1	−44.4	CF	6

* D, day only; N, night only; D+N, day and night.

† $b_{20} = (TS - 20 \log \bar{L})$.

‡ CF, caged-fish experiment; Comp, comparison with independent data; DB, *in situ* dual beam.

§References: 1, Edwards *et al.* (1984); 2, Armstrong (1986); 3, Foote *et al.* (1990); 4, Pieper (1979); 5, Jefferts *et al.* (1987); 6, Arnaya *et al.* (1989b).

**Dorsal mantle length.

calculate the value of **b** which best fits their data. This approach has the merit of simplicity, and it is supported by studies which have shown that altering the value of m does not have a large effect on the computed fish density. In view of the considerable scatter in the experimental data, little precision is lost by assuming a fixed value for m.

The experimental results for clupeoids and gadoids at 38 kHz are summarized by the target strength Equations 6.4 and 6.5 respectively. However, the no-swimbladder category includes species whose physiology, shape and size are very different. Results have been reported for crustaceans as well as fish, notably the Antarctic krill, *Euphausia superba* (Foote *et al.*, 1990). One equation is unlikely to be adequate to describe the target strengths of all these animals.

Target strength data from the extensive literature on this subject are summarized in Tables 6.2 to 6.4. We have not included measurements on immobilized fish, nor experimental results published before 1979. It is only in recent years that work with live fish and modern calibrated equipment has provided reliable numeric values for the target strengths of fish and other aquatic animals.

The target strength data and size dependences described above may be used to give a first indication of the values required for the interpretation of acoustic surveys. Random variation of the individual-fish target strength about the expected value has little effect on the estimate of fish density. However, systematic changes in target strength will bias the survey results. Such changes might occur if the behaviour and the condition of the observed fish were unusual. There is a continuing need for target strength data to be collected on fish which are representative of those to be surveyed, and on other species which may be present so that their contribution to the echo integration may be deduced accurately.

Chapter seven

Practical acoustic surveying

7.1 INTRODUCTION

The acoustic estimation of fish abundance has much in common with other survey methods used in fisheries research. It involves the collection of data from the area inhabited by the species of interest, and the analysis of acoustic and other records to provide whatever information is required about the fish in the area at the time of the survey. The usual intention is to determine the size of a particular 'stock' of fish. The stock is defined by the needs of fishery management, and it is not the same as the biological concept of a 'population' (Gulland, 1983). Stocks are often defined in terms of the fish resident within artificial boundaries, e.g. national fishery limits, thus excluding any part of the relevant population that is located elsewhere.

Several practical problems have a bearing on the reliability of survey results. The accuracy required of the acoustic instrumentation is close to the limit of present-day technology. The instruments must be used correctly, with a good understanding of the factors which determine their performance. It is particularly important to calibrate the acoustic equipment regularly in accordance with the procedures recommended in Chapter 3. Furthermore, the area to be surveyed is often very large, while the volume of water sampled by the acoustic beam is small. There are random errors and biases associated with the sampling strategy, since it is necessary to extrapolate from the observations to estimate the fish abundance and distribution over the entire area of interest. Similar statistical problems arise in surveys by other methods such as trawling.

The purpose of any survey is to provide useful information about the stock of interest. The sampling strategy and the protocols for data collection should be designed to provide the most accurate information that can be obtained within the limits set by the available resources. The first step is to be clear about the objectives of the work. Is the intention to determine the fish abundance in absolute terms, or is it the change since the previous survey that is of most interest? Is it necessary to determine the geographical distribution of the stock, or only an estimate of the total abundance? How much error can be accepted in the abundance estimate, and how should the confidence limits be assessed? These questions must be answered before deciding on the detailed plan for the survey.

The data collected during the survey will come from a number of sources, notably the echo integrator and the echogram. The examination of fish catches will indicate the proportion of different species present, and the size distribution of each species, which is important in taking account of the length dependence of the target strength. Independent measurements of the target strength of fish *in situ* may be collected if equipment such as the split-beam echosounder is available. All these data contribute to the analysis which provides quantitative information about the stock, the total abundance and perhaps the structure in terms of the number of individuals at each age.

In this chapter, we discuss the equipment required to conduct the survey, and practical questions about its use. The physical calibration of the acoustic equipment has already been described (Chapter 3), but here we cover two additional methods for measuring the overall performance in the field: the live-fish calibration and the inter-ship comparison. We go on to consider how surveys should be planned to meet the objectives set in advance, especially the design of the cruise track and the sampling strategy to make the best use of the time available. The illustrations include examples of cruise tracks which show how the theoretical principles are applied in practice to the design of surveys for open water or areas bounded by irregular coastlines.

The analysis of survey data is considered here only to the extent that it has a bearing on the survey design and methods of data collection. A more complete description of the methods used to analyse survey data will be found in the next chapter.

List of symbols

a_1, a_2	Side lengths of a rectangular area element
a'	Width of an area element normal to the cruise track
A	Size of the area to be surveyed
A_r	Area of the rth region
C	Time for calibrating the acoustic instruments
CV	Coefficient of variation
d	Degree of coverage, equal to $D/\sqrt{A}$

D	Total length of the cruise track
D_r	Length of track allocated to the rth region
F	Fishing time
H	Time for hydrographic stations
L	Time for loading and unloading the ship
M	Time for travelling to and from the survey area
N_r	Average number of transects passing through an area element
P	Proportion of the day when echo integration is useful
r_i	Random number in the interval 0 to 1
s	Slope of a functional regression
S	Average distance between successive transects
S_r	Value of S in the rth region
T	Total time available for surveying and related activities
v	Speed of the survey vessel

7.2 EQUIPMENT FOR ACOUSTIC SURVEYS

The choice of equipment for a survey may be constrained by whatever is available to the surveyors, however good or bad it may be. It is often a case of making the best of whatever acoustic instruments and fishing gear are to hand. The important point is to ensure that the available equipment is adequate to meet the objectives of the survey and that it is used correctly. The essential requirements include a good echosounder, fishing gear to catch representative samples of the detected targets, and of course a suitable ship or boat.

Echosounder

Successful acoustic surveying calls for an echosounder with accurate time-varied gain and stable electronics, whose performance is consistent and predictable. This kind of instrument is usually described as a 'scientific' echosounder. When combined with an echo counter or an echo integrator, it provides numerical records which can be logged on-line by a computer or manually on paper. These records constitute the primary acoustic data of the survey. The conventional single-beam echosounder is adequate for this purpose. However, if there is a requirement to measure *in situ* target strengths during the survey, the echosounder should be of the dual-beam or split-beam type, which is capable of resolving and measuring individual targets. The different kinds of echosounder which may be used on acoustic surveys have been described in Chapter 3.

Echosounders which operate on AC power are best suited to research vessels or fishing boats in the medium-to-large size range. Some manufacturers produce portable, battery-powered equipment which can be used

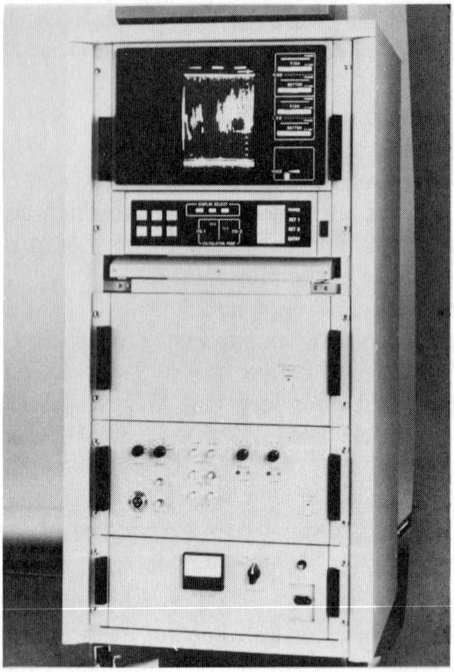

Fig. 7.1 Furuno FQ-70 single-beam echosounder with independent processors for echo integration and echo counting; dual frequency 50 and 200 kHz; colour CRT display. Reproduced with permission from Furuno Electric Co. Ltd, Nishinomiya City, Japan.

successfully on small boats. Figures 7.1 to 7.3 and Plate 8 illustrate some instruments developed specifically for acoustic survey work.

Transducer

Two kinds of transducer may be supplied with the echosounder, according to whether the device is made from ceramic (piezo-electric) material or metal plates which operate by the magneto-strictive effect (Urick, 1983). The ceramic transducers are more reliable and consistent in performance.

The transducer may be attached to the hull of the survey vessel or mounted in a towed body (Lawrie, 1984). There are arguments for and against both methods. When the transducer is a permanent fixture on the hull, electrical faults are less likely because the cable connections are not subjected to continual flexing. The hull mounting is an easy option in many respects, and once installed, the transducer is permanently available for use. However, when the transducer is in a separate body which is towed off the side of the ship, better results can be expected in bad weather, for

Fig. 7.2 Simrad EK-400 38 kHz single-beam echosounder, with paper recorder and ancillary echo-integration equipment, on the research ship *Scotia*. Crown copyright reserved.

two reasons. Firstly, aeration around the transducer is more likely on the hull of the vessel, and when this occurs, the acoustic measurements are severely degraded. Secondly, a towed body that is correctly adjusted does not pitch and roll to the same extent as the vessel. The body is towed at a few metres depth, below the immediate influence of the surface waves, and it provides a relatively stable platform for the transducer. Some research vessels are fitted with a stabilized platform on the hull which compensates for the angular motion of the vessel, but this is an expensive solution.

The towed body is simple to install, even on a commercial fishing boat with limited facilities for scientific instruments. Thus it is the obvious choice when the survey vessel does not have a hull-mounted transducer, or at least one which is good enough for the purpose. The towed body need not be particularly large (Figs 7.4, 7.5). Furthermore, it is easier to calibrate a transducer in a separate body, which is more accessible than a fixture on the hull, and the calibration may be done in relatively shallow water. We therefore recommend that the transducer should be deployed in a towed body whenever practicable.

The operating position of the towed body should be one or two metres off the ship's side, not too far aft where propeller noise or the wake

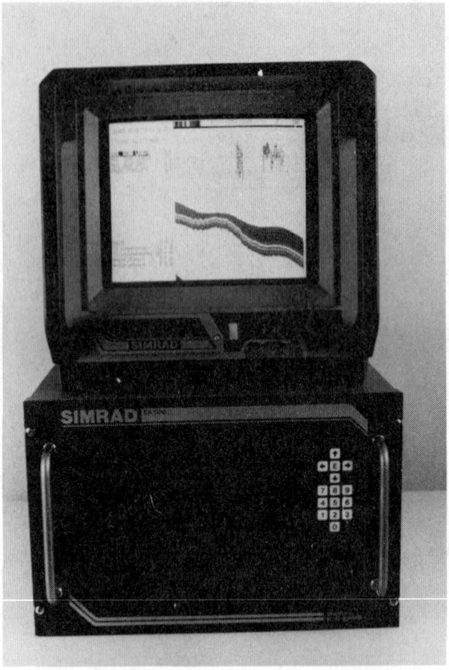

Fig. 7.3 Simrad EK-500 38 kHz split-beam echosounder with colour CRT display and controlled by selecting options from menus on the display. Reproduced with permission from Simrad A/S, Horten, Norway.

might cause problems, and yet far enough from the towing point to avoid excessive acceleration caused by the pitching and rolling of the vessel. Thus the body should be approximately on the stern quarter and towed from a forward boom (Fig. 7.6). It is convenient to use a separate lifting wire to raise and lower the body, while the electrical connections are in the fixed length of towing cable. This avoids the complication of transmitting signals through slip-rings on a winch which would be necessary if the same cable were used for lifting and towing.

Survey vessel

Ship time is expensive. In most cases the running cost of the vessel will dominate the economics of the survey. Large ships cost more than small boats, but they can work in worse weather conditions, so the cost is not the only factor to be considered. There may be little or no choice about the vessel to be used for the survey – it may simply be a question of having to use whatever is available. But if there is a choice, some thought

Fig. 7.4 Towed body (weight 60 kg) for housing the transducer shown separately on the left, as used by the Marine Laboratory, Aberdeen, UK. Crown copyright reserved.

should be given to the size of the vessel and the facilities required in relation to the objectives and the location of the survey. Table 7.1 gives an indication of the type of survey it might be reasonable to undertake with different kinds of vessel. For work in exposed areas of ocean, the larger the ship the better it will be for the comfort of personnel, apart from anything else. On the other hand, if the stock includes fish that are close inshore, a large ship might be unsuitable in view of the need to navigate in confined areas near the coast.

The survey vessel must be capable of fishing to catch samples of the echo traces, with sufficient towing power if this is to be done by trawling. Dry accommodation is required for the electronic equipment, which must be protected from the sea and, in the tropics, from humidity as well. The

Fig. 7.5 Towed body (paravane) for housing a small high-frequency transducer, shown here with the Biosonics 420 kHz dual-beam echosounder and echo processor. Reproduced with permission from Biosonics Inc., Seattle, USA.

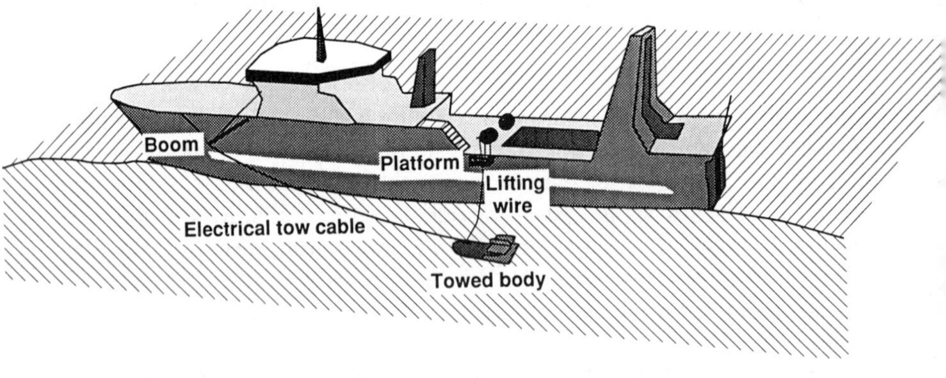

Fig. 7.6 Handling arrangement for the towed transducer body. Electrical signals are transmitted via conductors in the fixed length of towing cable. The body is towed from the forward outrigged boom, but is raised and lowered by the separate lifting wire, which is slack while towing.

Table 7.1 Practical limits on acoustic surveys according to the type and size of vessel

Vessel type	*Maximum survey duration*	*Maximum wind force (Beaufort scale)*	*Area*
Open boat	a few hours	2	Small lake
Cabin cruiser (6–8 m)	24 hours	4	Lake or inshore
Small (10–15 m)	a few days	5	Sheltered sea
Medium (30–40 m)	1–2 weeks	6–7	Offshore sea
Large (60 m or more)	3 weeks	8	Open ocean

electrical power supply must be of the correct voltage and frequency for the instruments, unless they are self-powered by batteries.

Fishing gear

The collection of biological samples is an essential part of any acoustic survey. It is done to determine the species composition and the size distribution of targets detected by the echosounder. The samples are collected by fishing with a trawl or other type of gear. General information on the design and operation of fishing gear will be found in FAO (1972), Strange (1981) and von Brandt (1984).

It is not necessary to catch a large quantity of fish. More importantly, the size and species composition of the catch should be representative of the fish population in the area. Ideally, the gear should have the same efficiency in catching different species and sizes of fish, that is to say, it should be 'non-selective'. Unfortunately, all fishing gears are selective to some extent (ICNAF, 1963).

Drift nets and hooks on lines are highly selective in the size of fish taken. The use of selective gear is generally encouraged in commercial fishing, to reduce the mortality of young fish, but the opposite is required of the gears used for research surveys. For this reason, lines and drift nets should only be used to collect samples when no other method is available, for example if the vessel is not equipped to trawl.

The purse seine is selective in a different way. With the aid of sonar, it can be used to capture an entire school. It is an effective method for the study of school structure. The catch determines the total biomass, which can then be compared with acoustic measurements of the school shape and density (Misund and Øvredal, 1988). However, the purse seine is not well suited to the sampling requirements of echo integrator surveys. Several species may be present in the area, whereas each school will normally consist of one species only. Thus a few purse seine catches will not give a good indication of the species composition of the population at large.

In some fisheries the purse seine is used in conjunction with a FAD (fish aggregating device), and again the catch will be some unknown selection from the local population. Furthermore, purse seining is a highly skilled business. On no account should it be attempted as the primary sampling method on a vessel whose crew have no experience of working this gear.

The pelagic trawl (Fig. 7.7) is commonly used to sample echo traces, and it is the recommended method provided that the survey vessel has sufficient towing power and is equipped for this type of fishing. In particular, it is essential to have a netsonde. This is an echosounder whose transducer is attached to the net, to detect fish in the vicinity and to determine the gear depth by reference to the surface or bottom echoes. Suppose that it is decided to sample a particular school which has been detected by the shipboard echosounder. The survey stops, the ship turns and the trawl is shot, towing back along the survey track. The depth of the trawl is adjusted by reference to the netsonde echogram until it matches that of the school. This procedure is known as aimed trawling. It is most effective when searchlight or multi-beam sonar is used to detect the school ahead of the vessel, since the fish may have moved off the survey track. Pelagic trawling is not an effective method for sampling schools when it is done without the aid of acoustic instruments. There is too much empty water between the schools and the chances of taking a representative catch by blind fishing are extremely slim.

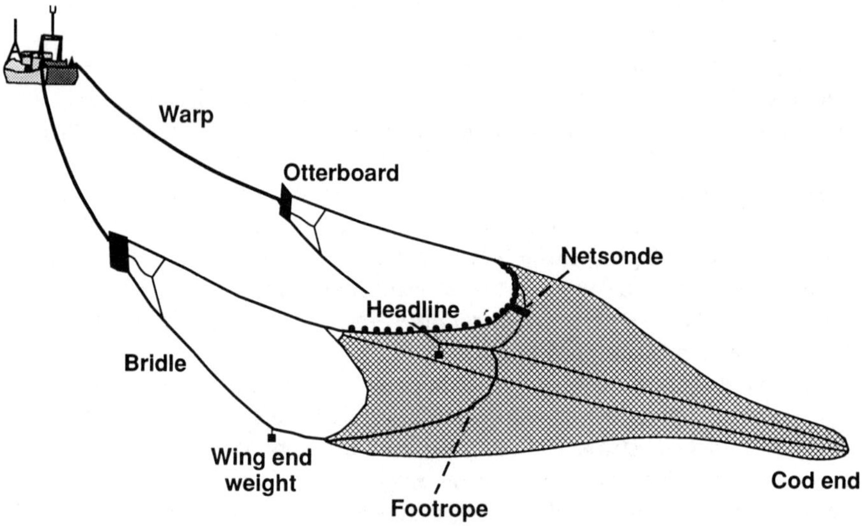

Fig. 7.7 Single-boat pelagic trawl. The netsonde is an echosounder, attached to the headline, which monitors the vertical net opening and the net depth. Netsonde signals are transmitted to the ship via an acoustic link or a separate cable (not shown).

The pelagic trawl must not be too large in relation to the towing power of the vessel, since it must be towed at a speed of 4 knots ($2.6\,m\,s^{-1}$) or more. If the towing speed is too low, the larger fish (which can swim faster) may escape the net while the smaller ones are caught more efficiently. On the other hand, the trawl must not be too small. It should have a mouth opening of at least 10 m in the fishing condition. Small nets may be unsuccessful because the fish do not have far to swim in order to escape (Wardle, 1983). The pelagic trawl may not be the best choice for boats with engines smaller than 600 hp which would be unable to tow a large enough gear at the required speed. However, in the case of surveys conducted at night when the fish are dispersed, escape motivated by vision is less important and a small trawl towed at low speed may then provide adequate samples.

Different sampling problems arise in the case of surveys for very fast swimmers such as adult salmon. These fish cannot be caught by trawl, except perhaps for the occasional straggler. Samples may be taken by angling or drift netting, but these methods are highly selective. The bag net or the beach seine is a better choice for the sampling of migrating fish close to the shore. However, when the survey is to be done in a small lake, the population structure may be well known from other biological studies of the area, or it may be deduced from the catches taken by commercial or sport fishermen.

Essential requirements

Success in acoustic surveying depends upon instruments and other equipment whose performance is good enough to meet the objectives of the survey. The essential requirements are as follows.

1. A scientific echosounder and signal-processing instruments (echo integrator or echo counter).
2. Equipment for calibrating the acoustic instruments (standard target and suspension apparatus).
3. A vessel of suitable size for the area to be surveyed, capable of fishing and with adequate facilities for scientific work (power supply, accommodation).
4. Fishing gear for the collection of biological samples.

7.3 PERFORMANCE TESTS

In this section we discuss two methods for testing the complete acoustic measurement system as installed on the survey vessel – live-fish calibration and inter-ship comparison. These techniques are not a substitute for the physical calibration procedures described in Chapter 3. However, they do allow the overall performance of the equipment to be tested on signals from live targets. Thus the acoustic properties of the fish are included in the measurement. The performance test should not be regarded as a precise cali- bration, but it could be useful in revealing a large change in the detectability of fish, due to an equipment failure perhaps, or differences between the performance of two vessels which otherwise might not be discovered.

Live-fish calibration

Johannesson and Losse (1977) first described this technique in which the performance of the equipment is measured while a known quantity of fish is insonified. The fish are caught locally and then placed in a cage which is suspended beneath the transducer. In principle, the one measurement determines the conversion factor for estimating the fish density from the echo integrals, taking account of the target strength of the fish as well as the physical parameters of the echosounder. The technique is similar to the caged fish method for measuring target strength (Chapter 6).

At first sight, live-fish calibration may appear to be a simple technique which obviates the need for the more complicated physical calibration of the acoustic instruments. However, we consider that live-fish calibration is only useful in providing an indication of the target strength of the fish. It is not a substitute for the recommended calibration procedure. The signal from the fish may vary for behavioural and physiological reasons to a much greater extent than the physical performance of the equipment. If live-fish calibration is the only measurement made, it will not be possible to say whether a substantial change in the sensitivity is caused by the fish targets, or by the equipment. Furthermore, it is doubtful whether the echoes received from captive fish of uncertain physiological condition will be representative of those from free-swimming fish in the wild. We do not recommend live-fish calibration as part of the normal survey procedure, unless there is no other means of estimating the target strength of the species of interest.

Inter-ship comparison

Acoustic surveys sometimes involve more than one ship. Each ship surveys part of the area occupied by the stock, and the results are combined to produce one estimate of the abundance. A common practice on this kind of survey is to conduct an inter-ship comparison when the two ships are operating in the same area. This technique is not an absolute calibration, but it is a useful means of checking that the acoustic systems on two ships are working correctly.

The inter-ship comparison works best when it is done in an area where there are substantial quantities of fish in layers, or dispersed aggregations of varying density. The comparison is performed by the two ships moving in formation, with one in the lead and the other about 400 m astern and far enough to the side to be clear of the leader's wake (Fig. 7.8). The two ships should take the lead in turns, exchanging position at the end of each transect (Rottingen, 1978).

Both ships record data from their echo integrators during the comparison. Normally a period of 2–6 hours is required to collect sufficient data, depending on the spatial distribution and the density of fish in the area. It is important to synchronize the integration periods on the two ships, so that they relate to the same portions of cruise track. Thus the following ship must record the echo integrals over time periods which are delayed relative to those of the leader. Constant radio communication between the ships is essential to ensure satisfactory cooperation and comparability of the results.

The two ships do not move on coincident tracks, because the following ship would then suffer interference from the wake of the lead vessel, but it is assumed that they will nevertheless detect the same density of fish on average. Once the data have been recorded, the two sets of echo integrator records, expressed in terms of fish density or other comparable units, are plotted against each other. Since we cannot assume that one ship is more liable to error than the other, the functional regression technique is used to fit a straight line to the results (Ricker, 1973). If the equipment on both ships is performing correctly, and if they have detected similar fish concen-

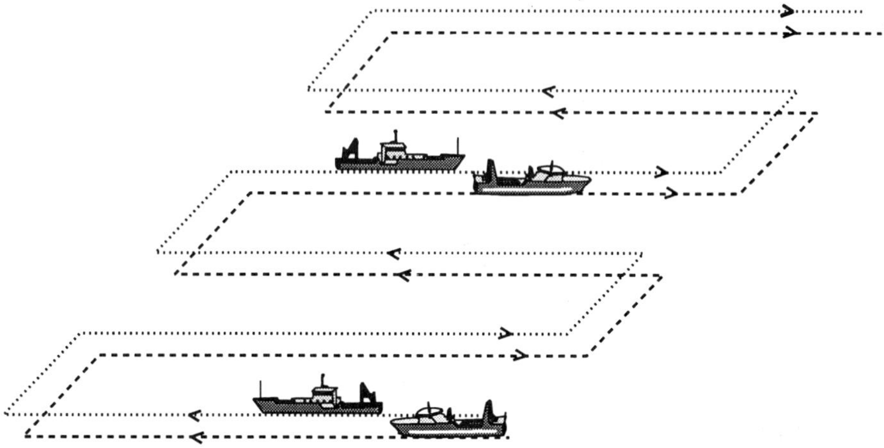

Fig. 7.8 The inter-ship comparison. Two ships move in formation while operating their acoustic instruments. The lead is taken by each vessel in turn, changing after each transect.

trations, then the regression line should have a slope *s* close to unity, any difference being explained by random sampling error. MacLennan and Pope (1983) have shown how the confidence limits on *s* may be estimated. If the 95% confidence limits do not include unity, as in the example shown in Fig. 7.9, this suggests that one ship is producing biased results, and the reasons for the discrepancy should be investigated.

The advantage of inter-ship comparison is that it will demonstrate any gross difference in acoustic performance. It is not a substitute for the proper calibration of the acoustic equipment, which should be carried out on both ships beforehand. If the comparison nevertheless reveals a large difference, the equipment on both ships should be recalibrated as soon as possible, in accordance with the procedures described in Chapter 3.

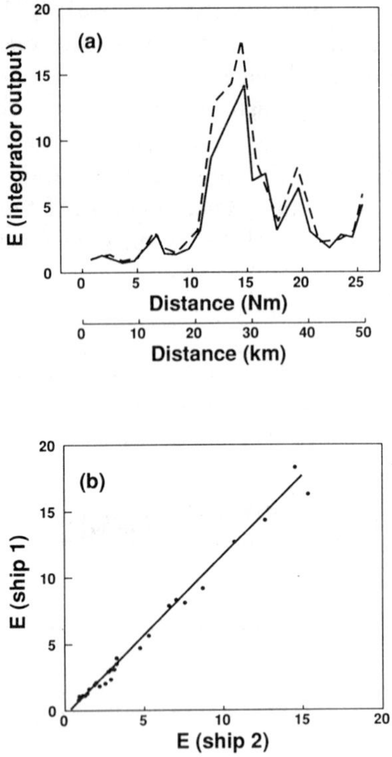

Fig. 7.9 Results of an inter-ship comparison. *E* is the mean echo integral per 1.852 km (=1 nautical mile, Nm) of cruise track, units arbitrary but the same on both ships; broken line, ship 1; solid curve, ship 2. (a) *E* v. distance sailed; (b) comparison of measured E values (points) and regression line *E* (ship 1)=1.21 *E* (ship 2)−21.3. Redrawn from Foote *et al.* (1987).

7.4 SURVEY DESIGN

Acoustic surveys are conducted to investigate a large volume of water. In practice, owing to the limited time available to perform the survey, only a small proportion of this volume can be observed acoustically. Thus the acoustic measurements are samples which are assumed to be representative of the wider fish distribution. The survey design is the planning of the cruise track, which needs to be considered well in advance of the survey itself. The object is to ensure that the resources available are adequate for the job to be done, that the resources are used to best effect, and that all the information required for the subsequent analysis is collected. The survey is designed by working through a set procedure along the following lines.

1. Define the geographical area to be covered, or if it is intended to employ an adaptive strategy (p. 227), decide on the principles to be applied in adjusting the coverage during the survey.
2. Estimate the resources required for an adequate survey of the defined area, principally the ship time in relation to the density of sampling.
3. Calculate the time available for the survey itself, making due allowance for other activities such as fishing.
4. Decide on the sampling strategy and the type of cruise track to be followed, i.e. the shape of the survey grid (triangular, rectangular, etc.).
5. Lay out the calculated length of cruise track on a map, ensuring that representative samples will be collected from all parts of the area as far as practicable.

If the survey is well designed, it will be much easier to analyse the data later and to produce satisfactory results. It is not a good idea to begin a survey with no strategy or plan as to how it should be conducted. This would very likely result in wasted time and inefficiency.

Acoustic surveys are similar in some respects to the sighting surveys which are conducted to count animals by visual observation. In both cases, the data are collected while the observer moves along the lines of a grid. The theory of sighting surveys corrects the population count for the probability of detection, which depends upon the range and the size of the observed animals (Quinn, 1985). However, this theory has little relevance to acoustic surveys, because the size and range of targets is taken into account by the automatic processing of the signal before the echo is integrated. There are statistical problems to be considered in the interpretation of acoustic data, but they have more to do with classical sampling theory (Cochran, 1977).

Geographical area

The area to be covered by the survey may be obvious, for example when it is bounded by land or national borders. In other cases, the boundaries may be decided by prior knowledge of the likely fish distribution and migration behaviour, or by reference to the water depth. The survey should be confined to the area where there is some chance of encountering the species of interest. If the surveyed area includes large regions of empty water, the coverage (sampling intensity) of fish concentrations will be less than optimum.

When the geographical limits of the stock distribution are uncertain, it may be decided to adopt an adaptive strategy, which allows the cruise track to be adjusted during the survey according to where concentrations of fish are observed. In that case, the area to be covered is not specified in detail beforehand, but it is still necessary to consider the range of possible options, so that the available time can be used to the best effect. All regions where fish may be found need to be surveyed to some extent, even though the coverage of some may be minimal.

Working time

The next step is to determine the time available for the collection of acoustic data, which we call the track time. Ideally, this should be decided on the basis of the sampling intensity needed to map the stock with acceptable precision. In practice, acoustic sampling is often constrained by the availability of the ship or other resources. Thus we begin with T, the total period within which the survey must be completed. The track-time is calculated by deducting from T the time required for other activities, such as loading and unloading the ship (L), and any travelling between the embarkation point and the survey area (M). It is also necessary to allow some time for calibrating the acoustic instruments (C), fishing to identify the echo traces (F) and taking hydrographic stations (H). Furthermore, the track time may be restricted to part of each day. If the fish migrate vertically in a diurnal cycle, the survey must be done during the hours when the fish are in mid-water. Some surveys will be restricted to the daylight hours when the fish are concentrated in schools. Others may be done only at night when the targets are dispersed. The important point is to ensure that the behaviour of the detected fish is consistent throughout the track time. The planning in advance of the survey must provide for all the ancillary activities as well as the collection of acoustic samples. It is not essential to conduct hydrographic stations if the environmental conditions are known well enough from other sources.

If v is the ship speed and P is the proportion of each day that can be used for echo integration, the total length of the cruise track is calculated as

$$D=(T-L-M-C-F-H)Pv \qquad (7.1)$$

It is necessary to decide in advance on a general scheme for allocating time among the different activities. For example, hydrographic data might be collected at selected positions along the cruise track. Each station might occupy an hour or so, depending on the water depth to be covered and the type of instrumentation available. Calibrations should not be performed in haste, and several hours must be allowed to do the job properly. The need for fishing is more difficult to predict, since fish samples are required only to partition the acoustic data between species and size groups of fish. However, it may be decided to allow for a certain number of trawl stations each day on average, and then to fish as and when there is doubt about the identity of the echo traces. As a rough guide, some 20% of the working time might be allowed for fishing on echo traces. It may require 1–3 hours to complete a trawl station on a large ship, and perhaps 30 minutes when using a light gear deployed from a small boat.

Strategic considerations

There are two fundamentally different approaches to the design of acoustic surveys, according to whether the cruise track is decided in advance (pre-planned) or adjusted on the basis of observations made during the survey (adaptive). The adaptive strategy has the advantage that it allows the geographical limits of the fish distribution to be explored. However, the non-random sampling which is implied by the adaptive approach may result in bias and uncertain statistical error in the abundance estimate. In most cases, it is better to plan the cruise track in advance, allowing changes during the survey only for practical reasons such as bad weather. The adaptive survey technique should only be considered if the main objective is to detect concentrations of fish which might be exploited by commercial fishing, and it is less important to obtain a good estimate of the stock abundance. The remainder of this section is concerned with pre-planned surveys.

The distribution of fish over the area to be surveyed may not be stationary in the statistical sense, meaning that the probability of finding a particular fish density is not the same everywhere. We may know that there are very few targets in some places, but expect dense concentrations in others. This topic is discussed further in section 8.4. If the broad pattern of the fish distribution is unknown, the best plan is to design the cruise track so that the whole area is covered evenly. On the other hand, if there is some indication as to where fish are most likely to be found, inshore as opposed to offshore for example, it is better to divide the area into regions, and to allocate more time and effort to those regions expected to contain the most fish. This is the technique of stratified sampling. For the fish densities observed on acoustic surveys, the variance usually increases with the mean. In that case, it can be shown that if the sampling effort is allocated between regions in proportion to the expected concentration of targets therein, the variance of the total abundance estimate is reduced (Cochran, 1977).

The general principle to be applied in dividing an area into regions is that the regions should be the largest divisions within which the distribution of the stock is believed to be stationary. If the area is divided into too many regions, this merely complicates the analysis with no improvement in precision. About five regions are as many as should be considered in most cases.

Prior information about the fish distribution may be obtained from commercial fishing activity, other research surveys or knowledge of fish behaviour. The important point is to think about the scale at which non-stationarity becomes evident; this determines the smallest region that needs to be considered. A region need not have a regular shape. For example, if the area is bounded by a long coast on one side and deep ocean on the other, as on the continental shelves off Morocco or Peru, the distribution of the stock perpendicular to the shore can be thought of as non-stationary on a smaller scale than that along the coast. In waters which shelve more gradually, such as the south-western part of the South China Sea, trends in the distribution are unlikely to be evident even on a scale of 50 km or more. At the other extreme, fish may concentrate along the edges of banks or other topographical features of the sea-bed, in which case the non-stationarity may occur on a scale of only one or two kilometres, as is found sometimes in the North Sea between the British Isles and Continental Europe.

A general rule which can be applied in deciding whether or how to stratify an area is to do this only when it is believed that the mean fish density in two potential regions will be different by at least a factor of two. Not more than three different track spacings should be used in allocating the cruise track between regions. What is required is a broad match between the intensity of sampling and the expected fish distribution. For example, suppose we believe that about two-thirds of the stock will be found within one-half of the area to be surveyed. It would then be reasonable to divide the area into two regions and to allocate two-thirds of the cruise track to the region containing more fish.

In the absence of prior information, the fish may be considered to be randomly distributed over the survey area, but this assumption no longer holds once concentrations have been located. It is necessary to be clear about the implications of this for the subsequent analysis of the acoustic data. For example, fishing is done to identify the echo traces. This activity will naturally occur in areas of higher-than-average density. Acoustic data may be collected along additional unplanned sections of cruise track while the ship manoeuvres to fish in the correct position. Some possibility of bias is introduced by the uneven sampling. To avoid this difficulty, the data collected along the unplanned sections of cruise track may simply be ignored in the analysis. However, if the fish are so mobile that there is no correlation between the densities recorded along coincident cruise tracks covered at different times, then all the data may be regarded as independent samples which may be used in estimating the stock abundance. There are other complications arising from the migration of fish to which we shall return later.

Sampling intensity

The next stage of the survey design is to decide how the cruise track of length D should be positioned to ensure the best coverage of the area to be surveyed. Some proportion of D is allocated to each region which is to be sampled at uniform intensity. The cruise track normally consists of a number of straight lines, called transects or legs, joined at the ends. Suppose that A_r is the area of a region to which the track length D_r has been allocated, and it is to be covered by a regular survey grid. The transect spacing is approximately

$$S_r = A_r / D_r \tag{7.2}$$

At this point it is necessary to consider how the acoustic data will later be used to estimate the stock abundance. This is commonly done by calculating the mean fish density in rectangular elements of area. Each region consists of a number of contiguous elements. The cruise track must pass through each element at least once to obtain useful data. It is supposed that the observed fish densities are representative of the element within which the data are collected. Thus the element must be small enough to ensure that the expected density is constant wherever the track goes. On the other hand, the width of the element cannot be smaller than the spacing of the transects, otherwise some elements would not be sampled at all. For example, suppose that the transects within a region are 15 km apart. This means that the smallest useful element would be a square of 15 km side. Within each square, it would be assumed that the conditions determining the aggregation of fish are stationary, so that the data collected from a cruise track crossing any part of the square would be representative of the element as a whole. More generally, the element might be a rectangle of sides a_1 and a_2, and area (a_1a_2). If the transect direction is arbitrary, the average number of transects passing through an element is

$$N_r = \sqrt{(a_1a_1)}/S_r \tag{7.3}$$

If the elements are squares of side a, this formula reduces to $N_r = a/S_r$, and the number of transects per element does not depend upon the direction of the cruise track. In the case of rectangular elements and transects which are nearly parallel, N_r is estimated more precisely as a'/S_r, where a' is the width of the element in the direction normal to the cruise track.

N_r should never be less than one. If it is, and if there is no possibility of allocating more cruise track to the region concerned, then the size of the element must be increased. If N_r is more than 1.5, consideration should be given to reducing D_r in that region, to allow better coverage elsewhere.

When designing the cruise track it is a good idea to have some understanding of how the precision of the results depends upon the transect spacing. This statistical problem has been addressed by Aglen (1983, 1989) (section 8.5). He suggests that the coefficient of variation (*CV*, the standard deviation divided by the mean) of the abundance estimate depends upon the degree of coverage, defined as $d = D/\sqrt{A}$ where *A* is the size of the area being surveyed. The larger *d* is, the smaller is *CV* although it depends also on the type of fish distribution. Better results can be expected when the fish are randomly distributed as individuals than when they are concentrated in isolated schools. For planning purposes, we may assume that $CV = 0.5/\sqrt{d}$. Thus if the required precision is $CV = 0.25$, *d* must be at least 4. However, this rough calculation of *CV* is not a substitute for the proper estimation of confidence limits after the survey (section 8.5).

Cruise track designs

Within each region, the allocated length of cruise track is laid along a series of straight lines joined at the ends. These lines are the transects, which normally run between points on opposite boundaries of the region. Several styles of track design may be adopted within this broad concept. The transects may be parallel to one another (Fig. 7.10(a)) or they may zigzag in a triangular pattern (Fig. 7.10(b)). Another consideration is whether the

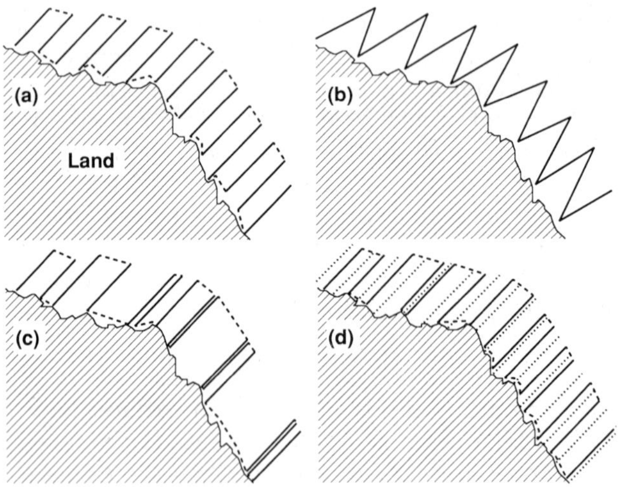

Fig. 7.10 Four types of cruise track. (a) Systematic parallel transects (broken lines join sections of track); (b) systematic triangular transects; (c) completely random design; (d) partly random design – one transect is placed randomly within each zone (dotted lines, zone boundaries).

track design should be systematic, with constant spacing between adjacent transects, or randomized in some way.

There are differing views as to whether the cruise track should be designed as a systematic or a randomized grid. Some maintain that the systematic approach is more useful in providing representative coverage of the whole area (Nickerson and Dowd, 1977; Kimura and Lemberg, 1981). Others have argued that randomly located, parallel transects are essential to satisfy well-established principles of statistical sampling theory (Jolly and Hampton, 1990). There are arguments for and against both views. We consider that the best design of cruise track depends upon what is known about the distribution and behaviour of the target species, and the kind of information which is required of the survey. The choice is not clear cut.

If one of the survey objectives is to determine the geographical distribution of the stock, a uniformly spaced track is better because the sampling effort is distributed evenly over the area of interest. The problem with the randomized track is that the sampling intensity is variable and large tracts of water may not be surveyed at all. If the fish are likely to be found in a few localized concentrations with much empty water in between, there is a risk that a survey on a randomized track will not detect major concentrations of fish, or features of their geographical distribution which could be important for the future development of a commercial fishery. Whether justified or not, the survey results may be criticized on the grounds that commercially important fish concentrations might have been located in those parts of the area not covered by cruise track. Gaps in the coverage are most noticeable when the transect positions are completely random (Fig. 7.10(c)). Each transect begins at a point which may be anywhere on the coastal boundary. Alternatively, there is the partly randomized design, in which the area is first divided into equally spaced zones. Exactly one transect is randomly positioned within each zone (Fig. 7.10(d)). This is the method recommended by Jolly and Hampton (1990). It spreads the coverage more evenly while maintaining the important features of the random sampling strategy.

If the expected value of the fish density is the same everywhere within a region, then the abundance estimate will be unbiased whether the track is systematic or random. An evenly spaced track has the advantage of providing better qualitative information about the geographical distribution of the stock. However, it is more difficult to determine the variance of the abundance estimate when the cruise track is systematic, and indeed this can only be done if some assumptions are made about the spatial distribution of the stock. For example, if we assume that the fish are distributed randomly, with no correlation on the scale of the transect spacing, then the regularity of the survey grid is not important. The densities observed along successive sections of track may then be considered as independent random samples. Unfortunately, the spatial distribution of fish is often more complicated than this. The distribution may be patchy, with strong correlation between the densities observed at nearby points.

If the main requirement is to obtain the most precise estimate of abundance, then we believe that in most cases a systematic track design will give the best results. It may not be possible to obtain a reliable indication of the precision, but if the geographical distribution of the stock is also required, this will be better understood. On the other hand, if it is considered to be more important to know the stock size within reliable confidence limits, then we suggest that the transects should be parallel and partly randomized as described above. We do not recommend the completely random design in any circumstances.

The above discussion refers to the calculation of variance from the results of one survey taken in isolation, that is to say based on the observations made along a continuous cruise track. Alternatively, the variance might be determined from the results of several surveys covering the same area, conducted by different vessels concurrently or one vessel at different times. In that case, the result of each survey is an independent estimate of the abundance.

The randomized track should always be based on parallel transects. The systematic grid may be parallel or triangular. The latter is not recommended when the transects are very long compared to their average separation, because the transect spacing (and the sampling intensity) becomes uneven near the turning points on the boundaries. We suggest that the transects should be parallel whenever the length exceeds five times the spacing, and conversely that the triangular grid is the better design when the transect length is shorter than this limit. A mixture of parallel and triangular transects might be used in a region which is wedge-shaped so that the length/spacing ratio is variable.

Transect direction

The survey grid might be constructed from transects laid on east–west lines, in which case the general progression of the cruise track is from north to south (or vice versa). The transects might equally be laid in any other direction, and the best choice is not always obvious. Several factors need to be taken into account, as listed below.

1. If the area to be covered is much longer than it is wide, then the transects should be parallel to the shorter axis. This minimizes the time between successive transects.
2. If the stock exhibits a density gradient in one direction only, between inshore and offshore regions for example, then the transects should run in the direction of the density gradient.
3. If the fish are known to be migrating in a particular direction, then the transects should be parallel to this direction. Thus successive transects

will run with and against the movement of the stock, which reduces the bias in the abundance estimate (section 8.6).

4. The presence of islands or the shape of the regional boundary may restrict the choice of transect direction. Another practical consideration is the position of the start and end points of the cruise track; these should preferably be close to the shore base of the survey.

In practice, the several factors to be considered may not suggest the same transect direction. It is necessary to decide on the compromise that will give the best results.

Mapping the cruise track

Once the type of transect to be followed is decided, the calculated length of track is drawn on a map of the area to be surveyed. When there are several regions to be sampled at different intensities, the mapping calculations need to be performed separately for each region, with the start and end points arranged to ensure continuity of the track at the boundaries.

In the case of the systematic triangular grid, the transects run between turning points on opposite sides of the region. The distance between the turning points along a boundary is twice the average spacing of the transects (Fig. 7.10(b)). If the sampling intensity has been decided, say to be twice that of some other region, this will determine the transect spacing and therefore the distance between the turning points. The turning points are plotted on the map at staggered positions on the opposing boundaries and joined by straight lines to form the grid. If the transect spacing has not been determined explicitly, but a particular length of track has to be fitted into the region, then it is simply a question of adjusting the distance between the turning points (and the number of transects) to achieve the desired result. The number of transects is given approximately by the track length divided by the average width of the region.

The systematic parallel grid is mapped in much the same way as regards calculating the number and spacing of transects, but there is now the added complication of how to deal with the short portions of track joining the ends. One view is simply to continue each transect to the regional boundary and to ignore data collected along the joining track as not being part of the survey design. This procedure may appear to be wasteful, but it simplifies the analysis of the results. The extent to which the intertransect data can be usefully employed in the analysis depends upon the fish distribution at the regional boundary. If the fish density is likely to change rapidly with position, as might occur at a boundary on a coastline, the inter-transect data should be ignored. On the other hand, if the boundary is in open water, it may be reasonable to take account of the observations along the joining section. In that case, the transects should not be continued all the way to the boundary. They should end at points short of the boundary by half the transect spacing. This is done to ensure that the centre and the edges of the region are sampled at the same intensity.

If the transects are to be randomized, first calculate the average spacing *S*, which is equal to that of a systematic parallel grid having the same total track length. We begin by dividing the area into zones of width *S*. A sequence of numbers r_1, r_2, etc. is written down, each number having a random value from a uniform distribution on the interval 0 to 1. These numbers can be obtained from mathematical tables or the random-number generator which is provided with most computers. The one transect in the *i*th zone is placed at the distance r_iS from the zone boundary. Thus an adjacent pair of transects may be very close together or as much as 2 *S*

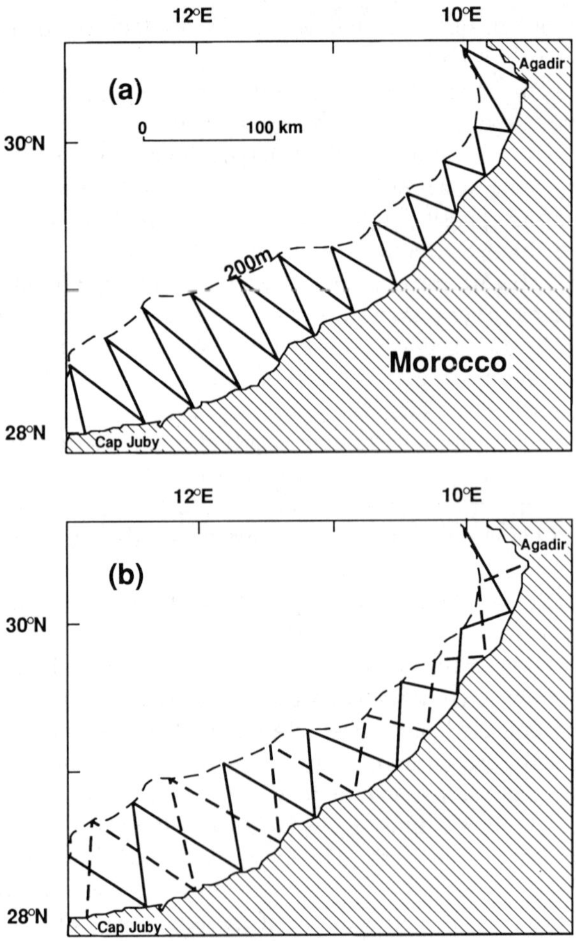

Fig. 7.11 Example cruise tracks: continental shelf off Morocco, survey from coastline to 200 m depth contour. (a) Triangular grid in one direction; (b) return grid of interleaved transects.

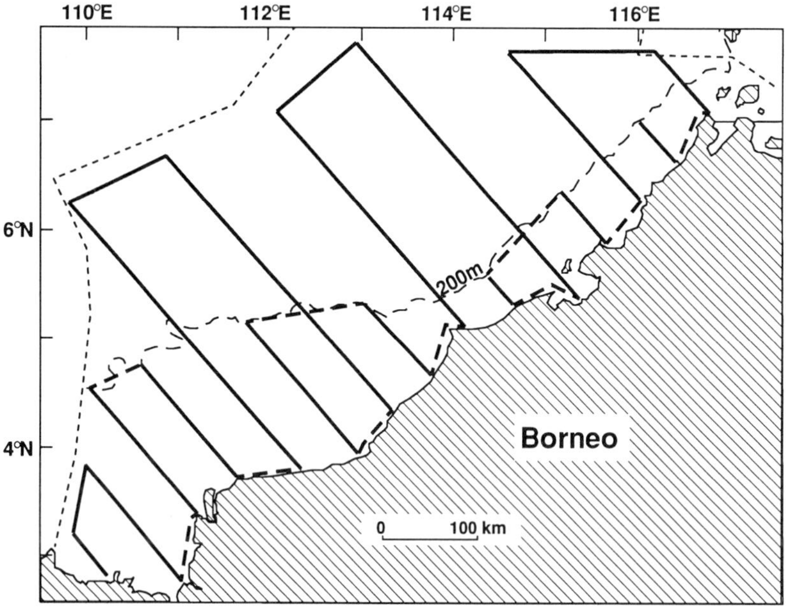

Fig. 7.12 Example cruise track: South China Sea, shelf and deep-water regions. The transect spacing is closer on the shelf where fish are more likely to be found.

apart. It is wasteful of survey time to include a pair of transects which are nearly coincident, and furthermore the two samples thus obtained may be highly correlated. This problem can be avoided by constraining the transect separation to be no less than some minimum, say 10% of the average spacing, which may be done by deleting from the random number series any of the r_i which are less than 0.05 or greater than 0.95. This procedure could result in bias if the fish distribution were non-stationary, but in practice it is most unlikely that non-stationarity would be evident on the scale of 10% of the transect spacing.

Specimen examples of cruise tracks are shown in Figs 7.11 to 7.14. These illustrate some of the practical problems which have to be solved in survey design, to take account of (a) geographical features such as islands, coastlines and enclosed areas with an irregular boundary, (b) the need for different sampling intensities from place to place and (c) the migration behaviour of the stock.

The EDSU

The elementary distance sampling unit (EDSU) is the length of cruise track along which the acoustic measurements are averaged to give one sample. The survey is conducted by collecting a series of samples from contiguous

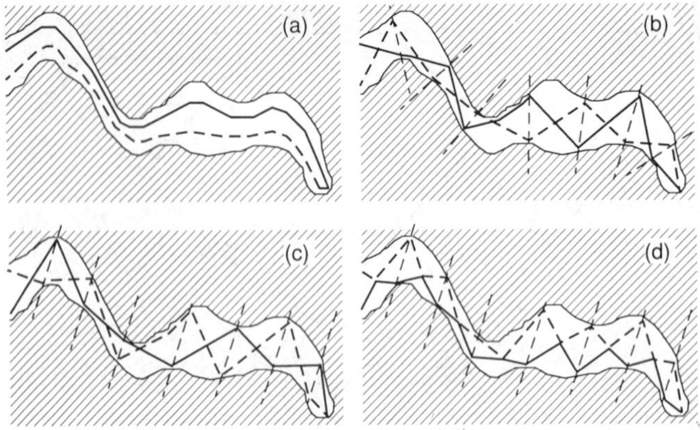

Fig. 7.13 Example cruise tracks: long narrow fjord or lake. (a) In one side, out the other; even area coverage but sensitive to crossways migration of fish. (b) Interleaved triangular track in directions determined by normals to the coastline; incorrect coverage on corners. (c) This problem is reduced with regular spacing of the track. (d) Separate triangular grids give the best area coverage but are sensitive to crossways migration.

sections of track 1 EDSU long. Each sample is considered to be representative of the fish density at the centre of the corresponding EDSU.

The optimum length of the EDSU must be decided at an early stage of the survey design. If the EDSU is too large, potentially useful information about the geographical distribution of the stock will be lost. If it is too small, successive samples will be correlated, in which case it will be difficult to determine the confidence limits on the stock-abundance estimate. As a general rule, the EDSU should be just large enough so that in regions where fish are observed, the correlation between pairs of successive samples is acceptably small. In this context, 'acceptably small' means that the error limits (at the 95% confidence level) on the correlation coefficient estimated from the observed fish densities should encompass zero (MacLennan and MacKenzie, 1988).

The data analysis is performed by calculating the abundance in elementary units of area. There should be several samples within each area element, whose width should therefore be much larger than the EDSU. The size of the element may be restricted by other considerations, in which case the need for adequate sampling may require an EDSU which is too short to avoid the serial correlation. Various analytical techniques have been proposed to overcome this problem (section 8.5), but there is nevertheless doubt as to whether reliable confidence limits on the abundance estimate can be determined from a series of samples which is serially correlated. It is much better to avoid the problem by choosing the EDSU to be large enough so that the acoustic data may reasonably be considered as a set of independent random samples.

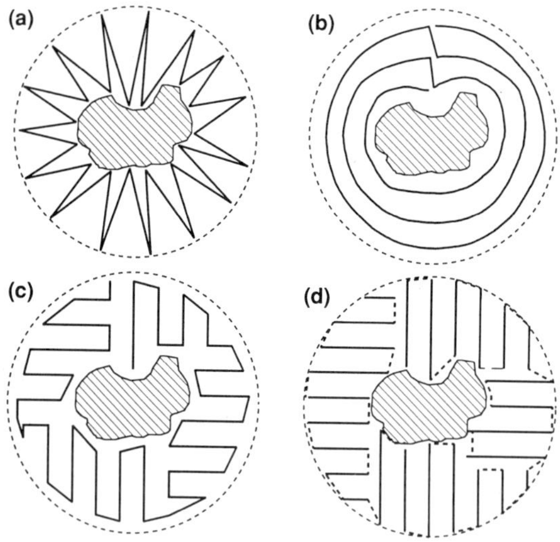

Fig. 7.14 Example cruise tracks: ocean around an island. (a) Triangular, greater sampling intensity inshore; (b) circular grid, even intensity but sensitive to the migration of fish in any direction; (c) rectangular, all data used; (d) parallel, joining sections (broken lines) not used.

The optimum length of the EDSU may be known from previous surveys of the same area. If not, it may be decided on the basis of normal practice on surveys of similar areas elsewhere. The EDSU could be as short as 0.1 miles (1 mile = 1853 m), which would be appropriate to dense schools within a fjord, or as much as 10 miles in the case of species that are widely distributed over large areas of ocean. More usually the EDSU might be in the range 1–5 miles. (Note that the nautical mile is larger than the statute mile, 1609 m, with which it must not be confused.)

It is often convenient to organize the data collection within intervals of time rather than distance. If the vessel travels at 10 knots, then 1 mile of track is covered in 6 minutes. If it had been decided that the EDSU should be 1 mile, then the samples may be recorded as the average fish density time and the distance travelled may not be exact, if the vessel speed is uncertain, but this is not an important factor in the recommended analysis procedure, which treats the EDSU as a nominal distance.

7.5 ADAPTIVE SURVEYING

So far we have considered only the pre-planned survey in which the cruise track is decided on the basis of prior knowledge and decisions. There are circumstances in which it might be desirable to adjust the cruise track as the work proceeds, for example if it is important to locate commercially exploitable fish concentrations. It may be decided that areas of high concentration should be surveyed more intensively than elsewhere. However, the adaptive survey is not appropriate when the principal objective is to determine the stock abundance. The problem is that the acoustic data cannot be considered to be random samples, because the position at which each measurement is made has been determined to some extent by the earlier observations, and so the abundance estimate may be biased. It is not easy to allow for the bias in the analysis, and whether this can be done at all depends upon assumptions about the fish distribution which may be difficult to validate.

We shall consider three kinds of adaptive survey. One technique is to begin with an outline survey, which is a rapid investigation of a large area using widely spaced transects. This is followed by more intensive examination of particular regions where fish concentrations have been detected. Another approach is to fix the transect spacing in advance, but to allow the length of the legs to be changed during the survey. Thirdly, the transect lengths may be decided beforehand whereas the spacing is varied according to the observations made at the time.

There are two important considerations to be borne in mind when making the real-time decisions required during an adaptive survey. Firstly, when the sampling intensity is reduced, the coverage must still be sufficient to provide good enough information to decide the subsequent sampling strategy. Secondly, when the plan is to return to regions where large fish concentrations have been observed, success depends upon the assumption that the fish distribution does not change with time, so that the concentrations can be found again without difficulty.

Outline survey

The survey is conducted in two stages. First, the vessel covers the area of interest on a widely spaced grid, to detect regions of high fish density. This stage should occupy no more than 25% of the time available. The vessel then returns to the regions where fish have been observed, and the remainder of the time is spent in surveying these regions more intensively (Fig. 7.15). This technique is not useful if the fish are likely to migrate or disperse in the time between the initial sweep of the area and the return visit. Furthermore, if the initial sweep is too widely spaced, some localized concentrations may not be detected at all. The outline survey works best when the area to be examined is not too large, within a fjord for example, and the fish are believed to be concentrated in a few large and static schools.

Variable transect length

This technique may be applied when the spatial distribution is well defined in one direction. For example, suppose there is a coastline along one edge of

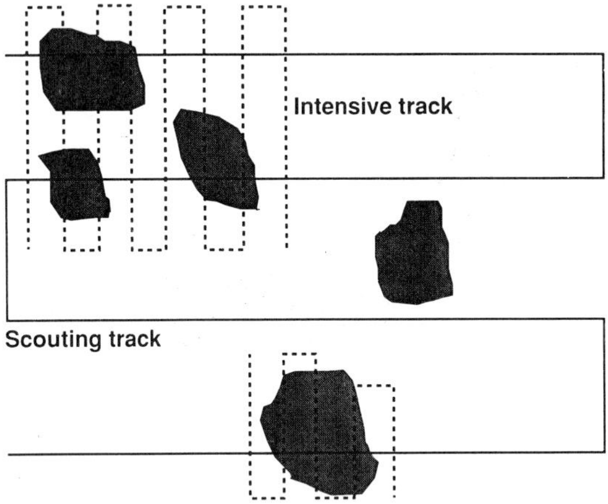

Fig. 7.15 The outline survey. Fish concentrations (dark patches) detected by the initial scouting grid (solid line) are later surveyed more intensively (broken lines), but some concentrations may not be detected at all.

the area to be surveyed, and the stock is located mainly in the shallow water near the coast. The survey is designed initially as a grid of transects running between turning points on the inshore and offshore boundaries. During each run in the offshore direction, it may be decided to terminate the transect once the observed fish density has declined to a small proportion of that observed near the coast (Fig. 7.16). The acoustic data may be analysed in the normal way, by calculating the abundance in elements of area, on the assumption that negligible quantities of fish would have been observed along the abandoned parts of the cruise track. To facilitate the analysis, once the decision to turn has been taken, the transect should nevertheless be continued to the edge of the current area element.

Variable transect spacing

Suppose that the fish are expected to occur in local aggregations, but in regions which are unknown in advance, as clusters of migrating schools for example. The general plan is to increase the sampling of any region where the observed fish density is much higher than the average, by reducing the transect spacing (Fig. 7.17). The transects continue to run for the full length to avoid gaps in the coverage.

The transect spacing should be decided on the basis of objective criteria. A simple technique is to observe the mean fish density along each transect, $\bar{F}$ say, and to make the spacing to the next transect proportional to $(1/\bar{F})$, subject to the calculated spacing being contained within practical limits.

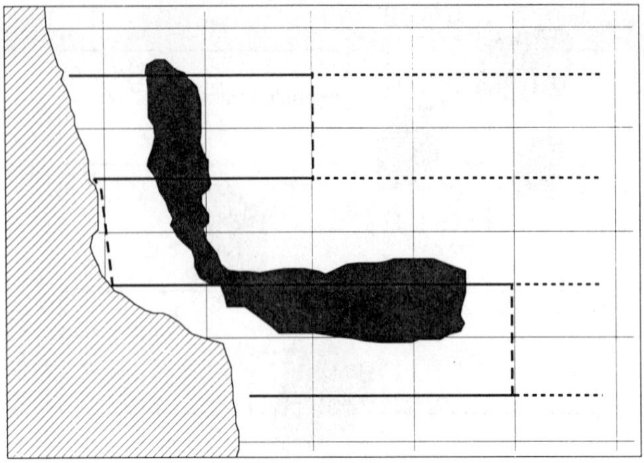

Fig. 7.16 Adaptive survey, variable transect length. Squares (light line) are the analysis area elements; the dark line is the cruise track (start, top left). On each offshore run, the transect ends when the fish density is consistently low and the edge of a square is reached.

Alternatively, the variance rather than the mean of the density measurements might be used to determine the spacing, as proposed by Stolyrenko (1988), on the grounds that precision is improved by sampling more intensively in regions of high variance. When few fish are observed, the variance is also small and the effect is to increase the coverage of the main

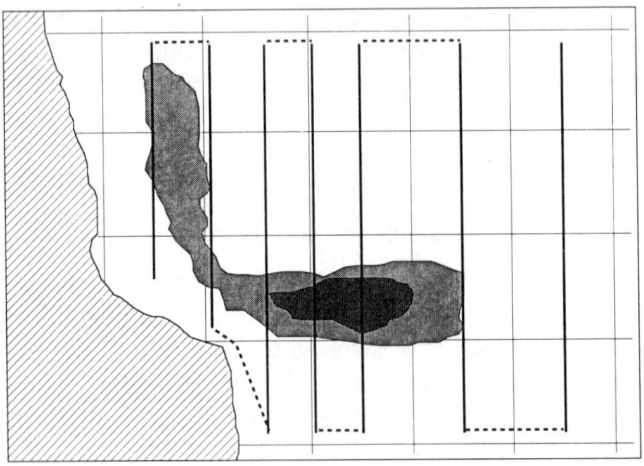

Fig. 7.17 Adaptive survey, variable transect spacing. Squares (light lines) are the analysis area elements; the dark line is the cruise track (start, top right). The grey scale is the fish density: when high densities are observed, the transect spacing is reduced, and vice versa.

concentrations. The two methods may not be much different in practice. According to Aglen (1989), it is characteristic of natural fish distributions that the standard deviation (square root of the variance) is proportional to the mean of the fish density.

Another approach is to design the survey grid in advance with a fixed transect spacing, but ensuring that the pre-planned track does not require all the time available for the survey. The spare time is used to cover extra transects in regions of high density. Whenever the observed density exceeds some limit, one extra transect is traversed half-way between those of the pre-planned grid, so that the sampling intensity in that region is doubled.

Randomized adaptive transects

The adaptive methods discussed above all suffer from the problem that the abundance of the stock cannot be estimated reliably, because of the bias introduced by locating the transects with reference to the observed fish densities. The bias may be reduced by randomizing the transect spacings.

We begin with a pre-planned set of parallel transects randomly placed in zones (p. 224), leaving some time for extra transects to be inserted during the survey. When it is decided to increase the sampling intensity, a new transect (also randomly placed) is included within the current zone and those ahead. As before, the new transects may be constrained to be at least 10% of the average spacing from any other. This procedure results in the first additional transect sometimes being out of the normal sequence, when the track progression must reverse to run the extra transect. However, all

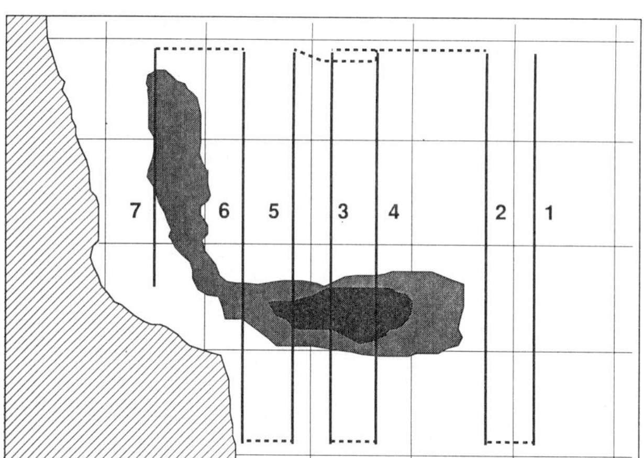

Fig. 7.18 Randomized adaptive survey. Light lines denote zone boundaries; the dark line is the cruise track. The grey scale is the fish density: when high densities are observed, an extra transect is randomly positioned in each zone. Transects are numbered in the order run.

the transects in each subsequent zone may be taken in the normal progressive sequence. When it is decided to reduce the sampling intensity, all the transects placed in the current zone must be completed first. Alternatively, the transects within each zone may be taken in random order, in which case the sampling intensity may be reduced at any time. Both these procedures are free of bias provided that zones with different sampling intensities are treated as separate regions in the analysis.

Figure 7.18 illustrates the randomized adaptive method. The survey initially progresses from east to west. When the fish concentration is observed on transect 3, the sampling intensity is increased. By chance, transect 4 is back to the east, but the later transects are taken in the normal sequence. The method may appear complicated at first sight, but it is simple to implement in practice and provides an estimate of the abundance with minimal bias and without the need for doubtful assumptions about the spatial distribution of the stock. There is a theoretical possibility of bias (p. 225), but it is unlikely to be important in practice.

Chapter eight

Data analysis

8.1 INTRODUCTION

After an acoustic survey has been conducted in accordance with the practical guidelines described earlier, the analysis is the final stage in which, all being well, useful results are derived from the collected data. Depending on the objectives of the survey, we may have to estimate the abundance of the species of interest, the geographical distribution of fish concentrations and/or the age structure of the population. In addition, some indication of the accuracy of the results is necessary.

The abundance may be estimated as the quantity of fish in the stock, or as an index which shows how the stock has changed relative to some previous estimate. It is preferable to know the absolute quantity, but if the error in this estimate is very large, the relative index may be more useful since it can be determined more accurately.

The information available from the survey or other sources may include any or all of the following items.

1. Acoustic data: echo integrals or echo-counts accumulated along sections of cruise track, relating to one or more depth channels, and the corresponding echograms;
2. Calibration results: measurements of the transducer sensitivity, the beam pattern and the time-varied gain;
3. Size and species composition of biological samples collected by fishing;
4. Target strength and its dependence on size (length or weight) for each species detected acoustically;
5. Hydrographic data: water temperature and salinity at various depths;
6. Geography of the area surveyed, e.g. location of coastlines and islands, and the cruise track.

In this Chapter, we describe how the analysis is done, through a series of calculations based on the information collected during the survey, to obtain numerical estimates of the abundance and other parameters of interest. The first step is to identify the source of echoes and the contribution of different species to the acoustic data. Once this is done, the surveyed area is divided into regions which are homogeneous as regards the expected density and size distribution of targets. The echo integrator conversion factor is calculated separately for each region, taking account of the effective target strength therein. These factors are used to convert the acoustic data to estimates of the fish density along the cruise track. The spatial distribution of the stock is described by constructing contour maps from the observed densities, or alternatively by numerical analysis of the acoustic data on a rectangular grid. We discuss the Kriging technique, which can be used to reveal spatial correlation in the observations. These preliminary considerations lead to a discussion of methods for estimating the total abundance of the stock. We go on to consider the precision of the abundance estimate, namely the statistical uncertainty associated with the sampling strategy. The importance of fish behaviour is discussed, notably problems due to migration, diurnal behaviour rhythms and the reactions of fish to the survey vessel. Several other sources of error are considered, and finally we discuss the overall accuracy of the abundance estimate and how it can be assessed in practice.

Figure 8.1 is a flow diagram which illustrates the various steps of the analysis and the order in which the calculations are performed. These procedures are similar in general terms to the analysis methods described by Johannesson and Mitson (1983), Dalen and Nakken (1983) and Simmonds and MacLennan (1988).

List of symbols

a, b	Constants in the formula for CV as a function of the degree of coverage
a_x, b_x	Constants in formulas relating target strength to the fish length ($x = i$, n or w)
a_f, b_f	Constants in the weight–length relationship of fish
A	Size of the surveyed area
A_b	Extra attenuation due to the bubble layer
A_k	Area of the kth elementary statistical sampling rectangle (ESSR)
A_p	Probability factor in the equation for D_p
A_r	Area of the rth region
c	Speed of sound
c_1	Assumed value of c
c_i	Echo-integrator conversion factor for species i
CV	Coefficient of variation
d	Degree of coverage ($D/\sqrt{A}$)
D	Total length of the cruise track

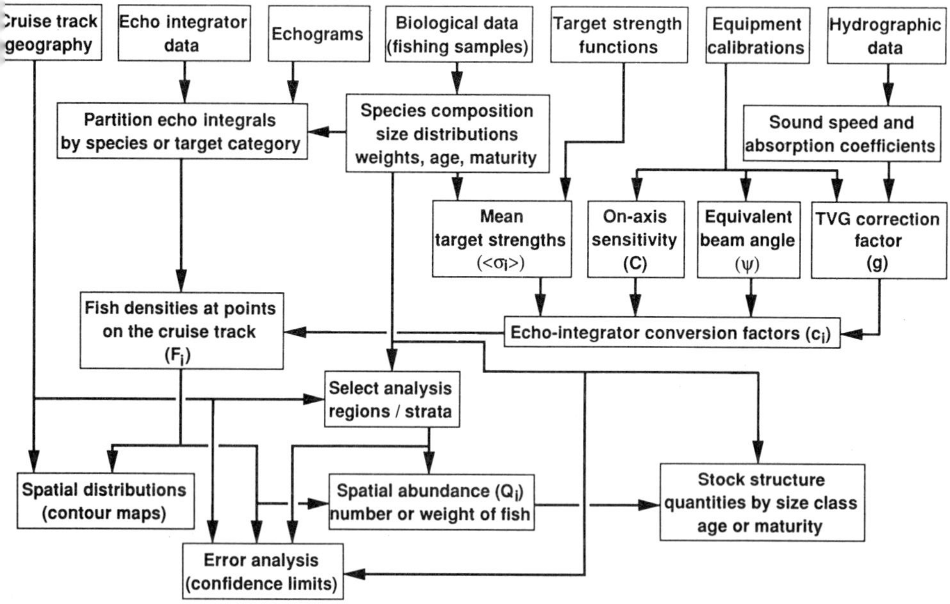

Fig. 8.1 Flow diagram of the analysis procedure.

D_I Minimum length of the elementary distance sampling unit (EDSU) for independent samples
D_{max} Maximum difference between two cumulative probability distributions
D_p Value of D_{max} exceeded with probability P
E Echo integral, usually the mean of many measurements
E_i Partitioned echo integral of species i
E_m Echo integral of a species mixture
E_k Mean echo integral near trawl-station k
E_{ik} Mean echo integral of species i near trawl-station k
$E(\hat{Q})$ Expected value of $\hat{Q}$
f Frequency of the echosounder transmission
f_{ij} Mean frequency of length-class j for species i
F Density of fish per unit area of the integration channel
F_i Density of species i, or the ith density observation
$\bar{F}_k$ Arithmetic mean density for stratum k
$\tilde{F}_k$ Arithmetic mean density for transect k
$\bar{g}$ Hydrographic error factor
$G_N(u)$ Function used to estimate the mean and variance of log-normal data
h Distance between two sampled locations
i, j, k Index numbers

K	Physical calibration factor of the echo integrator
L	Fish length
L_j	Fish length at the midpoint of size class j
ΔL	Interval between successive size classes
$L(\lambda)$	Likelihood function used in the Box–Cox test
m	Sample mean (i.e. the arithmetic average)
M	Number of trawl stations, regions, etc.
M_i	Number of catches containing species i
n	Number of fish, samples, etc
n_{ijk}	Number of species i in length-class j caught at station k
N	Number of surveys, samples, etc
N_{ik}	Number of species i caught at station k
p	Probability of observing zero density
P	Probability (Kolmogorov–Smirnov test)
$P(F)$	Probability density function (PDF) of F
q_{ik}	Quantity of species i caught at station k
Q	Quantity (abundance) of fish
$\hat{Q}$	Estimate of the true abundance Q
Q_k	Quantity of catch at station k
r	Rotation rate of the transducer
R_0	Mean depth of the integration channel
s^2	Sample variance
S	Residual sum of squares
SE	Standard error (standard deviation of the estimated mean)
SNR	Signal-to-noise ratio
t	Multiplier of SE in the calculation of confidence limits
t_k	Time spent fishing at station k
T	Water temperature
TS	Target strength
TS_{n}	Target strength of one fish
TS_{w}	Target strength of unit weight of fish
u	General variable
v	Cruising speed of the survey vessel through the water
v_{f}	Migration speed of fish
v_{s}	Speed at which the survey progresses in the direction of migration
v_{w}	Wind speed
$\hat{V}$	Estimated variance of $\hat{Q}$
w_i	Weighted proportion of species i in fish samples
W	Weight of one fish
W_{t}	Total weight of fish sample
x	General variable
X_i	Variable in the error analysis
ΔX_i	Error in X_i
Y	Transformed density, first kind ($Y_i = F_i^{\lambda}$)

Z	Transformed density, second kind ($Z_i = (F_i^{\lambda} - 1)/\lambda$)
β	Acoustic absorption coefficient (in nepers m^{-1})
β_1	Assumed value of β
ε_i	Standard deviation of $(\Delta X_i / X_i)$
λ	Power of F_i in the transformed density
λ_m	Most likely value of λ for the transformed data to be normally distributed
λ_1, λ_2	95% confidence limits on λ_m
γ	Spatial correlation parameter used to construct the variogram
μ	Population mean (in the statistical sense)
ϕ	Angular rotation of targets relative to the transducer
ϕ_a	Value of ϕ due to pitching and rolling
ϕ_f	Value of ϕ due to linear movement of the transducer
ψ	Equivalent beam angle of the transducer
θ_{3dB}	Angle from the acoustic axis at which the transducer sensitivity is 3 dB below the maximum
ρ_L	Serial correlation coefficient of paired samples at positions $(i, i+L)$ in the sequence of observations
σ	(1) Acoustic cross-section (2) Square root of the population variance
$\langle\sigma_i\rangle$	Mean acoustic cross-section of species *i*

8.2 SPECIES COMPOSITION

The first step is to examine the echograms to see whether any of the marks can be identified as having come from a particular species. For each species present, it is also necessary to determine the size distribution of individual animals. In general, it is necessary to rely on fishing samples for the identification of targets, as described in the following subsection. However, it may be known that certain species cause distinctive marks on the echogram, or they may occur at different depths. This prior knowledge may allow at least some of the marks to be positively identified (Midttun and Nakken, 1977). The echo integrals corresponding to known marks should be treated separately from the remainder of the acoustic data. It may even be possible on some occasions to classify marks by the size composition of individual fish as well as the species, for example if older and younger fish inhabit different depths. Suppose there were two year classes in the population whose marks could be identified in this way. The respective marks should be considered in the same way as different species, since they represent discrete components of the stock.

Although it may be possible to identify only a small proportion of the marks from the acoustic records alone, it is worthwhile performing a separate analysis of the identified records which will thus not be subject to errors in the interpretation of fishing samples. However, this must not be taken as an excuse to avoid fishing. The collection of biological samples remains an essential part of the survey procedure.

Analysis of fishing samples

Here we consider the interpretation of samples collected by trawling, but the same principles would apply to the catches taken by any other kind of gear. Trawling is conducted at a number of positions around the surveyed area, wherever there is doubt as to the identity of echo traces. The object is to determine the species and size composition of the local population. These parameters may not be the same over the whole of the surveyed area. Within any part of the area where uniform conditions are believed to apply, there will probably be results from several hauls made at different locations. The compositions indicated by hauls at different positions have to be combined to estimate the mean composition relevant to the whole area. There are several methods for doing this.

Suppose there are M_i hauls in which the catch of species i is sufficient to establish a reasonable length distribution. N_{ik} is the total number of species i in haul k and n_{ijk} is the number in length class j. When the catch is large, it is normal practice to sample a manageable proportion, in which case N_{ik} and n_{ijk} are the observed counts raised to estimates for the whole catch. In the analysis for each species, it is better to ignore any haul in which N_{ik} is too small to establish a reasonable length distribution. Thus M_i may not be the same for all species. The duration of haul k is t_k. Ideally, the average composition should be calculated by weighting the results from each haul according to the quantity of fish it represents, or the effective area sampled, but it is difficult to say what this is. The catching efficiency of the trawl is highly variable, especially in pelagic fishing, and the observed catch-rates may be quite unrepresentative of the true population densities.

When the catches are moderately large, so that a good length distribution can be established for each haul, the best plan is to give equal weight to each distribution, taking no account of differences in the total catch. The mean frequency of species i and length class j is f_{ij}. In this case (Method 1) the estimation formula is

$$f_{ij} = \sum_{k=1}^{M_i} (n_{ijk} / N_{ik}) / M_i \tag{8.1}$$

If the catches are small and the length distributions are poorly defined for any one haul, it is better to combine the samples, giving equal weight to each fish but taking account of the haul duration, since it is supposed that, on average, longer hauls will produce more fish. This is Method 2, in which the frequencies are estimated as

$$f_{ij} = \sum_{k=1}^{M_i} (n_{ijk}/t_k) \Bigg/ \sum_{k=1}^{M_i} (N_{ik}/t_k) \tag{8.2}$$

A third possibility (Method 3) is to weight the results in proportion to the echo integrals recorded in the vicinity of the fishing position. This should not be attempted unless each catch consists of one species only, because the echo integrals give no information as to the species composition of mixed targets. If E_{ik} is the mean echo integral for species i in the vicinity of haul k, supposed to be proportional to the fish density, the frequencies are estimated as

$$f_{ij} = \sum_{k=1}^{M_i} (n_{ijk} E_{ik} / N_{ik}) \Bigg/ \left(\sum_{k=1}^{M_i} E_{ik} \right) \tag{8.3}$$

Method 3 is not useful unless the catches are large enough to establish good length distributions. In addition, the frequency estimates may be biased if there is a systematic relationship between fish density and fish size, for example if the young fish are more dispersed than the adults. For most applications we recommend that Method 1 or Method 2 should be used to average the fishing results, depending on the sample sizes and the quality of the length distributions obtained for each haul.

Partitioning

We now come to the problem of dividing (partitioning) the echo integrals into components associated with particular species or, more generally, types of target. This is done by comparing the echogram and the integram at corresponding times. The integram is a graph of the accumulated echo integral against time. The slope of the integram depends upon the density of detected targets; the greater the density, the steeper is the slope. Figures 8.2 to 8.4 show examples of the acoustic records collected on surveys which we shall use to illustrate how the partitioning is done. In addition to the acoustic data, of course, the partitioning also takes account of the fishing samples, which are required to identify the echo traces.

Figure 8.2 comes from a survey of the South China Sea. It shows two schools in shallow water (a, b) followed by a small group of schools (c) in deeper water. The more diffuse marks are due to plankton. The integration has been performed in two channels whose limits are marked on the echogram, firstly the pelagic channel which extends from 5 m depth to the line 2 m above the bottom, and secondly the bottom channel which covers the last 2 m of water.

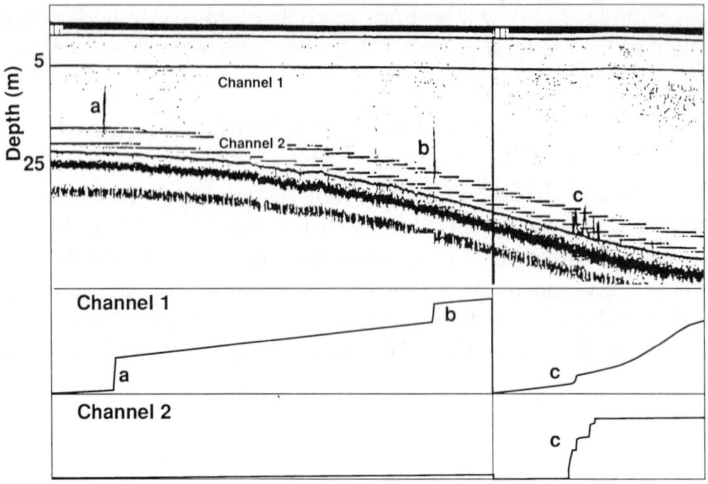

Fig. 8.2 Acoustic records showing two large schools (a,b) and a cluster of small schools (c) on the echogram (above) and the integram (below) at corresponding times. The integration is done in two contiguous depth channels. The vertical line between (b) and (c) marks an integrator reset. The trace is 15 minutes (4 km) long. Recorded on 13 June 1989, 11.00 hrs, South China Sea.

The shallow slope of the Channel 1 integram is due to the plankton and this component of the echo integral is ignored. The sharp deflections coincident with the school-marks a and b are clearly associated with the fish. Thus the component of the echo integral attributable to fish is the sum of the sharp deflections coincident with the schools, while the remainder is attributed to the plankton. In the case of the more extended marks at *c*, the Channel 2 integram is less steep but nevertheless the discontinuous jump is identified as the contribution of the fish. The schools have caused deflections in both channels, which are summed to give the fish component of the echo integral for the entire water column.

Thus it is not difficult to partition the echo integrals between plankton and schooling fish. Partitioning between different fish species is more problematical. Referring again to Fig. 8.2, the appearance of the school marks would suggest that a and b are one species while *c* is another, but beyond that the identification must depend upon catching samples from these or similar marks in the vicinity. In this example, the fishing results did not allow precise identification, but c is believed to be anchovy, *Engraulis* sp., while a and b are either *Sardinella aurita* or *Rastrelliger* sp.

Figure 8.3 illustrates another example commonly found in practice, again from the South China Sea, where the fish are seen to be in clusters of small schools. Again, the fishing information did not allow the schools to be identified to species level. They are attributed to a mixture of some six species

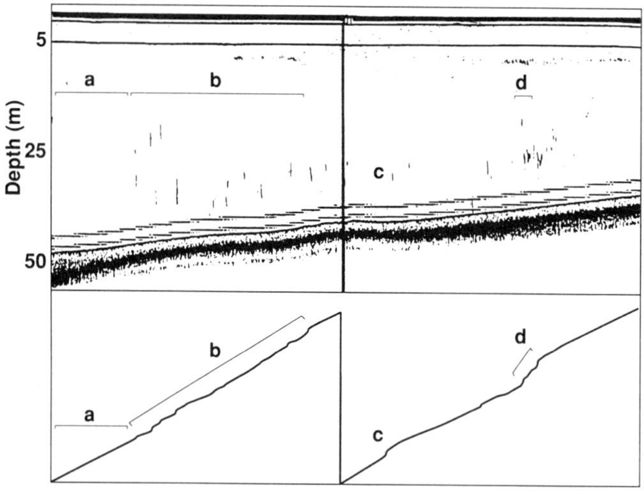

Fig. 8.3 Acoustic record showing plankton (a), scattered fish schools (b,c) and clustered schools (d) on the echogram (above) and the integram (below) at corresponding times. The vertical line between (b) and (c) marks an integrator reset at 05.45 hrs. The record is 30 minutes (8 km) long. Recorded on 14 June 1989, 06.00 hrs, South China Sea.

dominated by *Carangoides malabaricus* and *Decapterus* sp. The same schools could be seen in mid-water during darkness, and in daylight much closer to the sea-bed, while their structure as clusters of small aggregations did not depend upon the time of day. Most of the schools do not give clear deflections on the integram, but considering the aggregation as a whole, there is a change of slope (b, d) compared to the period when no fish are seen on the echogram (a). To partition the echo integral, we assume that the plankton density giving rise to the background slope a is the same within the cluster. In the absence of fish, the integram would have been a straight line extrapolated from that in a. Thus the echo integral due to the fish is the difference between the end of the extrapolated line and the actual integram. The method is illustrated in Fig. 8.4.

There is a large subjective element in the identification of fish species from the appearance of echo traces, but the procedure should be made objective as far as possible. The surveyors have to use their experience as well as the data to hand in deciding how to partition the echo integrals. They should establish a library of echo traces that have been positively identified by fishing, together with other relevant information such as the location, season and time of day. This will help to ensure that when different people have to identify the echo traces, as might apply when several ships are engaged on the same survey, the partitioning is based on consistent rules. The echo integration technique depends on the proportion of the echo

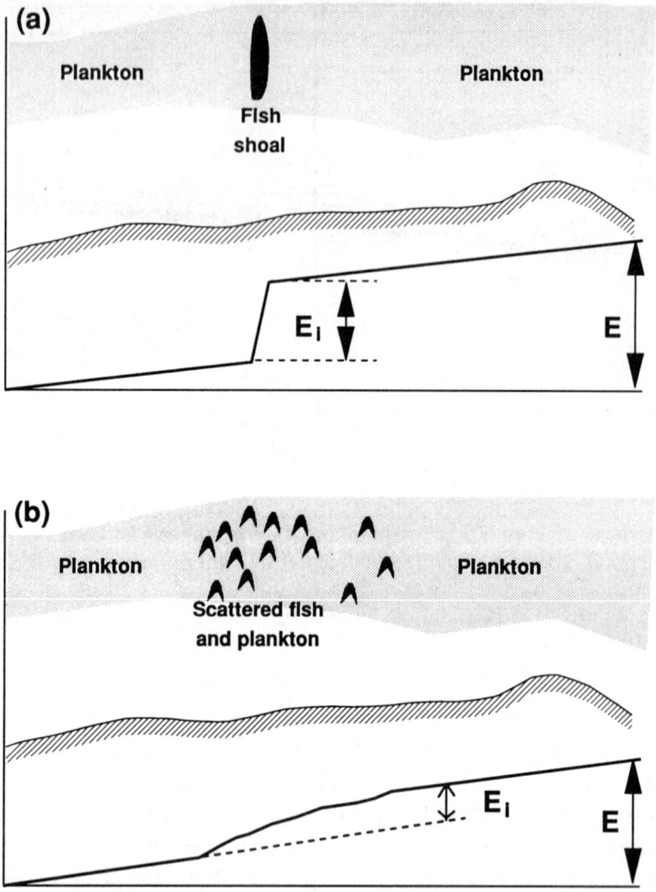

Fig. 8.4 Partitioning the echo integral. E is the total deflection of the integram and E_i is the part due to fish. (a) Diffuse plankton cause the slowly rising parts of the integram; one large fish school causes a sharp jump whose height is E_i. (b) A low-density fish aggregation is identified by the discontinuous slope; E_i is the deflection from the background slope due to plankton.

integrals due to each species in a mixed population being assessed accurately. This is particularly difficult in areas such as the South China Sea where there are many pelagic fish species and there is the further complication that the echo integrals are often dominated by plankton echoes.

Partitioning is not always a major problem. Off the coast of Morocco, for example, most of the schools are pilchard, *Sardina pilchardus* or the trumpet fish, *Macrorhamphosus scolopax*, the latter being found further from the shore. In the Arabian Sea and the northern part of the Indian Ocean, deep scattering layers of the lantern fish, *Benthosema pterotum*, are easily identified on the echogram, while in shallow waters, *S. aurita* and *Trachurus* sp. may be distinguished from anchovies (but not from each other) by the appearance of school-marks on the echogram. The anchovies are found in less dense schools and higher in the water column compared to the other species, but only during daylight. The schools dissociate at night, when the fish are found almost uniformly distributed between the surface and the sea-bed in waters less than 20 m deep.

The partitioning of the echo integrals is not necessarily done at the species level, but is more generally a question of classifying targets according to their behaviour and acoustic properties. The subsequent analysis is performed separately for each category of target. For example, if different size classes of one species can be identified, then the abundances of each class can be estimated to show the age-structured stock. At the other extreme, several species may occur in mixed aggregations giving rise to clusters of similar marks which cannot be identified with certainty. In that case it would be appropriate to consider a broad category which is a mixture of species with similar behaviour, and only the total abundance of the mixture could be estimated. However, the composition of the mixture must be known well enough to determine the mean target strength. This is not a problem if all the species are assumed to have the same length-dependence of the target strength.

Once the partitioning has been done, the result for each species or category is a set of mean echo integrals, and each mean refers to one EDSU (elementary distance sampling unit, p. 225) along the cruise track.

8.3 ECHO-INTEGRATOR CONVERSION FACTOR

The next step is to estimate the density of targets from the observed echo integrals. This is done by means of Equation 5.15, which we repeat here in a slightly different form

$$F_i = (K/\langle\sigma_i\rangle)E_i \tag{8.4}$$

The subscript i refers to one species or category of target, K is a calibration factor, $\langle\sigma_i\rangle$ is the mean acoustic cross-section of species i, E_i is the mean echo integral after partitioning, and F_i is the estimated density of species i, or the quantity per unit area of the integration channel. The quantity is the number or weight of species i, depending on whether $\langle\sigma_i\rangle$ is the mean cross-section per fish or per unit weight. The echo-integrator conversion factor, $c_i = (K/\langle\sigma_i\rangle)$, is not the same for all species. Furthermore, c_i depends upon the size distribution of the insonified targets, and if this is not the same over the whole surveyed area, the calculated conversion factors must take regional variation into account.

K is determined from the physical calibration of the equipment (Chapter 3); it does not depend upon the species or biological parameters. Several calibrations may be performed during a survey. In general, the measured values of K will differ but should be within 10% of one another. If two successive measurements are very different, by more than 20% or so, the cause should be investigated because a malfunction of the equipment is indicated. Otherwise, K should be taken as the average of the two measurements before and after the relevant part of the survey.

Homogeneous regions

There are several reasons why the size distribution of the stock might not be the same over the large area covered in the typical acoustic survey. When the surveyed area is bordered by a coastline, for example, there may be differences in the age structure of the stock between inshore fish and those in deeper water. Or the fish may associate with others near the same age, so that the stock consists of clusters of fish in particular age groups. If the fishing samples indicate consistent differences between one region and another, c_i must be determined separately for each region within which the size distribution is considered to be homogeneous.

The observed size distributions will vary due to sampling error as well as changes in the population structure between the trawl stations. To identify the homogeneous regions, we need to consider whether the distributions obtained from any two hauls are significantly different. This might be evident from inspection of the numerical data if there are clear differences between clusters of trawl stations. Alternatively, a more objective approach to the problem is to apply the Kolmogorov–Smirnov (KS) test to the observed size distributions (Campbell, 1974). This test is sensitive both to the position of modes and to the shape of the distributions. The result of the test is a number P in the range 0 to 1 (Appendix E). $P=0$ when the distributions are identical, and $P=1$ when there is no similarity. If the distributions come from the same population, subject only to sampling error, P is expected to be between 0.1 and 0.3.

It is not suggested that the KS test should be applied as an automatic procedure. A more thoughtful approach is necessary. We might begin by assigning the trawl stations to groups within which P for each pair is less than 0.2, and see whether this suggests a sensible division of the surveyed area into homogeneous regions. The calculations are then repeated for similarity thresholds $P=0.15$ and $P=0.25$. If the grouping of the trawl stations is sensitive to the threshold value of P, this indicates that the differences in size distribution are small enough to be ignored, and only one region (the surveyed area) needs to be considered. Otherwise, the boundaries between the homogeneous regions are determined by the condition that any part of the surveyed area is in the region corresponding to that of the closest trawl station.

Table 8.1 shows an example of the *P*-values obtained by comparing the samples collected at 28 trawl stations, displayed as a matrix. In Table 8.2, the distributions are grouped according to value of *P*, and the clear separation into four groups suggests that the surveyed area should be divided into that number of regions. The boundaries of the regions in relation to the trawl stations are shown in Fig. 8.5.

The KS test can also be used to compare the species mix in different parts of the surveyed area. In this case, the proportion of each species in a sample is treated in the same way as the frequency of a length class in the previous analysis, and the value of *P* indicates whether or not the species mix is similar between pairs of trawl stations.

It is not useful to divide the surveyed area into a large number of regions. About four or five are as many as should be considered. Furthermore, there should be more than one trawl station in each region.

Single species

The mean cross-section $\langle\sigma\rangle$ may be determined directly from *in situ* measurements of target strength made during the survey. More usually it is derived from a function which describes the length dependence of the target strength, normally expressed in the form

$$TS = a_i + b_i \log L \tag{8.5}$$

a_i and b_i are constants for the *i*th species, assumed to be known from experimental evidence (Chapter 6). The equivalent formula for the cross-section is

$$\sigma = 4\pi\, 10^{\{(a_i + b_i \log L)/10\}} \tag{8.6}$$

The mean cross-section is calculated as the average σ over the size distribution of the insonified fish. Thus

$$\langle\sigma_i\rangle = 4\pi \sum_j f_{ij}\, 10^{[(a_i + b_i \log L)/10]} \tag{8.7}$$

L_j is the midpoint of the *j*th size class and f_{ij} is the corresponding frequency as deduced from the fishing samples by the method described earlier (section 8.2). The echo-integrator conversion factor is immediately determined as $c_i = K/\langle\sigma_i\rangle$. The calculation is repeated for any species whose target strength function is known.

Note that it is the cross-section that is averaged, not the target strength. The arithmetic average of the target strengths calculated from Equation 8.5 gives a different result which is incorrect. The term 'mean target strength' is often seen in the literature, but this is normally the target strength equivalent to $\langle\sigma_i\rangle$, namely $10 \log(\langle\sigma_i\rangle/4\pi)$.

Table **8.1** Comparison of the length distributions of catches at paired trawl stations. The reference numbers (bold print) identify the trawl stations corresponding to each row and column. Values shown are the *P*-values from the KS test, which compares the length distributions

221	0.54	0.14	0.11	0.15	0.18	0.62	0.36	0.26	0.42	0.11	0.29	0.41	0.09
	223	0.41	0.48	0.65	0.65	0.88	0.79	0.67	0.14	0.64	0.69	0.74	0.59
		225	0.08	0.26	0.31	0.72	0.48	0.35	0.28	0.25	0.37	0.51	0.20
			226	0.22	0.27	0.64	0.44	0.29	0.34	0.22	0.30	0.43	0.16
				227	0.05	0.47	0.22	0.10	0.54	0.16	0.14	0.27	0.10
					228	0.48	0.21	0.11	0.55	0.15	0.15	0.27	0.12
						230	0.28	0.37	0.85	0.64	0.34	0.21	0.56
							231	0.14	0.72	0.36	0.14	0.08	0.29
								233	0.57	0.27	0.04	0.18	0.19
									234	0.53	0.58	0.69	0.48
										235	0.30	0.42	0.09
											237	0.16	0.23
												238	0.34
													239

Mixed species

Sometimes two or more species are found in mixed concentrations such that the marks on the echogram caused by each species cannot be distinguished. From inspection of the echogram, the echo integrals can be partitioned to provide data for the mixture as one category, but not for the individual species. However, further partitioning to species level is possible by reference to the composition of the trawl catches (Nakken and Dommasnes, 1975). Suppose E_{m} is the mean echo integral of the mixture, and w_i is the proportion of the *i*th species, calculated as a weighted average of the catch compositions (see below). It is necessary to know the target strength function and the size distribution of each species, from which $\langle\sigma_i\rangle$ is determined as before, and $c_i = K/\langle\sigma_i\rangle$. The echo integral contributed

Table 8.1 (Continued)

0.55	0.35	0.23	0.79	0.31	0.85	0.04	0.79	0.52	0.25	0.16	1.00	0.99	0.99
0.84	0.77	0.55	0.65	0.67	0.55	0.54	0.45	0.10	0.53	0.57	1.00	0.98	0.98
0.65	0.48	0.33	0.75	0.41	0.81	0.13	0.74	0.39	0.31	0.28	1.00	0.99	0.99
0.57	0.44	0.25	0.76	0.33	0.79	0.08	0.72	0.45	0.23	0.20	1.00	0.99	0.99
0.40	0.22	0.18	0.83	0.16	0.90	0.19	0.86	0.60	0.15	0.13	1.00	0.99	0.99
0.41	0.20	0.14	0.80	0.17	0.86	0.22	0.84	0.59	0.12	0.11	1.00	0.99	0.99
0.18	0.31	0.39	0.89	0.32	0.96	0.64	0.96	0.85	0.41	0.48	1.00	1.00	1.00
0.20	0.07	0.25	0.85	0.13	0.92	0.41	0.91	0.74	0.26	0.23	1.00	1.00	1.00
0.30	0.15	0.17	0.82	0.08	0.89	0.28	0.86	0.62	0.15	0.13	1.00	0.99	0.99
0.81	0.72	0.52	0.67	0.60	0.58	0.40	0.49	0.13	0.46	0.49	1.00	0.98	0.99
0.56	0.35	0.24	0.81	0.32	0.88	0.15	0.86	0.59	0.26	0.16	1.00	0.99	0.99
0.27	0.14	0.16	0.80	0.06	0.87	0.31	0.86	0.64	0.16	0.14	1.00	0.99	0.99
0.14	0.14	0.20	0.80	0.12	0.87	0.44	0.86	0.69	0.22	0.27	1.00	0.99	0.99
0.48	0.29	0.16	0.73	0.24	0.80	0.13	0.77	0.53	0.18	0.08	1.00	0.96	0.97
240	0.22	0.32	0.88	0.24	0.95	0.57	0.95	0.80	0.35	0.40	1.00	0.99	0.99
	241	0.22	0.81	0.12	0.88	0.40	0.88	0.72	0.26	0.24	0.99	0.98	0.98
		242	0.66	0.14	0.72	0.26	0.70	0.50	0.08	0.11	1.00	0.96	0.97
			243	0.79	0.51	0.81	0.47	0.59	0.72	0.72	0.67	0.48	0.49
				244	0.86	0.34	0.82	0.62	0.14	0.17	1.00	0.99	0.99
					245	0.87	0.10	0.45	0.79	0.78	1.00	0.97	0.97
						247	0.81	0.51	0.27	0.20	1.00	0.99	0.99
							248	0.37	0.72	0.73	0.98	0.94	0.95
								250	0.48	0.52	1.00	0.97	0.97
									251	0.12	1.00	0.98	0.99
										252	1.00	0.97	0.98
											253	0.55	0.56
												254	0.04
													255

by each species is proportional to the product of w_i and $\langle\sigma_i\rangle$. Thus the partitioned echo integrals are

$$E_i = w_i\langle\sigma_i\rangle E_{\mathrm{m}} \Bigg/ \left(\sum_j w_j\langle\sigma_j\rangle\right) \tag{8.8}$$

The w_i are expressed as the proportional number or weight of each species, according to the units used for $\langle\sigma_i\rangle$ and c_i. Consistent units must be used throughout the analysis, but the principles are the same whether it is the number of individuals or the total weight that is to be estimated. Suppose there are M trawl stations in a homogeneous region. q_{ik} is the quantity of the ith species caught at station k, and $Q_k = \sum_i q_{ik}$ is the total catch. The w_i are determined by combining the catch compositions, in a similar manner

Table 8.2 Comparison of the length distribution of herring catches at 28 trawl stations of NE Scotland. There are four sets of similar distributions, identified as areas I to IV using results of the KS tests from Table 8.1

	Area I, haul no.				*Area II, haul no.*				*Area III, haul no.*				*Area IV, haul no.*			
Length (cm)	*253*	*254*	*255*	*Mean*	*243*	*245*	*248*	*Mean*	*223*	*234*	*250*	*Mean*	*221*	*225*	*226*	*227*
15.5			0.01	0.00												
16.0	0.01		0.01	0.01												
16.5	0.07	0.00	0.00	0.03	0.00			0.00								
17.0	0.13	0.01	0.01	0.05												
17.5	0.13	0.02	0.00	0.05	0.01			0.00								
18.0	0.17	0.04	0.04	0.08	0.03			0.01								
18.5	0.17	0.07	0.08	0.11	0.05			0.02								
19.0	0.15	0.18	0.14	0.16	0.08			0.03								
19.5	0.08	0.18	0.20	0.15	0.09			0.03								
20.0	0.03	0.23	0.22	0.16	0.06		0.00	0.02								
20.5	0.04	0.14	0.15	0.11	0.07		0.00	0.02								
21.0	0.01	0.05	0.08	0.05	0.06		0.00	0.02								
21.5	0.00	0.03	0.03	0.02	0.02		0.00	0.01								
22.0	0.00	0.01	0.01	0.01	0.01		0.00	0.00								
22.5		0.00	0.00	0.00	0.02		0.02	0.01			0.00	0.00				
23.0		0.00	0.01	0.00	0.04	0.04	0.05	0.04			0.01	0.00				
23.5		0.00	0.01	0.00	0.04	0.09	0.07	0.06	0.00		0.01	0.00				
24.0	0.00	0.01	0.01	0.01	0.07	0.16	0.14	0.12	0.01	0.01	0.05	0.02				
24.5		0.00		0.00	0.06	0.24	0.16	0.15	0.05	0.04	0.09	0.06		0.00		
25.0		0.00	0.00	0.00	0.07	0.25	0.22	0.18	0.16	0.15	0.17	0.16	0.01	0.03	0.03	0.00
25.5			0.00	0.00	0.06	0.13	0.17	0.12	0.27	0.16	0.14	0.19	0.05	0.06	0.09	0.01
26.0					0.04	0.05	0.11	0.07	0.21	0.22	0.21	0.21	0.10	0.19	0.11	0.08
26.5		0.00		0.00	0.03	0.02	0.03	0.03	0.17	0.19	0.14	0.17	0.18	0.18	0.22	0.14

27.0		0.00		0.00	0.04	0.01	0.00	0.02	0.06	0.15	0.07	0.09	0.20	0.19	0.18	0.18
27.5					0.03	0.01	0.00	0.01	0.04	0.03	0.05	0.04	0.15	0.14	0.10	0.14
28.0					0.01	0.01		0.01	0.01	0.02	0.02	0.02	0.13	0.07	0.07	0.12
28.5					0.00		0.01	0.00	0.01	0.02	0.01	0.01	0.06	0.03	0.08	0.10
29.0					0.01		0.00	0.00	0.01	0.01	0.02	0.01	0.04	0.03	0.04	0.06
29.5					0.00			0.00		0.01	0.01	0.01	0.02	0.03	0.03	0.06
30.0											0.00	0.00	0.02	0.02	0.01	0.02
30.5											0.00	0.00	0.03	0.01	0.01	0.03
31.0														0.01	0.01	0.03
31.5														0.01	0.00	0.01
32.0															0.00	0.01
32.5													0.01			0.01
33.0																
33.5																
34.0																
34.5																
35.0																
Mean length (cm)	18.73	20.30	20.30	19.77	22.98	25.22	25.26	24.49	26.37	26.64	26.36	26.45	27.85	27.48	27.67	28.37
Mean weight (g)	47.87	64.14	64.06	58.69	106.42	138.54	139.93	128.30	162.48	168.87	162.90	164.75	198.69	189.46	194.34	213.18
TS per individual	−45.74	−45.03	−45.04	−45.26	−43.90	−43.16	−43.14	−43.39	−42.77	−42.68	−42.77	−42.74	−42.30	−42.41	−42.35	−42.13
TS per kg	−32.54	−33.11	−33.10	−32.94	−34.17	−34.57	−34.60	−34.47	−34.88	−34.96	−34.89	−34.91	−35.28	−35.19	−35.24	−35.42

Table 8.2 (*contd.*)

Length	*228*	*230*	*231*	*233*	*235*	*237*	*238*	*239*	*240*	*241*	*242*	*244*	*247*	*251*	*252*	*Mean*
	Area IV, *haul no.* (*cont.*)															
15.5																
16.0																
16.5																
17.0																
17.5																
18.0																
18.5										0.01						0.00
19.0																
19.5										0.01						0.00
20.0																
20.5																
21.0																
21.5																
22.0											0.00					0.00
22.5								0.00								0.00
23.0								0.01			0.02				0.00	0.00
23.5								0.01			0.03				0.01	0.00
24.0								0.01			0.02			0.01	0.01	0.00
24.5						0.00	0.01	0.01	0.00	0.01	0.04		0.00	0.02	0.01	0.01
25.0	0.02			0.00	0.01	0.01	0.01	0.02	0.00		0.03	0.02	0.01	0.04	0.03	0.01
25.5	0.03		0.01	0.02	0.02	0.03	0.04	0.05		0.02	0.04	0.03	0.02	0.05	0.05	0.03
26.0	0.06		0.03	0.07	0.06	0.04	0.03	0.07		0.04	0.06	0.07	0.14	0.11	0.10	0.07

26.5	0.12	0.02	0.04	0.11	0.14	0.09	0.04	0.10	0.02	0.01	0.08	0.08	0.20	0.11	0.08	0.10
27.0	0.13	0.03	0.11	0.13	0.20	0.15	0.09	0.19	0.06	0.09	0.07	0.10	0.22	0.10	0.12	0.13
27.5	0.17	0.04	0.14	0.13	0.25	0.11	0.08	0.13	0.06	0.19	0.09	0.09	0.15	0.06	0.11	0.12
28.0	0.16	0.11	0.14	0.10	0.15	0.09	0.12	0.15	0.11	0.11	0.11	0.12	0.11	0.07	0.14	0.11
28.5	0.09	0.14	0.10	0.11	0.07	0.14	0.09	0.07	0.15	0.13	0.07	0.09	0.05	0.08	0.10	0.09
29.0	0.06	0.09	0.09	0.06	0.06	0.06	0.10	0.07	0.18	0.12	0.10	0.11	0.03	0.07	0.05	0.08
29.5	0.07	0.12	0.10	0.05	0.01	0.07	0.09	0.03	0.09	0.07	0.09	0.08	0.03	0.08	0.05	0.06
30.0	0.05	0.10	0.04	0.06	0.01	0.07	0.06	0.03	0.10	0.05	0.04	0.08	0.02	0.05	0.05	0.05
30.5	0.02	0.09	0.06	0.04	0.01	0.02	0.06	0.02	0.06	0.03	0.04	0.04	0.00	0.05	0.03	0.03
31.0	0.01	0.08	0.06	0.03	0.00	0.04	0.09	0.01	0.04	0.03	0.03	0.04	0.01	0.05	0.03	0.03
31.5	0.02	0.06	0.02	0.03		0.05	0.03	0.00	0.04	0.03	0.01	0.03	0.01	0.02	0.01	0.02
32.0	0.01	0.07	0.03	0.01	0.00	0.01	0.03	0.01	0.04	0.03	0.01	0.01	0.01	0.03	0.01	0.02
32.5	0.00	0.02	0.01	0.01		0.00	0.02		0.02	0.01	0.00	0.00		0.01		0.01
33.0	0.00	0.02	0.01	0.02		0.00	0.02		0.00		0.01		0.00	0.00		0.01
33.5			0.00	0.00							0.00					0.00
34.0		0.00					0.00									0.00
34.5																
35.0						0.00										0.00
Mean length (cm)	28.34	30.13	29.17	28.74	27.94	28.79	29.32	27.92	29.59	28.86	28.18	28.75	27.78	28.42	28.12	28.50
Mean weight (g)	212.23	264.93	236.11	225.80	200.29	225.83	241.58	201.49	247.69	228.39	211.02	224.29	196.95	216.58	207.40	217.70
TS per Individual	−42.14	−41.61	−41.89	−42.00	−42.27	−42.00	−41.84	−42.27	−41.77	−41.97	−42.18	−42.01	−42.32	−42.11	−42.20	−42.09
TS per kg	−35.41	−35.84	−35.62	−35.54	−35.29	−35.54	−35.67	−35.31	−35.71	−35.56	−35.42	−35.52	−35.26	−35.46	−35.37	−35.47

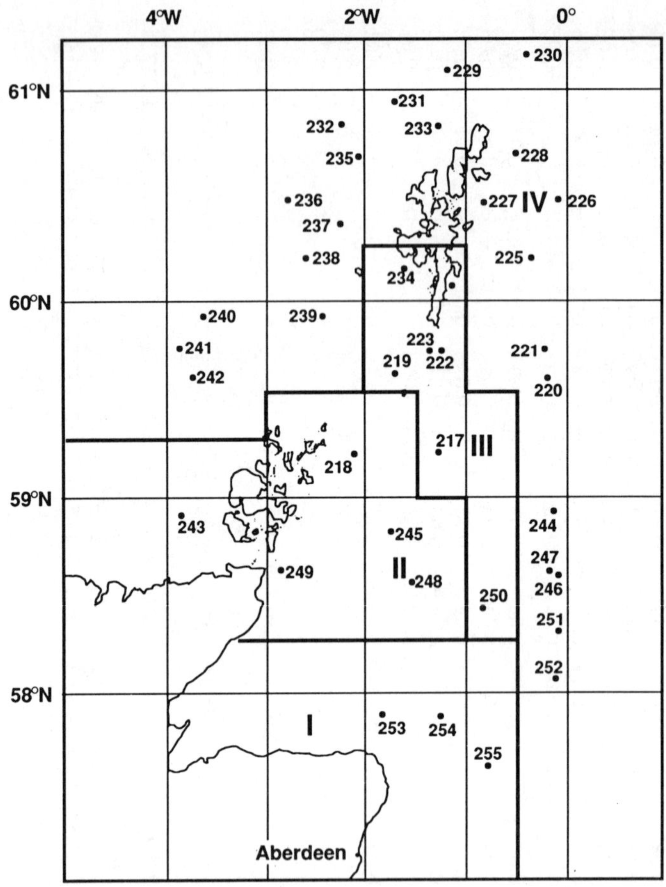

Fig. 8.5 Locations of trawl stations (●) off NE Scotland, numbered to correspond with Tables 8.1 and 8.2, and the boundaries of homogeneous regions (I–IV) identified by the KS test (heavy lines).

to the treatment of length frequencies (section 8.2). Again, there are three methods to be considered which differ in the weight given to each sample, and the respective formulas are presented below.

Method 1: equal weight is given to the proportions in each catch

$$w_i = \left(\sum_k q_{ik} / Q_k \right) \Big/ M \tag{8.9}$$

Method 2: equal weight is given to each catch-rate

$$w_i = \left(\sum_k q_{ik} / t_k \right) \Big/ \sum_k Q_k / t_k \tag{8.10}$$

Method 3: The proportions are weighted by the echo integrals E_k observed in the vicinity of the trawl stations.

$$w_i = \left(\sum_k E_k q_{ik}/Q_k\right) \bigg/ \left(\sum_k E_k\right) \quad (8.11)$$

In each case, the sums are over all the catches from $k=1$ to $k=M$. We recommend Method 2 as being the most generally applicable. Method 1 is better if the catches are taken from a few schools by aimed fishing (p. 210), when the catch rate may bear no relation to the mean density. Method 3 might be expected to give the best result when the echo integrals are highly variable, and again the catch rate might not be a reliable indicator of the density.

Pelagic trawling is a poor method of sampling fish densities, and substantial errors may arise in estimating the proportions of species in mixed aggregations. If there is any possibility of partitioning the echo integrals to species level from examination of the echograms, this should be attempted in preference to the technique of Nakken and Dommasnes (1975). Even if the interpretation of the echogram is uncertain, the error in acoustic partitioning may well be smaller than that of the catch analysis.

Weight–length relationships

The abundance is expressed either as the total weight or as the number of fish in the stock. When considering the structure of the stock, it is convenient to work with the numbers at each age, whereas an assessment of the commercial fishing opportunities would normally be expressed as the weight of the optimum yield. Consistent units must be used throughout the analysis. Thus if the abundance is required as a weight while the target strength function is given for individual fish, the latter must be converted to compatible units. This is done by reference to the weight–length relationship for the species in question.

For a fish of length L, the weight W is variable but the mean relationship is given by an equation of the form

$$W = \mathrm{a_f}\, L^{\mathrm{b_f}} \quad (8.12)$$

where $\mathrm{a_f}$ and $\mathrm{b_f}$ are constants for one species. Suppose the target strength of one fish is given as

$$TS_\mathrm{n} = \mathrm{a_n} + \mathrm{b_n} \log L \quad (8.13)$$

The corresponding function TS_w, the target strength of unit weight of fish, has the same form with different constants

$$TS_\mathrm{w} = \mathrm{a_w} + \mathrm{b_w} \log L \quad (8.14)$$

The number of individuals in a unit weight of fish is $(1/W)$, so the constant coefficients are related by the formulae

$$a_w = a_n - 10 \log a_f \qquad (8.15)$$

$$b_w = b_n - 10\, b_f \qquad (8.16)$$

The weight–length relationship is non-linear. This must be taken into account when estimating the total weight from the numbers in discrete size classes. Suppose there are n_j individuals in the jth class, L_j is the mean length and ΔL is the interval between successive classes. An unbiased estimate of the total weight is

$$W_t = a_f \sum_j n_j \{(L_j + \Delta L/2)^{b_f+1} - (L_j - \Delta L/2)^{b_f+1}\} / \{(b_f+1)\Delta L\} \qquad (8.17)$$

8.4 ABUNDANCE ESTIMATION

The analysis so far has produced for each EDSU of the cruise track, and for each species and depth-channel considered, an estimate of the mean density of the insonified fish. The next step is to determine the total abundance in the surveyed area. To do this, it is necessary to make some assumptions about the density of fish in places where there are no observations, that is in the water between the legs of the cruise track which has not been insonified by the acoustic beam. We have to consider the extent to which the observed densities along the cruise track are representative of the surveyed area as a whole.

The abundance is calculated independently for each species or category of target for which data have been obtained by partitioning the echo integrals. In the following discussion, we shall confine attention to the data relating to one species only.

It is an unfortunate fact that the densities observed during an acoustic survey may include a few values much larger than most. It is tempting to ignore unusual observations on the grounds that they are unrepresentative in some way, but this is wrong. While it is reasonable to look for errors or equipment failure as a good cause for rejection, all valid observations should be included in the calculation of abundance, however unusual they may appear at first sight.

Spatial variation

Before proceeding, we must introduce a few statistical concepts. The fish density is a stochastic variable, one described by a statistical distribution whose mean is estimated as the average of a large number of measurements. It is often found that fish tend to concentrate in some localities while they are scarce in others. This kind of spatial distribution is said to be contagious, as opposed to the case of fish randomly and independently distributed in space, when the probability of a given density is the same everywhere. In either case, if the density at one position is observed over a period of time, the measurements will cover a range of values owing to the movement of fish within the local population.

Thus the observed densities are the result of a stochastic process which may depend upon the location and time of the observations. This leads to the concept of stationarity. The process is said to be stationary if all the samples are drawn from the same statistical distribution regardless of time or position. If the statistical distribution changes with time or position, it is then described as non-stationary.

We shall now use the term PDF (probability density function) for the statistical distribution which controls the individual observations, to avoid confusion with the spatial distribution, which refers to the location of fish relative to one another. When the fish are randomly and independently distributed, the density PDF is stationary. When the distribution is contagious, stationarity depends upon the mobility of the concentrations and whether they can occur anywhere. If the fish are in aggregations which can be found anywhere with equal probability, the density PDF is stationary. On the other hand, if the fish are expected to be concentrated in particular localities, inshore as opposed to offshore for example, and are expected to remain so for any realization (see below) of the stochastic process, the PDF is non-stationary. Paradoxically, when the fish concentrations are less mobile, the density PDF is less likely to be stationary (in the statistical sense).

It is important to understand what is meant by a 'realization' in this context. Suppose that surveys were to be conducted in 100 similar lakes. Each lake has the same topographical features and depth profiles, but the hydrographic conditions and the weather may change. There are identical fish populations in all the lakes, but their mobility ensures that the actual spatial distributions are different. Each survey generates a set of estimates of the fish density at a number of positions in one lake. This set is one realization of the stochastic process which describes the variability arising from the behaviour of the fish, environmental conditions, sampling error and measurement inaccuracy. The 100 surveys should be sufficient to reveal the statistics of this process, in particular whether it is stationary or not. The process is stationary if there is equal probability of finding fish in any part of a lake. However, if the largest concentrations are consistently found in one region but not another, in shallow but not in deep water for example, then the process is non-stationary.

The contagiousness of the distribution depends upon the spatial scale of the observations. If a small area is being considered, the mean density might be similar at all points, but this would not be true of an area that is large compared to the size of the fish concentrations (Fig. 8.6).

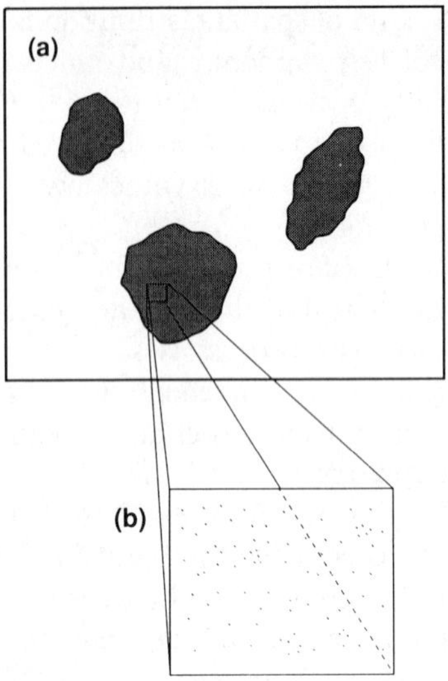

Fig. 8.6 Fish distribution in relation to the scale. In the large square (a), the fish are seen to be in a few isolated schools. This is a contagious distribution. When the small square is magnified (b), the fish within it appear to be randomly distributed.

The densities F_i recorded at positions along the cruise track are samples of the population in A, the whole area covered by the survey. $\bar{F}$ is the arithmetic average of all the observations. If there is no prior knowledge of the distribution, $\bar{F}$ is an unbiased estimate of the mean density applicable to A, and an estimate of the total abundance follows immediately as $\bar{F}$ times the area of A. However, when the distribution is contagious, $\bar{F}$ although unbiased is not the best estimator, that is to say it could be very different from the true mean density. To obtain the best estimate, the one likely to be closest to the true value, we must once again consider dividing the surveyed area into strata within which the abundance may be estimated with confidence. These strata are not to be confused with the regional divisions considered in the survey design, which allowed for some regions to be sampled more intensively than others, and those based on the composition of catches in the calculation of the conversion factors. While the stratification required for the abundance estimation should take account of the regional boundaries defined earlier, it is likely to be on a smaller scale. Each region may be further divided into several strata. Several methods have been employed to estimate the abundance by stratification of the surveyed area, as described below. As a general principle, the strata should comprise the largest non-overlapping contiguous sub-areas such that the density PDF within each sub-area is stationary.

Spatial correlation is another statistical feature that has a bearing on the analysis. When the distribution is contagious, the mean density shows strong local correlations over the surveyed area. The densities observed at two nearby points are more likely to be similar than those at positions far apart. Thus the sequence of measurements along the cruise track may be serially correlated, which means that the probability of observing F_i depends upon the actual densities recorded earlier. Serial correlation is easier to illustrate by considering the case of a highly contagious population, one in which the fish occur in a few dense concentrations with much empty water in between. The measurement sequence contains strings of consecutive zeros corresponding to the sections of cruise track crossing empty water. Now suppose that the survey is in progress and that a measurement has just been made. If the result is $F=0$, the probability that the next result will be the same is higher than the proportion of zeros in all the measurements. The presence of correlation does not affect the estimate of abundance, provided the cruise track has been designed to give samples which can reasonably be considered as representative of the area in question, but it is important in estimating the variance and the sampling error. We shall return to this problem in section 8.5.

Contour maps

The distribution of fish in the surveyed area may be displayed as a contour map. The contours are curves connecting points of equal density, and when they are drawn on a map it is easy to see where the major concentrations of fish are located (Fig. 8.7).

The contours are constructed by reference to the observed densities along the cruise track. Each observation is supposed to be made at the midpoint of the corresponding EDSU. Thus we begin with a number of observations at known positions on the map. Contouring is essentially a matter of interpolating between the point observations, to deduce the positions at which the density is expected to equal the contour level. The contours may be constructed by several methods which differ in the weight given to each observation and the treatment of stochastic measurement errors (Powell, 1974). Here we describe a simple method which is not precise but should be adequate for most purposes in the context of acoustic surveys.

The first step is to decide the contour levels. The observed densities will often span several orders of magnitude, so a logarithmic scale is useful with contours, say, at 10, 100 and 1000 units of density. The point observations are inspected to see where the contours lie with respect to the cruise track. The intersection of each contour with the track is calculated by interpolation. Suppose F_1 and F_2 are two adjacent observations, one above and one below the contour level F_c. The intersection is located at a distance $(F_c - F_1)/(F_2 - F_1)$ times the EDSU from the F_1 position. When this has been done for all the intersections indicated by the observations, the structure of the contours ought to become clear. Finally, the

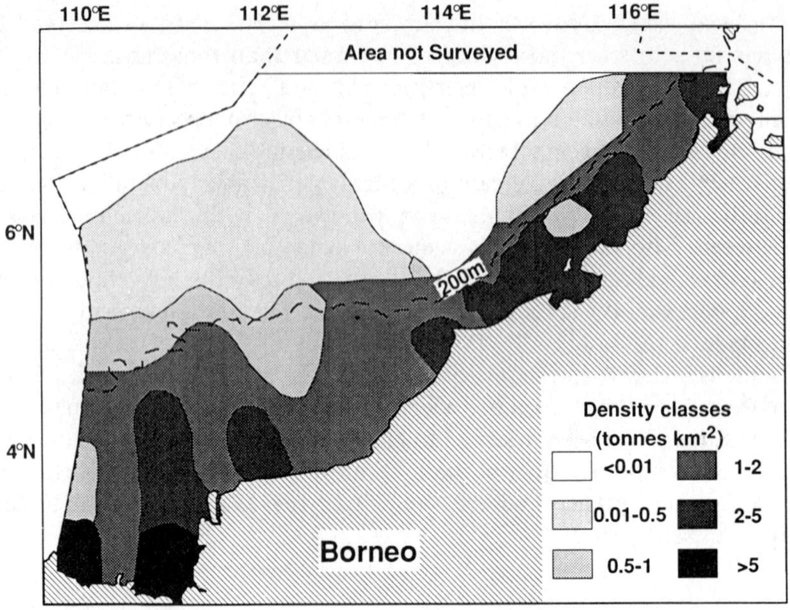

Fig. 8.7 Contour map of the fish density revealed by an acoustic survey of mixed pelagic species in the South China Sea.

contours are drawn as smooth curves joining the loci of the track intersections (Fig. 8.8).

In principle, the abundance can be determined from the contour map. The sub-area bounded by two contours is considered as one stratum. The area of each stratum is measured with a planimeter, and then multiplied by the average density to estimate the abundance. This is the post-sampling stratification method described by Johanesson and Mitson (1983). In essence, the echo integrator data are used to define the strata.

Post-sampling stratification is not recommended for the estimation of abundance. It suffers from fundamental theoretical problems to do with the fact that the same data are used to define the strata and to calculate the abundance. The measurements cannot be considered as random samples from a defined population because the boundaries of the strata are partly determined by the stochastic variation of the observed densities. Near the boundaries, the measurements are likely to be assigned to the stratum containing the values which are most similar. Thus the apparent variance within each stratum is less than it would be if the boundaries were fixed. Furthermore, the estimated abundance is sensitive to the choice of contour levels and may be biased. It is difficult if not impossible to determine confidence limits on the abundance estimates obtained by post-sampling stratification. To avoid these problems, it is essential that the strata

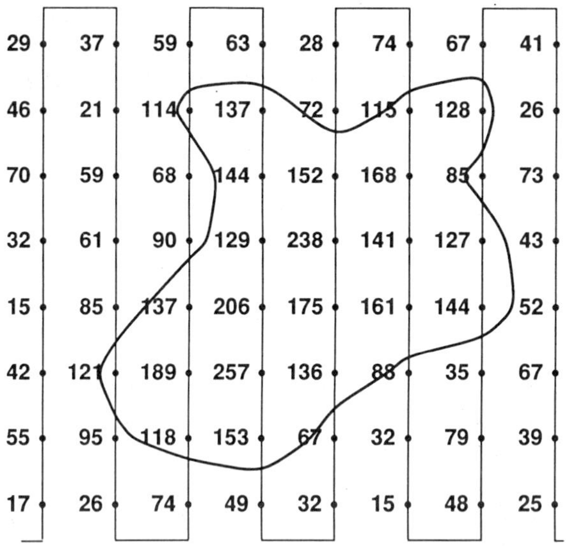

Fig. 8.8 Construction of contours. The numbers are the mean fish densities referred to the midpoint of each elementary distance sampling unit (EDSU) of the cruise track (light line). The contour for 100 density units (heavy line) is interpolated between the point samples.

should be selected by criteria that are independent of the acoustic measurements.

Contouring is most useful as a means of visualizing the important features of fish concentrations. It works best for contagious distributions in which the spatial variations are stronger than the random changes in time.

Rectangular grid

A simple method of stratifying the surveyed area is to divide it by equally spaced lines of latitude and longitude (Dalen and Nakken, 1983; Simmonds and MacLennan, 1988). The strata are the rectangles formed by the intersections of the latitudinal and longitudinal lines. Johannesson and Mitson (1983) refer to this type of stratum as the ESSR, or the 'elementary statistical sampling rectangle'.

It is assumed that the density PDF within each stratum is stationary, so the ESSR must not be too large. On the other hand, there must be at least one sample (and preferably more) within each ESSR, which dictates the minimum size. The cruise track must pass through each rectangle at least once. In the case of surveys for herring in the North Sea, it has been found that satisfactory results are obtained with a rectangular grid spaced at intervals of 0.5° in longitude and 0.25° in latitude. Thus the ESSR is approximately square with a side-length of 26 km, and there are about 100 ESSRs in the surveyed area.

One sample of the fish density is obtained from each EDSU along the cruise track. Suppose there are n_k samples in the kth ESSR ($k=1$ to M). F_{jk} is the jth sample and A_k is the area of the ESSR which may contain fish. A_k is not necessarily constant, because part of the ESSR may coincide with islands or the coast. The abundance Q is estimated from the average density within each ESSR multiplied by the corresponding area.

$$\hat{Q} = \sum_{k=1}^{M} \left\{ A_k \sum_{j=1}^{n_k} F_{jk} / n_k \right\} \tag{8.18}$$

The rectangular grid method has a particular advantage in that the geographical basis of the strata allows easy mapping, and the calculations are straightforward. A pocket calculator will suffice if a computer is not available.

Anderson and Kirkegaard (1985) have suggested that geographical strata might be bounded partly by depth contours instead of latitudes or longitudes. This approach should be advantageous if different fish densities are expected according to the water depth.

Transform methods

The PDF of the fish density is often positively skewed, which means that a large proportion of the observations yield small values, resulting in the characteristic histogram shown in Fig. 8.9. This type of PDF is very different from the symmetrical normal or Gaussian probability function on which much of sampling theory is based. If μ is the true mean and σ^2 is the true variance of F, the Gaussian PDF is

$$P(F) = \exp\{-(F-\mu)^2/2\sigma^2\}/\sqrt{(2\pi\sigma^2)} \tag{8.19}$$

For any stationary PDF, the arithmetic average $\bar{F}$ and the sample variance s^2 calculated from the observations are unbiased estimates of the true mean and variance respectively. But when the PDF is not Gaussian, these estimators although unbiased are not the most precise. They are subject to stochastic variation which may be very large.

More efficient estimators can be derived if the PDF is known. The principle behind this idea is that a new data set is conceived as a one-to-one transformation of the original observations, such that the new PDF is Gaussian. Statistical theory is applied to deduce new estimators for the mean and variance which are more accurate than $\bar{F}$ and s^2.

The first step is to determine the appropriate transformation. It is sufficient for practical purposes to consider only the class of power transformations $Z_i = (F_i^{\lambda} - 1)/\lambda$ for λ in the range 0 to 1. The limiting case $\lambda = 0$ is equivalent

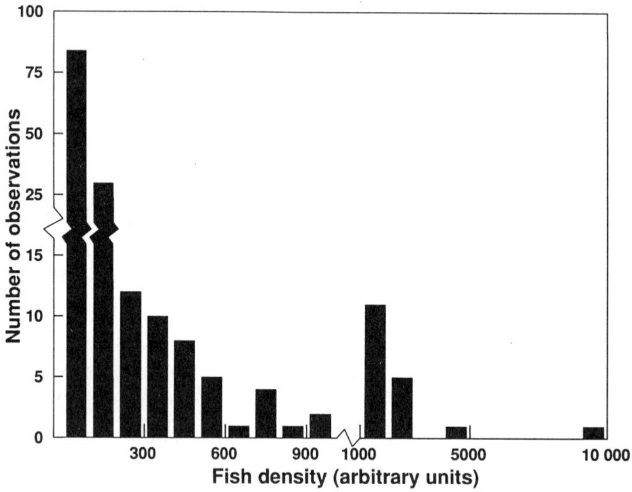

Fig. 8.9 Histogram of fish densities F recorded during a survey of herring in the North Sea. The class intervals are 100 for $F < 1000$ and 1000 for $F > 1000$. The distribution is positively skewed. Most of the observations are lower than the average.

to the log-transform $Z_i = \ln(F_i)$. The most likely value of λ may be determined from a test devised by Box and Cox (1964). Estimators for the special cases $\lambda = 0$, 1/6, 1/4, 1/3 and 1/2 have been described by MacLennan and MacKenzie (1988) (Appendix F). As far as we know, estimators have not been derived for an arbitrary value of λ, but in practice it is good enough to work with those from Appendix F for which λ is closest to the value indicated by the Box–Cox test.

The transform theory assumes that the samples are drawn from a stationary PDF which is zero for $F \leqslant 0$. Further complications arise when the fish distribution is contagious to the extent that there is a finite probability of observing $F = 0$. Aitchison (1955) and Pennington (1983) have considered this problem. It is supposed that the fish occur in patches with empty water in between, but the density PDF is stationary within each patch. Aitchison's method treats the zero values and the others as samples from different PDFs, and the estimators are modified to take account of the proportion of zeros in the data. The relevant formulae are in Appendix F.

In principle, the transform method should provide the best estimates of mean and variance, those most likely to be closest to the true values. However, the method depends upon the density PDF being known, and it must be unimodal. If the estimators for the wrong PDF are applied, the results will be biased to an uncertain extent. The contagion which is often a feature of the fish distribution is another practical problem. The transform method is not suitable for contagious distributions unless they conform to the assumptions of Aitchison's technique.

Kriging

Kriging is a powerful smoothing technique which can be used to calculate the surface which represents the spatial variation of the mean density. Clark (1979) has provided a simple explanation of the theory. Originally developed for surveys connected with mining operations, hence the term 'geostatistics' which appears in the cited literature, the use of Kriging in acoustic survey analysis has been proposed by Conan and Buerkle (1988), Guillard *et al.* (1990) and others. The technique is based on the variogram, which is a graph of the covariance γ of paired samples, plotted against the distance h between the sampled locations. If $\{F, F'\}$ is the set of paired samples for a given h, $\gamma(h)$ is the expected value of $\{(F-F')^2/2\}$. Strictly speaking, $\gamma(h)$ is the semi-variogram function, but the prefix is often omitted as we do here. Examples of variograms are shown in Fig. 8.10.

When the fish distribution is contagious, the covariance of observations made at closely spaced points is less than that of the data set as a whole.

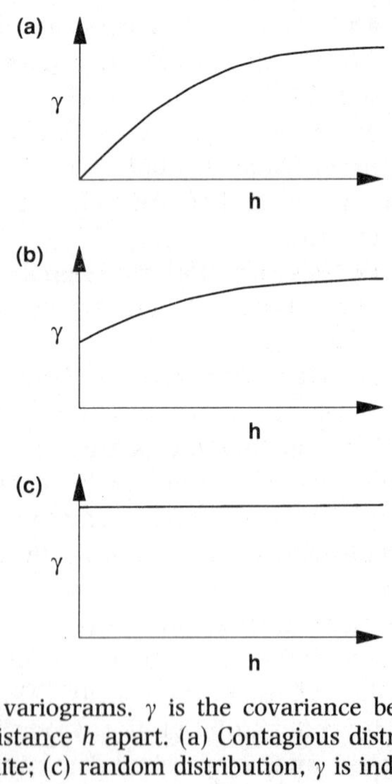

Fig. 8.10 Examples of variograms. γ is the covariance between the fish densities observed at locations distance h apart. (a) Contagious distribution, $\gamma(0)\rightarrow 0$; (b) the nugget effect, $\gamma(0)$ is finite; (c) random distribution, γ is independent of h.

Thus γ increases monotonically with h, and for large h it approaches the sample variance of all the observations. A curve is fitted to the points of the variogram, and the shape of this curve indicates the spatial characteristics of the fish distribution. If γ remains finite as h tends to zero, this is the 'nugget effect' which implies that there are two causes of variation, local measurement error and the spatial structure. The size of the nugget effect compared to the sample variance indicates which is the more important cause of variation. To estimate the abundance, the area represented by each sample is calculated, taking account of the other observations in the vicinity and their distance from the point in question. Kriging does not require the samples to be statistically independent. The effect of serial correlation is removed so that the variance of the abundance estimate is determined without bias.

At present, the Kriging technique is under development and it is not clear how useful it is for the analysis of acoustic data. The method works best for populations that have a spatial structure which does not change in time. Unfortunately, this is seldom true of pelagic fish. If the spatial distribution changes on the time scale of the survey owing to migration or random movement of the fish concentrations, the likely result is that the points on the variogram will show no consistent pattern, and the form of the fitted curve will be doubtful. As far as abundance estimation is concerned, the results obtained from Kriging are sensitive to the assumed form of the variogram function. It may be that future developments will demonstrate the advantages of Kriging more clearly, but for the present we do not recommend this technique for the analysis of acoustic surveys.

8.5 PRECISION OF THE ABUNDANCE ESTIMATE

The densities recorded during an acoustic survey relate to the fish detected by the echosounder. We suppose that the stock inhabits a certain volume of water, but only those fish which encounter the acoustic beam are detected. The proportion insonified of the total volume indicates the sampled fraction of the stock. This might be as small as 0.01% for a stock spread over a large area of ocean, and seldom exceeds 2% even for small lakes.

The error of the abundance estimate may be considered as the sum of two components, firstly the sampling error (precision) caused by the measurements being stochastic samples of the true mean density, and secondly the systematic error (bias) which effects all the observations equally, such as the error in the assumed value of the target strength. In this section we discuss the sampling error and how it can be determined. The systematic errors are discussed in section 8.7.

Repeated surveys

Suppose that a number of independent surveys have been made of the same stock, by using several ships at the same time or successive surveys of the same area by one ship. There are N surveys and the abundance estimates are $Q_1, Q_2 \ldots Q_N$. If these estimates are believed to be equally precise, the best estimate making use of all the data is the arithmetic average, $\bar{Q}$. s^2 is the sample variance and SE is the standard deviation of $\bar{Q}$, known as the standard error.

$$\bar{Q}=\sum_{j=1}^{N} Q_j/N \tag{8.20}$$

$$s^2=\sum_{j=1}^{N} (Q_j-\bar{Q})^2/(N-1) \tag{8.21}$$

$$SE=s/\sqrt{N} \tag{8.22}$$

It is convenient to describe the precision in terms of confidence limits on the best estimate. The result is given as a statement that Q is expected to lie in the interval $(\bar{Q}-t\cdot SE)$ to $(\bar{Q}+t\cdot SE)$. The multiplier t depends upon N and the confidence level. If the PDF of Q_j is Gaussian (normal), t is the appropriate percentile from Student's t-distribution. There are $(N-1)$ degrees of freedom, and the values given in Table 8.3 are used to determine the confidence limits. It is reasonable to assume normality in this instance, since each abundance estimate is obtained from the average of many density measurements. The central limit theorem provides that whatever the PDF of

Table 8.3 Percentile values from Student's t-distribution for various degrees of freedom and levels of confidence

Degrees of freedom	*Confidence level*		
	90%	*95%*	*99%*
1	6.31	12.71	63.66
2	2.92	4.30	9.92
3	2.35	3.18	5.84
4	2.13	2.78	4.60
6	1.94	2.45	3.70
8	1.86	2.31	3.35
10	1.81	2.23	3.17
15	1.75	2.13	2.95
20	1.72	2.09	2.84
30	1.70	2.04	2.75
∞	1.64	1.96	2.58

one observation, the average of many has a PDF much closer to the Gaussian function.

There are few reports in the literature of the precision being determined by repeated surveys. Stromme and Saetersdal (1987) have described work on tropical stocks of *Sardinella* spp. The results of independent surveys were generally within 10% of the average. Bailey and Simmonds (1990) compared two surveys conducted by different research vessels working independently but concurrently in the same area of the North Sea. The abundances of herring found by the two vessels differed by 10%. These results indicate that the precision of a properly designed acoustic survey can be extremely good. However, the cost of repeating the survey several times is often prohibitive. It is normally the case that only one survey of an area can be accomplished in the time available. The precision must then be estimated by reference to the data collected during the one survey, by analysis of the variations observed along the cruise track.

Stratified random transects

The analysis of precision is greatly simplified when the survey has been designed as randomly positioned parallel transects. Jolly and Hampton (1990) recommend the random transect approach because simple statistical theory can be applied without too many assumptions, and in particular there is no need to worry about spatial correlation. Shotton and Bazigos (1984) and Shotton (1981) support this view. The surveyed area will have been stratified (if necessary) into regions by the survey design, and the random positioning of transects is done independently for each region. The mean density observed along the whole length of a transect is considered as one sample. Suppose there are M regions. In the rth region whose area is A_r, there are n_r transects and the density samples are F_{jr} for $j=1$ to n_r. The abundance Q_r in this region and its variance V_r are estimated as

$$\hat{Q}_r = A_r \sum_{j=1}^{n_r} F_{jr} / n_r \qquad (8.23)$$

$$\hat{V}_r = \sum_{j=1}^{n_r} (A_r F_{jr} - \hat{Q}_r)^2 / [n_r(n_r - 1)] \qquad (8.24)$$

The total abundance and its variance are obtained by summing the results for each region.

$$\hat{Q} = \hat{Q}_1 + \hat{Q}_2 + \cdots + \hat{Q}_M \qquad (8.25)$$

$$\hat{V} = \hat{V}_1 + \hat{V}_2 + \cdots + \hat{V}_M \qquad (8.26)$$

The standard error of $\hat{Q}$ is the square root of $\hat{V}$. The confidence limits are once again determined by the standard error and Student's parameter t, the degrees of freedom in this case being the total number of transects less the number of regions.

Degree of coverage

Another approach is to look for a systematic relationship between the sampling error and the fraction of the stock which has been observed. It may be argued intuitively that the greater the sampling intensity, in terms of the length of cruise track in a particular area, the better the results should be. Aglen (1983, 1989) has investigated the precision of surveys made in different circumstances, in confined fjords and large areas of open sea, and in tropical as well as temperate waters. His results are illustrated in Fig. 8.11. As a measure of the sampling intensity, Aglen defined the degree of coverage to be $d=D/\sqrt{A}$, where D is the total length of the cruise track and A is the size of the surveyed area. For each survey, he determined the precision of the abundance estimate by comparing results derived from subsets of the transects. Thus every second transect might be used to obtain one estimate, and the intervening transects to provide another. In effect, the two interleaved sets of transects are considered as though they had been covered by different vessels, and the difference between the two results (assumed to be independent estimates of the same stock) is an indication of the sampling error. As a measure of the precision, Aglen uses CV, the coefficient of variation, which is the standard error of the abundance estimate divided by

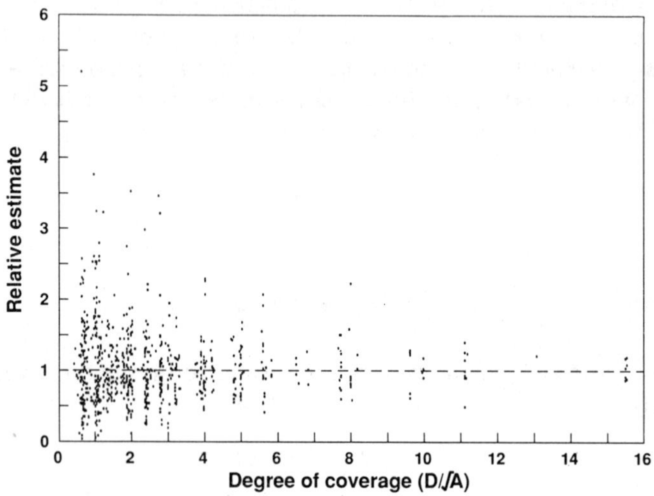

Fig. 8.11 Variation of the abundance estimate v. the degree of coverage ($D/\sqrt{A}$). Redrawn from Aglen (1989). Each point is the ratio of one estimate to the mean of several.

the mean. He suggests that CV depends upon the degree of coverage through a power law of the form

$$CV = a(D/\sqrt{A})^b \tag{8.27}$$

The coefficients a and b may be determined from the function which best fits the results obtained from many surveys. Aglen found that b is close to -0.5, the theoretical value predicted on the assumption that the transects provide independent samples. The value of a is more variable, between 0.4 and 0.8. It is reasonable to suppose that a depends upon the contagion of the fish distribution, a factor not considered explicitly by Aglen. Higher values of a would be appropriate in the case of fish concentrated in a few large schools, as opposed to the more uniform distribution of scattering layers.

Another useful conclusion from Aglen's work is that when the degree of coverage is more than 6, the sampling error in the abundance estimate has a PDF which is close to the Gaussian (Fig. 8.12). This simplifies the calculation of confidence limits when d is large enough.

Kimura and Lemberg (1981) support Aglen's use of $D/\sqrt{A}$ as the determinant of CV. Francis (1985) suggests that D/A is more appropriate. To compare the two measures, consider what happens if we combine the results from two surveys conducted in adjacent waters, each covering the same size of area and having the same track design. In the combined survey, D and A have each increased by a factor of two. Thus d has increased by $\sqrt{2}$, but D/A remains the same. If the two surveys have shown abundances Q_1 and Q_2, the result of the combined survey is $Q = Q_1 + Q_2$ and the variances are similarly additive. Thus the combined survey should produce results

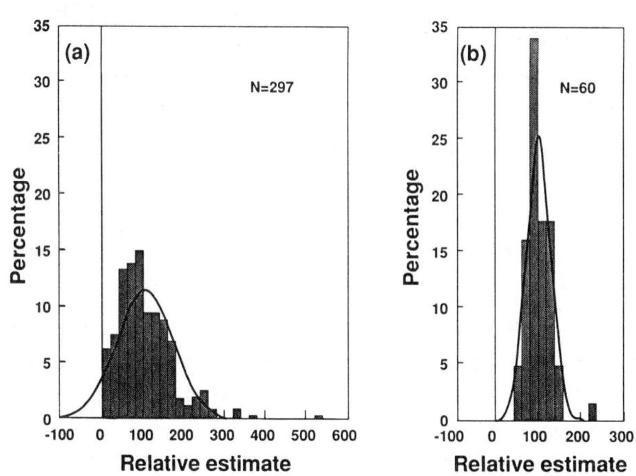

Fig. 8.12 Histograms of the estimated abundance as a proportion of the true value for degrees of coverage; (a) d < 2, (b) d > 6. The curves are fitted Gaussian functions; the fit is poor when d is small. Redrawn from Aglen (1989).

with a reduced CV. This suggests that $D/\sqrt{A}$ is the more useful measure, at least for the calculation of confidence limits, since any predictive formula requires a measure of the sampling intensity which changes under circumstances which alter CV.

According to Cochran (1977), the precision depends upon N, the number of independent samples collected from a population of size N_0, and the sampled fraction N/N_0 is unimportant. Aglen's interpretation of $D/\sqrt{A}$ as the sampling intensity may therefore be misleading. Alternatively, we can say that $N=D/D_I$, where D_I is the minimum EDSU which gives independent samples. D_I may depend upon A, because larger areas are more likely to exhibit contagious distributions. But whatever the explanation, Equation 8.27 is an empirical result which is helpful in designing the survey, since it indicates the track spacing necessary to achieve a given precision.

Other methods

In principle, any set of statistically independent samples of the fish density might be used to construct a confidence interval for the estimated abundance. The more samples we have, the more information is potentially available about the distribution and its variance. Earlier we considered the case of randomly positioned transects in which one sample is obtained from each transect (p. 265). The random positioning ensures the independence of the samples. In this context, independence means that for any two samples, the observation of one does not affect the PDF of the other.

When the samples are obtained from evenly spaced transects, or successive EDSUs within a transect, they might not be independent, in which case the sample variance is a poor indicator of the precision. Some test of independence is necessary. One method is to examine the serial correlation in the sample sequence considered as a time series (MacLennan and MacKenzie, 1988). If there are n samples in total, and there are n_L pairs separated by L observations in the sequence, the correlation coefficient for the lag L is

$$\rho_L=\{(n\text{-}1)/(n_L-1)\}\sum_{i=1}^{n_L}(F_i-\bar{F})(F_{i+L}-\bar{F})/\sum_{i=1}^{n}(F_i-\bar{F})^2 \qquad (8.28)$$

The samples are independent if the serial correlation is negligible. The variance of ρ_L is approximately $(1/n_L)$, and standard statistical tests will determine whether the observed ρ_L are consistent with the hypothesis that the expected value of ρ_L is zero.

When the sample sequence contains strings of zeros, the correlation test is likely to fail. If the non-zero sequences are found to have no internal correlation, Aitchison's method can be used to take separate account of the zero values (p. 289).

Referring to Equation 8.18, $\hat{Q}$ is the abundance estimate obtained from samples F_{jk} collected in M strata. Suppose first that within each stratum, the F_{jk} are independent samples. $\bar{F}_k$ is the arithmetic average of the n_k samples from stratum k, whose area is A_k, and the variance of $\hat{Q}$ is estimated as

$$\hat{V}=\sum_{k=1}^{M}\sum_{j=1}^{n_k} A_k^2(F_{jk}-\bar{F}_k)^2/\{n_k(n_k-1)\} \tag{8.29}$$

However, if the F_{jk} are not independent, the small-scale variations are difficult to interpret and Equation 8.29 is not valid. The observations might then be aggregated to give a smaller set which does satisfy the independence test, for example by considering each transect as one sample and one stratum. M is now the number of transects, $\tilde{F}_k$ is the mean density and A_k is the area represented by the kth transect. The abundance estimate and its variance are

$$\hat{Q}=\sum_{k=1}^{M} A_k\tilde{F}_k \tag{8.30}$$

$$\hat{V}=\{M/(M-1)\}\sum_{k=1}^{M}(A_k\tilde{F}_k-\hat{Q}/M)^2 \tag{8.31}$$

A third method, also relevant to non-independent data, is the cluster estimator described by Williamson (1982). Each transect is considered as a cluster of n_k observations, which may be correlated. $\hat{Q}$ is still given by Equation 8.30, but the variance estimator is weighted by n_k as well as the representative areas. The formula is

$$\hat{V}=\{M/(M-1)\}\sum_{k=1}^{M} n_k^2(MA_k\tilde{F}_k-\hat{Q})^2\Bigg/\left(\sum_{k=1}^{M} n_i\right)^2 \tag{8.32}$$

However, Equation 8.32 is an approximation which is invalid if the CV of the n_k is more than 0.2 (Kish, 1965), so the transects have to be similar in size. We prefer Equations 8.29 or 8.31 as being more generally applicable to the estimation of variance. Equation 8.29 is appropriate for non-stationary statistics when strata containing independent samples are easily defined. Equation 8.31 is preferred when the statistics are stationary and there is one independent sample per transect.

A different approach to the problem is the bootstrap or jack-knife technique (Efron and Tibshirani, 1986; Robotham and Castillo, 1990). The PDF indicated by the observed densities is used with a random number generator to produce new sets of simulated data. The abundance is estimated for each set and confidence limits are obtained by the method described for repeated surveys (p. 264). However, we consider that bootstrapping has limited relevance to acoustic surveys. It takes no account of the spatial distribution, and each observation is treated as a sample from a stationary PDF.

8.6 SOURCES OF SYSTEMATIC ERROR

So far we have restricted attention to the random error in $\hat{Q}$ caused by the sampling strategy. There are other sources of error to be considered before the overall accuracy of $\hat{Q}$ can be assessed. The non-sampling errors are systematic in that all the observations are similarly affected. Consequently, $\hat{Q}$ is a biased estimate of the abundance. Unlike the random error, the bias cannot be reduced by collecting more samples.

The target strength is an important source of systematic error. $\hat{Q}$ is inversely proportional to $\langle\sigma\rangle$. Thus any error in the assumed value of $\langle\sigma\rangle$ biases all the observed densities and the abundance. We shall say no more about target strength here, since the subject has been considered exhaustively in Chapter 6.

Equipment sensitivity

According to Equation 8.4, the estimated density is proportional to the calibration factor K which describes the physical performance of the echosounder and the echo integrator. K is determined by measuring the signal from a standard target (Chapter 3). The measurement error in K, and any change which occurs between the time of the calibration and the survey, has the same proportional effect on all the density observations and thus contributes to the bias in the abundance estimate.

If the calibration is performed carefully in accordance with the recommended procedure, the precision of K is very good. For scientific echosounders operating in the 30–50 kHz range, the measurement error in K should be less than $\pm 4\%$ (Foote *et al.*, 1987). As regards changes in the longer term, Simmonds (1990) has measured the performance of a Simrad EK400 echosounder over a 5-year period, and finds that the variation of K is within $\pm 7\%$.

Transducer motion

Because of the finite time between the transmitter pulse and the reception of echoes, any movement of the transducer degrades the amplitude of the received signal. The transducer moves forward with the vessel, and it also pitches and rolls, so the angular motion is more severe in bad weather. In consequence, there is some fluctuation of the received signal, but more importantly the mean amplitude is reduced. This bias occurs because of the change in the target direction as seen from the transducer (Stanton, 1982). In effect, the transmitting and receiving beam patterns have been misaligned by the angle ϕ, which is the change in the target direction between transmission and reception. For the same motion, the bias depends upon the beam pattern, being greater for narrow-beam transducers (Fig. 8.13).

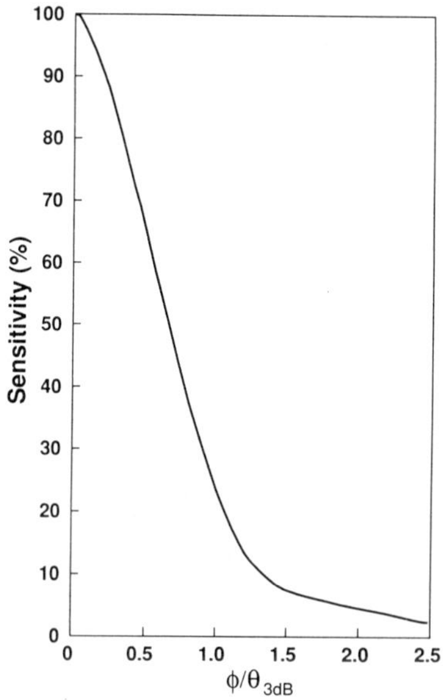

Fig. 8.13 Relative transducer sensitivity (echo intensity) resulting from angular movement ϕ of the transducer in pitching/rolling while insonifying randomly distributed targets. θ_{3dB} is the beam width of the transducer between 3 dB points.

In the case of pitch and roll, $\phi = \phi_a$ is simply the angle turned in the time delay of the echo. Thus ϕ_a increases with the range of the target and is proportional to the angular rate of rotation. The component owing to the forward motion, ϕ_f, does not depend upon the range. When the transducer is transmitting vertically, ϕ_f is a function of (v/c), the ratio of the cruising speed of the ship to that of sound.

$$\phi_f = 2 \tan^{-1}(v/c) \tag{8.33}$$

For example, if $v = 5.14\,\mathrm{ms}^{-1}$ (10 knots) and $c = 1500\,\mathrm{ms}^{-1}$, ϕ_f is 0.4°. For a transducer of 5° beam width, moving at 10 knots without pitching or rolling, the curve in Fig. 8.13 predicts a reduction of 1.5% in the received echo energy compared with the stationary case. Figure 8.14 illustrates the effect of pitching and rolling as a function of the rotation rate and the range. The total movement ϕ is somewhere in the interval $(\phi_a \pm \phi_f)$, depending on the direction of the rotation axis.

Surface bubble layer

Wind stress on the sea surface induces gas bubbles that increase the attenuation and scattering of sound waves. When the bubble layer extends below the transducer, the echo amplitudes are reduced by the factor A_b, the 'extra attenuation', which increases with the wind force. The motion of the

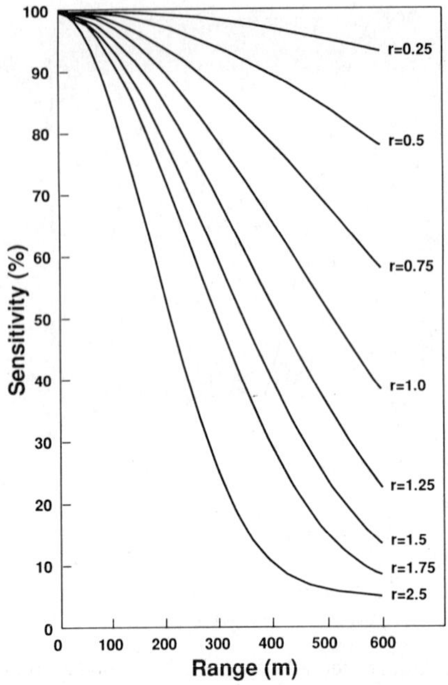

Fig. 8.14 Relative sensitivity (echo intensity) caused by pitching/rolling of the transducer without horizontal displacement, as a function of the target range and *r*, the rotation rate (degrees s^{-1}) divided by the beam width (degrees) between 3 dB points.

ship also induces bubbles, as evidenced by the foam seen in the wake. Again, the bubble density and hence the acoustic attenuation are likely to increase in bad weather.

Hall (1989) has reviewed the theoretical models which describe firstly the aeration due to wind stress, and secondly the propagation of sound through bubbly water. Weston and Ching (1989) have measured the extra attenuation in horizontal propagation at frequencies below 5 kHz. They attribute most of the attenuation to the surface waves; the losses caused by the bubble layer are uncertain. Weston (1989), having compared the observations with theory for frequencies up to 40 kHz, concludes that the predicted attenuation due to bubbles is typically half that observed. Novarini and Bruno (1982) have considered the effect of wind-induced aeration on vertical sound propagation at frequencies up to 150 kHz (see also Bruno and Novarini, 1983). They derive empirical equations which express A_b as a function of the wind speed v_w in knots (one knot $= 0.514\,\mathrm{m\,s^{-1}}$), the frequency f in kHz and the transducer depth z in metres. The equations are valid for wind speeds 6–30 knots, $z = 1$ m or more, and frequencies as follows:

Low frequencies, 8–60 kHz:

$$A_b = (0.9754 - 0.08957\ v_w + 0.002367\ v_w^2) f^{1.32} \times \exp(-1616.77\ z\ v_w^{-2.36}) \qquad (8.34)$$

High frequencies, 70–150 kHz:

$$A_b = 10^6 (5.94 - 0.551\ v_w + 0.0158\ v_w^2) f^{-2.451} \times \exp(-2756.5\ z\ v_w^{-2.579}) \qquad (8.35)$$

These equations are based on acoustic propagation theory and early measurement of bubble densities by Johnson and Cooke (1979). Dalen and Løvik (1981) have reported experiments with a transducer on the hull of a moving ship. They measured A_b in two ways, by reference firstly to the volume reverberation from the bubble layer, and secondly to the echo integral of the sea-bed echo. There is good agreement between the two methods, but the extra attenuation observed by Dalen and Løvik is much greater than that reported by others (Fig. 8.15). Comparison with theory (Weston, 1989) suggests that the extra attenuation is not all due to bubbles directly induced by the wind. The motion of the ship could be a contributory factor. Apart from the aeration in the wake, water displaced downwards by the ship could enhance the bubble density below a transducer on the hull. Dalen and Løvik note that A_b depends upon the wind direction relative to the ship's track, again suggesting some indirect effect through the weather-induced motion. They suppose that the observed attenuation would apply equally to a transducer in a towed body, but this is unlikely to be true of the component resulting from the ship motion.

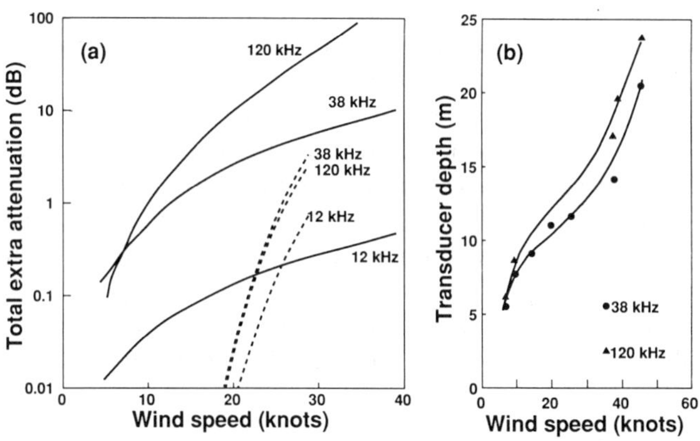

Fig. 8.15 Extra attenuation caused by wind-induced bubbles (1 knot = 0.514 m s^{-1}). (a) Attenuation at three frequencies; solid curves, measurements on a hull-mounted transducer 5 m deep (Dalen and Løvik, 1981); broken curves, predictions after Bruno and Novarini (1983). (b) Transducer depth required for attenuation ⩽0.3 dB.

We consider that the Novarini–Bruno equations are a good description of the extra attenuation caused by the natural environment, indicating the minimum level of A_b which will be greater in the survey application when additional aeration is generated by the ship. Furthermore, A_b will depend upon the weather during the previous few days, and the shelter afforded by the coast if it is nearby, as well as the wind speed at the time. If the survey must continue in bad weather, A_b should be measured as described by Dalen and Løvik (1981), to allow better compensation of this bias.

Hydrographic conditions

The intensity of the sound wave decreases as it propagates through the water because of beam spreading and absorption. These losses depend upon the sound speed c and the absorption coefficient β. In Chapter 2 we described how c and β are determined from empirical equations as functions of the water temperature, salinity and depth. Thus it is important to understand the hydrography of the area being surveyed. Ideally, the temperature and salinity should be measured at a number of stations, failing which we must rely upon other sources of information such as hydrographic maps.

The prevailing hydrographic conditions can change from place to place and from one season to another. In particular, the position and strength of the thermocline is important in vertical echosounding, causing the average sound speed over the transmission path to change with the depth of targets. The strongest thermoclines occur in tropical seas, where the temperature might be 30°C at the surface but only 12°C at 200 m depth. The change with depth occurs sharply at the thermocline. In the Persian Gulf, for example, the summer temperature can change by 16°C over just 20 m of depth, but the thermocline weakens in winter when the water is nearly isothermal. Thus local conditions and the seasonal factor need to be considered in estimating the likely values of c and β.

Suppose that the sound speed and absorption coefficient are assumed or calculated to be (c_1, β_1), which differ from the true values (c, β), the latter representing appropriate averages over the transmission path. These errors bias the observed fish density through the functional dependence of the equivalent beam angle ψ and the exact TVG function (MacLennan, 1990). ψ is approximately proportional to c^2. If the actual TVG function is implemented to be exact for the parameters c_1 and β_1, and R_0 is the depth at the middle of the integration channel, the echo integral is in error by the factor

$$\bar{g} = (c/c_1)^2 \exp\{2R_0(\beta - \beta_1 c_1/c)\} \tag{8.36}$$

Since $\hat{Q}$ is proportional to the ratio $(\bar{g}/\psi)$ (Equation 5.15), the variation of ψ with c compensates at least partly for the c^2 dependence of $\bar{g}$ through the first term of Equation 8.36. If the water temperature and the salinity are known, MacLennan (1990) suggests that at 38 kHz and $R_0 = 200$ m, the hydrographic error in $\hat{Q}$ is less than 3%. Larger errors may occur if the temperature and salinity are uncertain (Foote, 1981b). Figure 8.16 shows how the TVG error depends upon the range, the frequency and deviations in temperature.

In the case of an echo counter whose TVG is nominally 40 log R, $\bar{g}$ is proportional to $(c/c_1)^4$. Consequently, the hydrographic error in echo counting is greater than that applicable to echo integration.

Noise and reverberation

We use the term 'noise' to cover any unwanted signal produced by the echosounder, whether it is true electrical noise, acoustic reverberation or the merged echoes from plankton. Whatever the source, the noise obscures the smaller echoes that we wish to measure. When a signal threshold is applied to eliminate the noise, any echo smaller than the threshold is also ignored. The observed density is therefore biased because some proportion of the target population has not been detected. The bias depends upon the ratio of the signal and noise amplitudes (*SNR*).

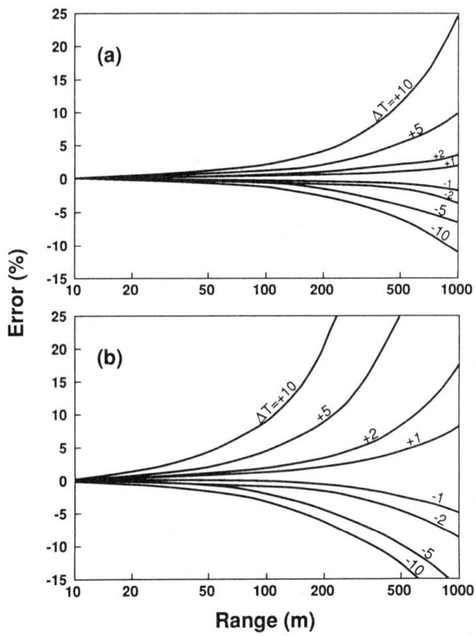

Fig. 8.16 Error in the estimated fish density caused by the TVG function being based on inaccurate hydrographic data, after Foote (1981b). The vertical scale is the percentage error between true and estimated densities. ΔT is the difference between the true and assumed water temperatures (°C). (a) 38 kHz; (b) 120 kHz.

Weimar and Ehrenberg (1975) studied the threshold bias by simulating the signals from an assumed target strength PDF. When *SNR* was 30 dB, the bias was about 25%. Foote *et al.* (1986) performed similar calculations but based on measurements of the noise and empirical target strength functions. They found no bias in the echo integration of large gadoids ($L > 50$ cm), but for smaller fish ($L = 10$–30 cm; $SNR = 15$–20 dB) the bias could be as much as 50%.

The threshold bias is important in echo counting as well as echo integration. Since the target strength is a stochastic variable, there is some probability of echoes below the threshold level, and they are ignored. Thus the observed echoes represent only part of the PDF, and the observed mean target strength is larger than it should be.

Nunnallee (1990) has proposed an alternative to the signal threshold for dealing with noise in echo integration. He measures the noise level at the echosounder output with the transmitter switched off. Subsequently, the noise contribution is subtracted from the echo integrals with no threshold applied. This procedure does not allow for reverberation, but it has the advantage that the threshold bias is eliminated. The non-noise echo integrals are subject to random errors, but they should average to zero over a long period of integration.

In section 8.2, we described a variant of Nunnallee's method, used to separate the school echo integrals from the planktonic background. This method gives good results when the schools or individual large fish produce clear marks on the echogram, in which case there is no need to use a threshold. On the other hand, if the plankton is so dense that the fish marks are obscured, an appropriate signal threshold may be the only way of rejecting the plankton echoes. Surveys in the tropics often suffer from strong plankton reverberation. It is less of a problem elsewhere, excepting the occasional short-lived bloom.

If the echo PDF of the target species is known, the threshold bias may be calculated and a correction may be applied to the estimated abundance. Unfortunately, the true PDF is seldom known well enough for this purpose. The problem is best avoided if possible.

Fish migration

It is well known that some fish migrate over long distances on an annual cycle (Harden Jones, 1968). One factor in the survey design is the timing in relation to the migration cycle, to ensure that the surveyed area includes the entire stock. But even if this condition is met, migration of the stock within the surveyed area can bias the abundance estimate. The extent of the problem depends upon the direction of migration in relation to the transects.

Suppose the fish are migrating at speed v_f, and v_s is the speed at which the survey progresses in the direction of migration. If v_s is positive, this means that the fish tend to follow the vessel as it travels along successive transects. If the cruise track were drawn on a map whose frame of reference moved with the fish, the transects would be closer together than those on the geostationary map. Thus the effective area applicable to the analysis is less than the actual area surveyed. The observed densities are unbiased, but since the abundance is the mean density multiplied by the effective area, the estimate $\hat{Q}$ is biased. The expected value of $\hat{Q}$ is

$$E(\hat{Q}) = Q[1-(v_f/v_s)] \tag{8.37}$$

Note that when the transects are long and perpendicular to the migration, v_s is much smaller than the cruising speed of the vessel. For example, if the cruising speed is 5 m s^{-1}, and the transect length is 10 times the spacing, then the survey progresses at $v_s = 0.5$ m s^{-1}, a value which could well be comparable with v_f. Harden Jones (1968) suggests that herring are capable of migration speeds up to 0.6 m s^{-1}. The swimming capability of fish depends upon their size, but adult herring and mackerel can sustain speeds around 1.0 m s^{-1} for long periods (He and Wardle, 1988; Lockwood, 1988).

The bias is greatly reduced if the transects run alternately with and against the migration, in which case

$$E(\hat{Q}) = Q\{1-(v_f/v_s)^2\} \tag{8.38}$$

and v_s is now equal to v. Taking the worst case as $v_f = 1$ m s^{-1} and $v_s = 5$ m s^{-1}, the bias is an underestimate of 4%. In practice, the transect direction may be decided by other factors such as the coastline or depth contours. If the transects must be perpendicular to the migration, we might consider traversing the surveyed area twice, in opposite directions with the outward and return transects interleaved (Fig. 8.17). Thus the survey begins and ends at the same place. This need not be too costly in ship time if the place concerned is the home port. As regards the analysis, the best plan is to treat the outward and return sections as if they were replicate surveys, and to estimate the abundance as the average of the two results.

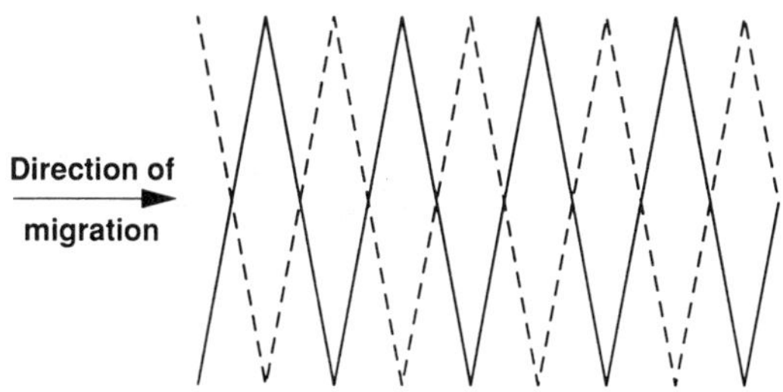

Fig. 8.17 Cruise track with transects running across the fish migration route. The outward section (solid line) is interleaved with the return section (broken line), and the track progresses equally with and against the migration.

Diurnal behaviour rhythms

The behaviour of many aquatic organisms changes with the time of day; the change is motivated at least in part by the light level. Several writers have discussed the implications for acoustic surveys (e.g. Diner and Masse, 1987; Levenez *et al.*, 1990; Thorne *et al.*, 1990). Foote (1981c) reports that the estimated abundance from repeated surveys of the same area shows consistent differences between those conducted at night and those by day. There are several reasons why such a bias might occur. Firstly, schools observed during the day are likely to disperse at night, changing the criteria for partitioning the echo integrals. Secondly, vertical migration could remove a variable proportion of the stock from the sampled volume, owing to fish rising above the transducer or descending too close to the sea-bed. Furthermore, the target strength may change as the swimbladder expands and contracts in response to the ambient pressure. Thirdly, the orientation of schooling fish may depend upon vision and visibility. Thus the mean target strength in daylight and in darkness may not be the same, indeed experimental evidence shows this to be a common occurrence (Nakken and Olsen, 1977; Sameoto, 1980).

The extent of day–night bias may be judged from the comparison of data collected at different times of day, in areas where the fish density is expected to be similar. More generally, it is important to recognize that the bias might suggest spurious changes in the spatial distribution, simply because some regions have been traversed by night and others by day. If the survey is conducted around the clock and all the data are to be used in the analysis, the bias in $\hat{Q}$ is reduced if the night and day samples can be considered as randomly positioned with respect to the fish concentrations, and the effective target strength is calculated as an appropriate diurnal average. Note that the sampling effort might be unevenly divided between night and day for operational reasons. For example, if fishing is conducted only when schools are seen on the echogram, and the schools disperse at night, there will be less time during the day for acoustic data collection.

There is no bias if the abundance is estimated with reference only to the acoustic data collected under similar light conditions. In high latitudes, this is a reasonable proposition during summer or winter. In the tropics, of course, there are around 12 hours each of daylight and darkness whatever the season. In these circumstances, separate analyses of the observations by day and by night might be considered, using different mean target strengths for the two periods.

Avoidance reactions

Another behavioural effect is the reaction of fish when they are startled by the survey vessel. Olsen *et al.* (1983a,b) showed that herring moved away from the approaching vessel, so that the density in the acoustic beam was less than the undisturbed value. Furthermore, television and photographic observations showed a change in the mean tilt angle of the reacting fish, thus the target strength would also change. K. Olsen (1969) believes that the noise created by the vessel is more important than the visual stimulus, and the rate of change of the noise level at a fixed point is especially important. This suggests that the avoidance reaction will depend upon the vessel speed and the depth of the fish, while the absolute level of the vessel noise may be less important. Diner and Masse (1987) studied the reactions of schools in the Bay of Biscay and the English Channel by using an omni-directional sonar to observe the horizontal movement of the schools around the vessel. Complex reactions were observed and few conclusions could be made, but the time of day seemed to be an important factor. Levenez *et al.* (1990) found that on dark nights, fish moved downwards in response to the lights of the survey vessel, although the mean echo integral was the same with the lights on or off.

Much of the experimental work has been descriptive and provides little quantitative evidence to assess the bias due to avoidance. However, Olsen *et al.* (1983b) observed herring at 40 m depth as a vessel approached at 10 knots ($5\,m\,s^{-1}$), using a Doppler sonar to measure the speed of targets (Holliday, 1977b). The fish descended at speeds up to $0.7\,m\,s^{-1}$. They suggested that the change of mean tilt angle in reacting fish was inversely proportional to the square of depth, but the results were not consistent between different vessels and species. In another quantitative experiment, Olsen (1990) compared the echo integrals from two stationary transducers suspended at 10 m depth, one close to the vessel track and the other about 20 m to the side (Fig. 8.18). much reduced density was observed immediately below the vessel.

Olsen's work is a notable contribution to the field, but it has mostly been done in sheltered fjords. It is uncertain to what extent his results are applicable to migrating schools in the open sea. Freon and Gerlotto (1988) have described methods for investigating the bias arising from the reactions of schooling fish. Gerlotto and Freon (1988) report observations on schools of tropical clupeoids. The structure of the school, notably the height and the fish density within it, changed as the vessel passed by. Schools of *Sardinella aurita* reacted by diving when encountered above 20 m depth, but those in deeper water did not respond. The reactions observed by Gerlotto and Freon are much less dramatic than those of herring described by Olsen. Gerlotto and Freon noted that their fish were unexploited, and they speculated that the stronger reactions observed elsewhere might be conditioned by previous encounters with fishing vessels, citing the work of Marler and Terrace (1984). We suspect that seasonal, geographical or environmental factors are more likely as an explanation.

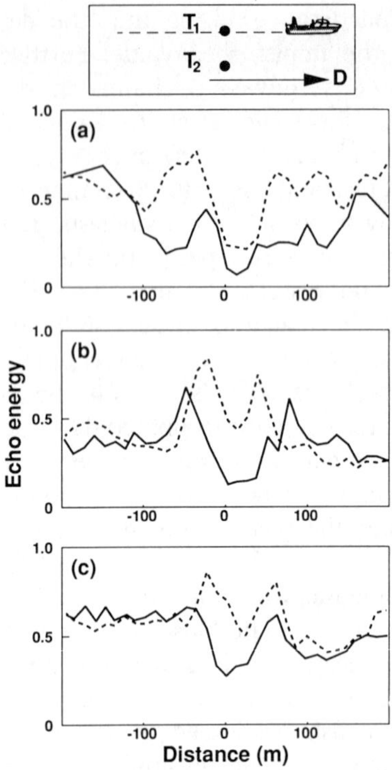

Fig. 8.18 Avoidance reactions of herring in a Norwegian fjord. Echo energy (arbitrary units) v. distance *D* of a ship passing stationary transducers at 9 knots ($4.6\,m\,s^{-1}$). Solid curve, transducer T_1 is on the ship's track; broken curve, T_2 is 20 m to the side; three different runs over fish at depths (a) 15–35 m, (b) 30–50 m, (c) 25–60 m. Redrawn from Olsen (1990).

When the fish react by diving, the target strength alters due to compression of the swimbladder as well as changes in the tilt angle (Harden Jones and Scholes, 1981; Hawkins, 1981; Ona, 1984). The swimbladder volume is proportional to the ambient pressure (Boyle's Law), and for the same distance descended, the target strength changes more in fish closer to the surface.

There is need for further work to quantify the effect of avoidance reactions. The avoidance bias is most likely to occur in shallow water and with non-migrating fish in sheltered areas, but little more can be said at present about the magnitude of the effect.

8.7 ACCURACY OF THE ABUNDANCE ESTIMATE

It can be very difficult to assess the accuracy of the abundance estimate, but it is essential to consider how good or bad the results are. There are two approaches to the problem. The intrinsic method is an error analysis of the survey itself. Alternatively, the acoustic abundance may be compared with independent estimates of the same stock, if any are available.

The abundance may be expressed as an absolute quantity of fish, or an index proportional to the quantity by a factor which is unknown but supposed to be constant from one survey to the next. The accuracy of the index depends mainly on the sampling errors discussed in section 8.5. The larger systematic errors, those associated with fish behaviour, may reasonably be expected not to change between surveys provided they are conducted in comparable circumstances, for example at the same time of year. Thus the accuracy of the index should be much better than that of the absolute abundance.

Intrinsic error analysis

The total error of the abundance estimate is considered as the combination of several individual errors whose magnitudes are known or may be calculated on reasonable assumptions. The many factors which contribute individual errors have been discussed in sections 8.5 and 8.6. In Table 8.4, we summarize these factors and indicate how much error might be expected under the conditions typically encountered on acoustic surveys, on the assumption that the survey has been well designed and competently conducted with properly calibrated scientific instruments. The errors are listed in two groups, (a) those which apply equally to the absolute abundance and the index, and (b) those applicable to the absolute abundance only.

Each factor contributes an error which is random, systematic or both. The distinction has been explained in section 8.6 (p. 270). The random errors may be reduced by collecting more samples, which means spending more time on the survey, but this is not true of the systematic errors, which bias all the observations equally.

To calculate the total error of estimate, we must first determine the individual errors. We suppose that the abundance estimate has been corrected for the biases in so far as they are known. Thus the individual errors are random variables, each of which has an associated PDF, assumed to be Gaussian, with an expectation of zero. Most of them are statistically independent. There is no reason to suppose, for example, that the equipment performance should have any bearing on the fish behaviour. Some errors are correlated; they require special treatment. Referring to Table 8.4, the only correlation which needs to be considered is that between the transducer motion and the bubble attenuation, since the associated errors are unidirectional biases, which increase in magnitude as the weather gets worse. There may be correlation between some of the behavioural errors, but they are poorly understood and the assump-

Table 8.4 Sources of error in acoustic surveys. The figures are an indication of the errors which might occur under typical survey conditions. List (a) applies to the acoustic index of relative abundance. List (b) is the additional error applying to the absolute abundance estimate

Source	*Random error*	*Systematic error (bias)*
(a) Acoustic index errors		
Physical calibration	±2%	±5%
Transducer motion*	–	0 to −30%
Bubble attenuation*	–	0 to −90%
Hydrographic conditions†	±2% to ±5%	–
Target strength	±5%	–
Species identification	–	0 to ±80%
Random sampling	±10% to ±40%	–
Fish migration	–	0 to ±40%
Diurnal behaviour‡	0 to 25%	–
(b) Absolute abundance errors		
Physical calibration	–	±3%
Hydrographic conditions†	–	0 to 5%
Target strength	–	0 to 50%
Avoidance reactions	–	uncertain

*Worst in bad weather.
†Worst at long range.
‡Only random if time of day is ignored.

tion of independence is as good as any. The errors within any correlated set are summed and the result is treated as one independent error in the subsequent analysis.

The individual errors are expressed as percentages and they have a multiplicative effect on the abundance estimate. The model is

$$\hat{Q} = X_1 X_2 \cdots X_n \tag{8.39}$$

Each variable X_i includes an error ΔX_i. There is a simple formula for the total error which is valid if $\Delta X_i \ll X_i$. While some $(\Delta X_i / X_i)$ may not be small, the more complicated theory required to deal with large but uncertain errors is unlikely to be worthwhile. If ε_i^2 is the expected value of $(\Delta X_i / X_i)^2$, or the variance of the proportional error, and the ΔX_i are independent, the variance of $\hat{Q}$ is estimated as

$$\hat{V} = \hat{Q}^2 \sum_{i=1}^{n} \varepsilon_i^2 \tag{8.40}$$

Comparison with other methods

The ultimate test of the acoustic survey technique is to compare estimates of the same stock obtained by acoustic and other methods. Suppose that methods A and B give the abundance estimates Q_a and Q_b with variances V_a and V_b respectively. The variances may be unknown, but the differences $(Q_a - Q_b)$ incorporate the combined error of both methods. If the errors are independent, the variance of $(Q_a - Q_b)$ is $V = V_a + V_b$. Given two series of comparable abundance measurements, from annual surveys of the same stock conducted over several years for example, V is estimated as the sample variance of $(Q_a - Q_b)$. The simple comparison does not indicate which method is more accurate, but the variance is a positive quantity and so V is an upper bound on the variances of both Q_a and Q_b.

Jakobsson (1983) compared acoustic abundances of the Icelandic summer-spawning herring against estimates from virtual population analysis (VPA), a retrospective method based on the catches taken in the fishery (Gulland, 1983). He found good agreement, but the comparison was not strictly valid because the acoustic data had been used to 'tune' the VPA, so that the most recent VPA and acoustic estimates were not completely independent. Nakken and Ulltang (1983) conducted a similar study of cod and haddock in the north-east Arctic. In this case the VPA and acoustic estimates were independent. When the cod and haddock were considered as one stock, the estimates agreed within $\pm 10\%$, but much larger differences were found in the results for each species. The problem appeared to be due to the trawl catches, which did not reflect the true proportion of cod and haddock in the sea.

In freshwater environments, it is often possible to count the run of migrating fish. Nunnallee (1983) has compared the count of sockeye salmon, *Oncorhynchus nerka*, observed passing a weir, against the population indicated by acoustic surveys of Cultus Lake in British Columbia. The acoustic estimates were obtained by a combination of echo counting and echo integration, surveying only at night when the juvenile sockeye were distributed in the upper part of the water column, mainly at depths 20–25 m. Several other species were present in the lake, but they did not migrate, so the sockeye population was estimated from the change in the acoustic abundance after the start of the sockeye run. The weir-count and the acoustic abundance differed by only 2.5%, much less than the 95% confidence interval suggested by intrinsic error analysis. Another approach is comparison with the fish densities indicated by trawl samples. Thorne (1983a) has discussed the accuracy of acoustic surveys of salmonid populations in North American lakes. He obtained independent abundance estimates from the catches of a large midwater trawl towed at night. The acoustic estimates were made by echo integration using either 105 or 120 kHz echosounders, giving 19 valid comparisons of juvenile sockeye salmon populations. Regression of the acoustic and trawl estimates showed a systematic difference of 4%, and a correlation coefficient of 0.73, which indicated a random sampling error of somewhat greater magnitude.

Bailey and Simmonds (1990) have compared the abundance of North Sea herring predicted by four methods – acoustic survey, the larval abundance index, the larval production index and VPA. When several independent measures of the same stock are available, the comparison should reveal any substantial difference in the accuracy of each method. Bailey and Simmonds found that the VPA estimate was the least reliable, and the acoustic estimate was superior to either of those based on larval sampling. In a series of five annual surveys, there were differences around 10% between concurrent acoustic and larval estimates of the stock. This result refers only to the random sampling error, because the larval indices do not provide absolute estimates of the spawning stock until they are calibrated against absolute estimates obtained by some other method.

Although it is not often that two assessment methods give satisfactory results for the same stock, the comparisons which have been made suggest that in many cases the acoustic technique is at least as good as, and probably better than, any other. Furthermore, the sources of error in acoustic abundance estimates have been more extensively investigated than appears to be the case for other methods, at least as regards the assessment of pelagic stocks in the sea. The most poorly understood errors are those related to fish behaviour and the partitioning of the echo integrals between species. These errors apply to the absolute abundance estimates, but in a well-designed series of surveys, they should not be important in the precision of the acoustic index. Again, the index may be converted to an absolute abundance by calibration against any independent measure of the stock that is considered to be reliable enough.

All survey techniques have their own advantages and disadvantages, and we do not claim that the acoustic method is the best in every case. What is important is that the sources of error in any method should be well-enough understood to judge the best approach to the problem of fish stock assessment. The acoustic survey has become well established as a useful technique in fishery research. There are many applications in which it is the only practical means of assessment available to fishery managers, but there are others in which no one method is satisfactory. Sometimes we have no choice but to apply different methods in parallel, to produce an average result whose confidence interval is acceptably small.

APPENDIX E The Kolmogorov–Smirnov Test

The Kolmogorov–Smirnov (KS) test compares two samples to determine whether they have been drawn from populations with different probability distributions. The test is sensitive to differences in the mean, variance and number of modes. Length frequency data from trawl samples can differ in all these features. A good description of the test and its limitations will be found in Campbell (1974).

The test statistic is D_{max}, defined as the maximum difference between the two cumulative probability distributions. If the two samples comprise N_1 and N_2 fish and the numbers in length class j are n_{j1} and n_{j2} respectively, the cumulative frequencies are

$$f_{i1} = \sum_{j=1}^{i} n_{j1} / N_1 \qquad \text{(E8.1)}$$

$$f_{i2} = \sum_{j=1}^{i} n_{j2} / N_2 \qquad \text{(E8.2)}$$

D_{max} is the largest value of $|f_{i1} - f_{i2}|$. On the null hypothesis that the samples have been drawn randomly from the same population, there is a small probability P that D_{max} will be larger than

$$D_p = A_p \sqrt{[(N_1 + N_2)/(N_1 N_2)]} \qquad \text{(E8.3)}$$

The factor A_p is 1.36, 1.63 or 1.95 for $P = 0.05$, 0.01 or 0.001 respectively. For example, suppose $N_1 = N_2 = 100$. If D_{max} is more than 0.2, there is less than 5% chance that the samples represent the same population.

Figure A8.1 shows the length frequencies of herring samples collected at three trawl stations to the north of Scotland, along with the corresponding cumulative probability distributions. N is close to 150 in each case. A separate test is done for each pair of stations with the following results.

Stations 1 and 2, $D_{max} = 0.13$

Stations 1 and 3, $D_{max} = 0.81$

Stations 2 and 3, $D_{max} = 0.86$

D_p is 0.24 for $P = 0.001$, so it is almost certain that the population at station 3 is different from the others, while those at stations 1 and 2 have the same size distribution.

Trawl samples are normally analysed in a manner which is incompatible with the strict rules of the KS test. In particular, the interval between length classes is required to be small so that no more than seven observations occur in any class. In the example illustrated in Fig. E8.1, there are 16 fish in the most frequent class. This results in D_{max} being underestimated by about 0.05 at the 5% probability level. Because of this uncertainty, we do not

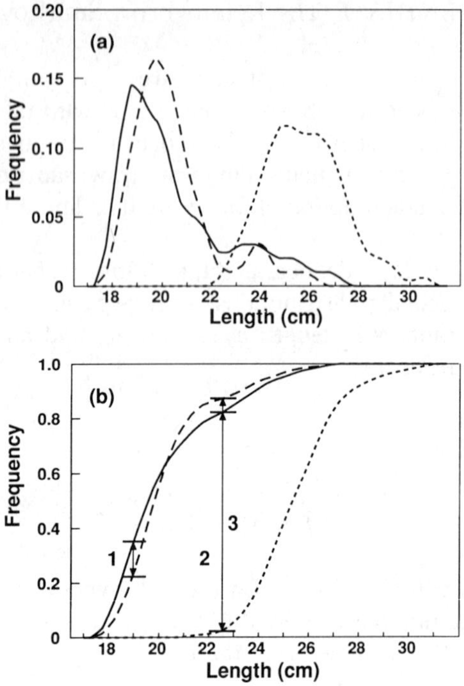

Fig. E8.1 Comparison of fish size distributions observed in three samples of herring. (a) Frequencies in 1 cm length classes; (b) cumulative frequencies showing D_{max}, the greatest difference between each sample pair, as vertical lines.

suggest that the KS test should be used as a completely objective procedure for classifying trawl samples in similarity groups defined by a pre-determined value of P. Instead, D_{max} is considered as an index of similarity and no more than that. Each sample is compared with all the others and the resulting values of D_{max} are displayed in a triangular matrix whose rows and columns are ordered by the station numbers. This matrix is examined for associations by comparing D_{max} with a threshold D_t. Samples from neighbouring stations for which $D_{max} < D_t$ are considered to represent the same population. D_t is set between 0.1 and 0.3, at a level where the associations are least sensitive to changes in the threshold. For the three samples described above, the two groups, (1, 2) and (3), are confirmed since they are consistently identified by any threshold between 0.13 and 0.81.

APPENDIX F Power Transformations

Consider the data set F_i consisting of N independent observations of the fish density ($i = 1, 2 \ldots N$). The F_i are all greater than zero and N is 2 or more. We suppose that the F_i are random samples with some probability distribution which is unlikely to be normal. It is convenient to work with normally distributed data, and this might be achieved by a transformation. In the case of the power transform, a new data set Y_i is obtained from the equations

$$\begin{aligned} Y_i &= F_i^{\lambda} \text{ for } \lambda > 0 \\ &= \ln(F_i) \quad \text{for } \lambda = 0 \end{aligned} \tag{F8.1}$$

We suppose that for some value of the parameter λ, the Y_i are normally distributed. The sample mean m and the residual sum of squares S are

$$m = \sum_i Y_i / N \tag{F8.2}$$

$$S = \sum_i (Y_i - m)^2 \tag{F8.3}$$

$\hat{F}$ is an estimate of the true mean density, and $\hat{V}$ is the variance of $\hat{F}$. The estimation formulas described below are applicable to a few discrete values of λ (MacLennan and MacKenzie, 1988).

Square-root normal ($\lambda = 1/2$)

$$\hat{F} = m^2 + S/N \tag{F8.4}$$

$$\hat{V} = \frac{4Sm^2}{N(N-1)} + \frac{S^2}{N^2} - \frac{S^2}{N^2-1}\left(1 - \frac{2}{N} + \frac{3}{N^2}\right) \tag{F8.5}$$

Cube-root normal ($\lambda = 1/3$)

$$\hat{F} = m^3 + 3mS/N \tag{F8.6}$$

$$\begin{aligned} \hat{V} = {} & \frac{9m^4S}{N(N-1)} + \frac{9m^2S^2}{(N-1)^2}\left[\left(1 - \frac{1}{N}\right)^2 - \frac{N-1}{N+1}\left(1 - \frac{4}{N} + \frac{5}{N^2}\right)\right] \\ & + \frac{3S^3}{N(N^2-1)(N+3)}\left(3 - \frac{6}{N} + \frac{5}{N^2}\right) \end{aligned} \tag{F8.7}$$

Fourth-root normal ($\lambda = 1/4$)

$$\hat{F} = m^4 + 6m^2 S/N + 3(N-1)(S/N)^2/(N+1) \tag{F8.8}$$

$$\begin{aligned} \hat{V} = {} & \hat{F}^2 - m^8 - \frac{m^6S}{N-1}\left(12 - \frac{28}{N}\right) \\ & - \frac{m^4S^2}{N^2-1}\left(\frac{210}{N^2} - \frac{180}{N} + 42\right) \\ & - \frac{m^2S^3}{(N^2-1)(N+3)}\left(36 - \frac{252}{N} + \frac{540}{N^2} - \frac{420}{N^3}\right) \end{aligned}$$

$$-\frac{S^4}{(N^2-1)\,(N+3)\,(N+5)}\left(9-\frac{36}{N}+\frac{126}{N^2}-\frac{180}{N^3}+\frac{105}{N^4}\right) \quad \text{(F8.9)}$$

Sixth-root normal ($\lambda=1/6$)

$$\hat{F}=m^6+\frac{15m^4S}{N}+45m^2\,\frac{N-1}{N+1}\left(\frac{S}{N}\right)^2$$

$$+\frac{15(N-1)^2}{(N+1)(N+3)}\left(\frac{S}{N}\right)^3 \quad \text{(F8.10)}$$

$$\hat{V}=\hat{F}^2-m^{12}-\frac{6m^{10}S}{N-1}\left(5-\frac{11}{N}\right)$$

$$-\frac{45m^8S^2}{N^2-1}\left(7-\frac{30}{N}+\frac{33}{N^2}\right)$$

$$-\frac{60m^6S^3}{(N^2-1)(N+3)}\left(23-\frac{147}{N}+\frac{315}{N^2}-\frac{231}{N^3}\right)$$

$$-\frac{225m^4S^4}{(N^2-1)(N+3)(N+5)}\left(11-\frac{92}{N}+\frac{294}{N^2}-\frac{420}{N^3}+\frac{231}{N^4}\right)$$

$$-\frac{270m^2S^5}{(N^2-1)(N+3)(N+5)(N+7)}$$

$$\left(5-\frac{55}{N}+\frac{230}{N^2}-\frac{490}{N^3}+\frac{525}{N^4}-\frac{231}{N^5}\right)$$

$$-\frac{45S^6}{(N^2-1)(N+3)(N+5)(N+7)(N+9)}$$

$$\left(5-\frac{30}{N}+\frac{165}{N^2}-\frac{460}{N^3}+\frac{735}{N^4}-\frac{450}{N^5}+\frac{231}{N^6}\right) \quad \text{(F8.11)}$$

Log-normal ($\lambda=0$)

$$\hat{F}=\exp(m)\,G_N[0.5S/(N-1)] \quad \text{(F8.12)}$$

$$\hat{V}=\hat{F}^2-\exp(2m)\,G_N[S(N-2)\,/\,(N-1)^2] \quad \text{(F8.13)}$$

The function $G_N(u)$ is computed from the following algorithm:

$$G_N(u) = 1 + (N-1)u/N \tag{F8.14}$$

$$x = (N-1)^3 u^2/[2N^2(N+1)]; \quad j=3 \tag{F8.15}$$

$$G_N(u) \rightarrow G_N(u) + x \tag{F8.16}$$

$$x \rightarrow x(N-1)^2 u/[Nj(N+2j-3)]; \quad j \rightarrow j+1 \tag{F8.17}$$

Expressions A8.19 and A8.20 are repeated until convergence, when x is very close to zero.

The Box–Cox test

The estimation formulae (F8.7 to F8.20) are unbiased only if the transformed variable has a normal distribution. The formulae then represent minimum variance unbiased estimators (MVUEs). It is therefore necessary to determine the value of λ most appropriate to a given data set. This can be done using a test devised by Box and Cox (1964). Equation F8.1 is not continuous at $\lambda=0$, and to overcome this difficulty the test is applied to another data set Z_i, defined as

$$\begin{aligned} Z_i &= (F_i^{\lambda} - 1)/\lambda \quad \text{for } \lambda > 0 \\ &= \ln(F_i) \quad \text{for } \lambda = 0 \end{aligned} \tag{F8.18}$$

The Box–Cox function is

$$\begin{aligned} L(\lambda) = &-(N/2)\ln\left\{\sum_{i=1}^{N}(Z_i - \bar{Z})^2/N\right\} \\ &+ (\lambda - 1)\sum_{i=1}^{N}\ln(F_i) \end{aligned} \tag{F8.19}$$

The maximum of $L(\lambda)$, at λ_m say, indicates the value of λ for which the transformed data are most nearly matched to the normal distribution. A 95% confidence interval on λ_m, λ_1 to λ_2 say, is obtained as the solution of

$$L(\lambda_1) = L(\lambda_2) = L(\lambda_m) - 1.92 \tag{F8.20}$$

Zero values

If the species of interest is absent from part of the surveyed area, some of the observed densities will be zero (empty water) while the others are stochastic samples taken within the regions where fish are found. The density is non-stationary (p. 255), but the mean and variance can be estimated by the method of Aitchison (1955) which treats the zero and non-zero data separately. Consider the probability distribution.

$$P_1(F)=p \quad \text{for } F=0$$
$$=(1-p)P(F) \text{ for } F>0 \qquad \text{(F8.21)}$$

where p is the finite probability that an observation will be exactly $F=0$. Assume that $P(F)$ is such that under the transformation $F(\lambda)\rightarrow Z$, the Z are approximately normally distributed. If E is the expectation operator, let

$$\hat{F}=E(F|F>0) \qquad \text{(F8.22)}$$

$$\hat{V}=\mathrm{Var}(\hat{F}|F>0) \qquad \text{(F8.23)}$$

Thus $\hat{F}$ and $\hat{V}$ are determined by the non-zero values only. For a sample of size M containing N values greater than zero, Pennington (1983) gives the following formulas for F_r, the estimate of the mean, and $\hat{V}_r$, the estimated variance of the mean.

$$\hat{F}_r=N\hat{F}/M \qquad \text{(F8.24)}$$

$$\hat{V}_r=N(M-N)\hat{F}^2/[M^2(M-1)]$$
$$+N(N-1)\hat{V}/[M(M-1)] \qquad \text{(F8.25)}$$

The estimate of p is N/M. The most appropriate estimators (according to the Box–Cox test) from Equations F8.4 to F8.17 are first used to obtain $\hat{F}$ and $\hat{V}$ from calculations with the non-zero data. Then Equations F8.24 and F8.25 provide the equivalent statistics for the whole sample, taking account of the regions with and without fish.

References

Abramowitz, M. and Stegun, I.A. (1964) Handbook of Mathematical Functions. *Natl. Bur. Stand. Applied Mathematical Series*, **55**, Govt. Printing Office, Washington DC.

Aglen, A. (1983) Random errors of acoustic fish abundance estimates in relation to the survey grid density applied. *FAO Fish. Rep.*, **300**, 293–8.

Aglen, A. (1989) Empirical results on precision-effort relationships for acoustic surveys. Int. Coun. Explor. Sea CM 1989/B:30, 28 pp. (mimeo).

Aitchison, J. (1955) On the distribution of a positive random variable having a discrete probability mass at the origin. *J. Am. Stat. Soc.*, **50**, 901–8.

Aksland, M. (1983) Acoustic abundance estimation of the spawning component of the local herring stock in Lindaaspollene, Western Norway. *FiskDir. Skr. Ser. Havunders.*, **17**, 297–334.

Aksland, M. (1986) Estimating numbers of pelagic fish by echo integration. *J. Cons. perm. int. Explor. Mer*, **43**, 7–25.

Anderson, K.P. and Kirkegaard, E. (1985) Sources and magnitude of random errors in acoustic fish abundance estimate. Int. Coun. Explor. Sea CM 1985/B:40, 17 pp. (mimeo).

Anderson, V.C. (1950) Sound scattering from a fluid sphere. *J. acoust. Soc. Am.*, **22**, 426–31.

Aplin, J.A. (1947) The effect of explosives on marine life. *Calif. Fish Game*, **33**, 23–30.

Armstrong, F. (1986) Target strength of sandeels. Int. Coun. Explor. Sea CM 1986/B:5, 6 pp. (mimeo).

Arnaya, I.N., Sano, N. and Iida, K. (1989a) Studies on acoustic target strength of squid II: effect of behaviour on averaged dorsal aspect target strength. *Bull. Fac. Fish. Hokkaido Univ.*, **40**, 83–99.

Arnaya, I.N., Sano, N. and Iida, K. (1989b) Studies on acoustic target strength of squid III: measurement of the mean target strength of small live squid. *Bull. Fac. Fish. Hokkaido Univ.*, **40**(2), 100–15.

Arnold, G.P. and Metcalfe, J.D. (1989) Acoustic telemetry: progress and potential in understanding fish behaviour. *Proc. Inst. Acoust.*, **11**(3), 97–103.

Arnold, G.P., Greer Walker, M. and Holford, B.H. (1990) Fish behaviour: achievements and potential of high-resolution sector-scanning sonar. *Rapp. P.-v. Réun. Cons. perm. int. Explor. Mer*, **189**, 112–22.

Au, W.W.L. and Martin, D.W. (1989) Insights into dolphin sonar discrimination capabilities from human listening experiments. *J. acoust. Soc. Am.*, **86**, 1662–70.

Au, W.W.L. and Moore, P.W.B. (1990) Critical ratio and critical bandwidth for the Atlantic bottlenose dolphin. *J. acoust. Soc. Am.*, **88**, 1635–8.

Au, W.W.L. and Pawloski, J.L. (1989) Detection of noise with rippled spectra by the Atlantic bottlenose dolphin. *J. acoust. Soc. Am.*, **86**, 591–6.

Au, W.W.L., Schusterman, R.J. and Kersting, D.A. (1980) Sphere-cylinder discrimination via echolocation by *Tursiops truncatus*, in *Animal sonar systems* (eds R.G. Busnel and J.F. Fish), Plenum, New York, pp. 859–62.

Bailey, R.S. and Simmonds, E.J. (1990) The use of acoustic surveys in the assessment of the North Sea herring stock and a comparison with other methods. *Rapp. P.-v. Réun. Cons. perm. int. Explor. Mer*, **189**, 9–17.

Bary, B.M. (1966) Backscattering at 12 kc/s in relation to biomass and numbers of zooplanktonic organisms in Saanich Inlet, British Columbia. *Deep Sea Res.*, **13**, 655–66.

Baxter, L. (1985) Mortality of fish subjected to explosive shock as applied to oil well severance on Georges Bank. pp. 119–35 in Greene *et al.* (1985), q.v.

Bazigos, G.P. (1981) A manual on acoustic surveys: sampling methods. Interregional fisheries development and management programme, Rep. CECAF/ECAF ser. 80/17 (En). Food and Agriculture Organisation, Rome, 137 pp.

Berka, R. (1986) The transport of live fish: a review. *European Inland Fish. Advisory Comm. tech. Pap.*, no. 48, 52 pp.

Beverton, R.J. (1990) Small marine pelagic fish and the threat of fishing; are they endangered? *J. Fish Biol.*, **37** (Supp. A), 5–16.

Blaxter, J.H.S. and Batty, R.S. (1984) The herring swimbladder: loss and gain of gas. *J. mar. biol. Ass. UK*, **64**, 441–59.

Blaxter, J.H.S. and Batty, R.S. (1990) Swimbladder 'behaviour' and target strength. *Rapp. P.-v. Réun. Cons. perm. int. Explor. Mer*, **189**, 233–44.

Blaxter, J.H.S. and Hunter, J.R. (1982) The biology of the clupeoid fishes. *Adv. mar. Biol.*, **20**, 1–223.

Blue, J.E. (1984) Physical calibration. *Rapp. P.-v. Réun. Cons. perm. int. Explor. Mer*, **184**, 19–24.

Bodholt, H. (1977) Variance error in echo integrator output. *Rapp. P.-v. Réun. Cons. perm. int. Explor. Mer*, **170**, 196–204.

Bodholt, H., Ness, H. and Solli, H. (1989) A new echo-sounder system. *Proc. Inst. Acoust.*, **11**(3), 123–30.

Borisenko, E.S., Gusar, A.G. and Goncharov, S.M. (1989) The target strength dependence of some freshwater species on their length-weight characteristics. *Proc. Inst. Acoust.*, **11**, 27–34. Institute of Acoustics, St Albans, England.

Box, G.E.P. and Cox, D.R. (1964) An analysis of transformations. *J. R. statist. Soc.*, **B**, **26**, 211–52.

Brandt, A. von: *see* von Brandt.

Breder, C.M., jun. (1976) Fish schools as operational structures. *Fish. Bull.*, **74**, 471–502. US Dept. of Commerce, Seattle, WA.

Bruno, D.R. and Novarini, J.C. (1983) High frequency sound attenuation caused by the wind-generated bubble layer in the open sea. *J. acoust. Soc. Am.*, **73**, 1064–5.

Burczynski, J. (1979) Introduction to the use of sonar systems for estimating fish biomass. *FAO Fish. tech. Pap.*, no. 191, 89 pp.

Burczynski, J.J. and Johnson, R.L. (1986) Application of dual-beam acoustic survey techniques to limnetic populations of juvenile sockeye salmon *Oncorhynchus nerka*. *Can. J. Fish. Aquat. Sci.*, **43**, 1776–88.

Burczynski, J.J., Johnson, R.L. and Kirchner, W.B. (1990) Acoustic estimation of high densities of fish in sea pens. *Rapp. P.-v. Réun. Cons. perm. int. Explor. Mer*, **189**, 54–64.

Campbell, R.C. (1974) *Statistics for Biologists*, 2nd edn, Cambridge University Press, 383 pp.

Carlson, T.J. and Jackson, D.R. (1980) Empirical evaluation of the feasibility of split beam methods for direct in situ target strength measurement of single fish. Seattle Applied Physics Laboratory, University of Washington (APL-WW 8006).

Castile, B.D. (1975) Reverberation from plankton at 330 kHz in the Western Pacific. *J. acoust. Soc. Am.*, **58**, 972–6.

Chapman, C.J. and Hawkins, A.D. (1973) A field study of hearing in the cod, *Gadus morhua*. *J. comp. Physiol.*, **85**, 147–67.

Chapman, C.J. and Johnstone, A.D.F. (1974) Some auditory discrimination experiments on marine fish. *J. exp. Biol.*, **61**, 521–8.

Chapman, C.J. and Sand, O. (1974) Field studies of hearing in two species of flatfish, *Pleuronectes platessa* and *Limanda limanda*. *Comp. Biochem. Physiol.*, **47A**, 371–85.

Chen, C.T. and Millero, F.J. (1976) Reevaluation of Wilson's sound-speed measurements for pure water. *J. acoust. Soc. Am.*, **60**, 1270–73.

Clark, I. (1979) *Practical Geostatistics*, Elsevier, London, 129 pp.

Clay, C.S. (1983) Deconvolution of the fish scattering PDF from the echo PDF for a single transducer sonar. *J. acoust. Soc. Am.*, **73**, 1989–94.

Clay, C.S. and Heist, B.G. (1984) Acoustic scattering by fish - acoustic models and a two-parameter fit. *J. acoust. Soc. Am.*, **75**, 1077–83.

Clay, C.S. and Medwin, H. (1977) *Acoustical Oceanography: Principles and Applications*, Wiley, New York, 544 pp.

Cochran, W.G. (1977) *Sampling Techniques*, 3rd edn, Wiley, New York, 428 pp.

Cole, R.H. (1948) *Underwater Explosions*, Princeton University Press, Princeton, NJ, 436 pp.

Conan, G.Y. and Buerkle, U. (1988) Geostatistical analysis of spatial distribution in a school of herring. Int. Coun. Explor. Sea CM 1988/D:21, 8 pp. (mimeo).

Coombs, S. and Fay, R.R. (1989) The temporal evolution of masking and frequency selectivity in the goldfish (*Carassius auratus*). *J. acoust. Soc. Am.*, **86**, 925–33.

Coombs, S. and Popper, A.N. (1979) Hearing differences among Hawaiian squirrelfish (Holocentridae) related to differences in the peripheral auditory system. *J. comp. Physiol.*, **132**, 203–7.

Cooper, G. and Temple, J.A.G. (1983) Calculations of acoustic scattering from ellipsoidal voids: birds, krill and fish. *Ultrasonics*, **21**, 171–6.

Costello, J.H., Peiper, R.E. and Holliday, D.V. (1989) Comparison of acoustic and pump sampling techniques for the analysis of zooplankton distributions. *J. Plankton Res.*, **11**, 703–9.

Craig, R.E. (1983) Re-definition of sonar theory in terms of energy. *FAO Fish. Rep.*, **300**, 1–3.

Craig, R.E. and Forbes, S.T. (1969) A sonar for fish counting. *FiskDir. Skr. Ser. Havunders.*, **15**, 210–19.

Crittenden, R.N., Thomas, G.C., Marino, D.A. and Thorne, R.E. (1988) A weighted duration-in-beam estimator for the volume sampled by a quantitative echosounder. *Can. J. Fish. Aquat. Sci.*, **45**, 1249–56.

Cushing, D.H. (1977) Observations on fish shoals with the ARL scanner. *Rapp. P.-v. Réun. Cons. perm. int. Explor. Mer*, **170**, 15–20.

Dalen, J. and Kristensen, K.E. (1990) Comparative studies of theoretical and empirical target-strength models of euphausiids (krill) in relation to field-experiment data. *Rapp. P.-v. Réun. Cons. perm. int. Explor. Mer*, **189**, 336–44.

Dalen, J. and Løvik, A. (1981) The influence of wind-induced bubbles on echo-integration surveys. *J. acoust. Soc. Am.*, **69**, 1653–59.

Dalen, J. and Nakken, O. (1983) On the application of the echo integration method. Int Coun. Explor. Sea CM 1983/B:19, 30 pp. (mimeo).

Dalen, J. and Smedstad, O.M. (1983) Abundance estimation of demersal fish in the Barents Sea by an extended acoustic method. *FAO Fish. Rep.*, **300**, 232–9.

Degnbol, P. and Lewy, P. (1990) Interpretation of target-strength information from split beam data. *Rapp. P.-v. Réun. Cons. perm. int. Explor. Mer*, **189**, 274–82.

Degnbol, P., Lassen, H. and Staehr, K.J. (1985) In-situ determination of target strength of herring and sprat at 38 and 120 kHz. *Dana*, **5**, 45–54.

Del Grosso, V.A. and Mader, C.W. (1972) Speed of sound in pure water. *J. acoust. Soc. Am.*, **52**, 1442–6.

de Munck, J.C. and Schellart, N.A.M. (1987) A model for the nearfield acoustics of the fish swimbladder and its relevance for directional hearing. *J. acoust. Soc. Am.*, **81**, 556–60.

Denbigh, P.N. and Smith, Q. (1987) Experiments with an acoustic technique for estimating fish number density not requiring fish target strength. Paper no. 28A, Int. Symp. Fisheries Acoustics, Seattle, WA, 22–26 June 1987, 12 pp. (mimeo).

Dickie, L.M., Dowd, R.G. and Bourdeau, P.R. (1983) On echo counting and logging system (ECOLOG) for demersal fish size distributions and densities. *Can. J. Fish. Aquat. Sci.*, **40**, 487–98.

Diercks, J.J., Trochta, R.T., Greenlaw, R.L. and Evans, W.E. (1971) Recording and analysis of dolphin echolocation signals. *J. acoust. Soc. Am.*, **49**, 1729–32.

Diner, N. and Masse, J. (1987) Fish school behaviour during echo survey observed by acoustic devices. Paper no. 30, Int. Symp. Fisheries Acoustics, Seattle, WA, 22–26 June 1987, 28 pp. (mimeo).

Do, M.A. and Surti, A.M. (1990) Estimation of dorsal aspect target strength of deep-water fish using a simple model of swimbladder backscattering. *J. acoust. Soc. Am.*, **87**, 1588–96.

Dragesund, O. and Olsen, S. (1965) On the possibility of estimating year-class strength by measuring echo-abundance of 0-group fish. *FiskDir. Skr. Ser. Havunders.*, **13**, 47–75.

Dragonette. L.R., Vogt, R.H., Flax, L. and Neubauer, W.G. (1974) Acoustic reflection from elastic spheres and rigid spheres and spheroids. II Transient analysis. *J. acoust. Soc. Am.*, **55**, 1130–7.

Duncan, P.M. (1985) Seismic sources in a marine environment. pp. 56–87 in Greene *et al.* (1985), q.v.

Edwards, J.I. and Armstrong, F. (1983) Measurement of the target strength of live herring and mackerel. *FAO Fish. Rep.*, **300**, 69–77.

Edwards, J.I. and Armstrong, F. (1984) Target strength experiments on caged fish. *Scott. Fish. Bull.*, **48**, 12–20.

Edwards, J.I., Armstrong, F., Magurran, A.E. and Pitcher, T.J. (1984) Herring, mackerel and sprat target strength experiments with behavioural observations. Int. Coun. Explor. Sea CM 1984/B:34, 21 pp. (mimeo).

Efron, B. and Tibshirani, R. (1986) Bootstrap methods for standard errors, confidence intervals, and other measures of statistical accuracy. *Statistical Science*, **2**, 54–77.

Ehrenberg, J.E. (1972) A method for extracting the fish target strength distribution from acoustic echoes. Conference 13–15 September 1972, Newport, Rhode Island, New York, USA. *Proc. 1972 IEEE Conf. Eng. Ocean Environ.*, 61–4. IEEE, New York.

Ehrenberg, J.E. (1973) Estimation of the intensity of a filtered Poisson process and its application to acoustic assessment of marine organisms. Univ. Wash. Sea Grant Publ. WSG 73–2, 135 pp.

Ehrenberg, J.E. (1974a) Recursive algorithm for estimating the spatial density of acoustic point scatterers. *J. acoust. Soc. Am.*, **56**, 542–7.

Ehrenberg, J.E. (1974b) Two applications for a dual-beam transducer in hydroacoustic fish assessment systems. Conference 21–23 August 1974, Halifax, Nova Scotia, Canada. Proc. Conf. Engineering in the Ocean Environment, 152–5. IEEE, New York.

Ehrenberg, J.E. (1979) A comparative analysis of in-situ methods for directly measuring the acoustic target strength of individual fish. *IEEE J. Oceanics Engineering*, Vol. OE–4, 141–52.

Ehrenberg, J.E. (1983) A review of in-situ target strength estimation techniques. *FAO Fish. Rep.*, **300**, 85–90.

Ehrenberg, J.E., Carlson, T.J., Traynor, J.J. and Williamson, N.J. (1981) Indirect measurement of the mean acoustic backscattering cross-section of fish. *J. acoust. Soc. Am.*, **69**, 955–62.

Enger, P.S. (1966) Acoustic threshold in goldfish and its relation to the sound source distance. *Comp. Biochem. Physiol.*, **18**, 859–68.

Evans, W.E. (1973) Echolocation by marine delphinids and one species of fresh-water dolphin. *J. acoust. Soc. Am.*, **54**, 191–9.

Everson, I. (1982) Diurnal variations in mean volume backscattering strength of an Antarctic krill (*Euphausia superba*) patch. *J. Plankton Res.*, **4**, 155–62.

FAO (1972) *FAO Catalogue of Fishing Gear Designs*, Food and Agriculture Organisation, Rome, 155 pp.

Fay, R.R. and Popper, A.N. (1980) Structure and function in teleost auditory systems, in *Comparative Studies of Hearing in Vertebrates* (eds A.N. Popper and R.R. Fay), Springer-Verlag, New York, pp. 1–42.

Fedotova, T.A. and Shatoba, O.E. (1983) Acoustic backscattering cross-section of cod averaged by sizes and inclination of fish. *FAO Fish. Rep.*, **300**, 63–8.

Fine, M.L., Winn, H.E. and Olla, B.L. (1977) Communication in fishes, in *How Animals Communicate* (ed. T.A. Sebeok), Indiana University Press, Bloomington, pp. 472–578.

Fish, M.P. (1964) Biological sources of sustained ambient sea noise, in *Marine Bio-acoustics* (ed. W.N. Tavolga), Pergamon Press, New York, pp. 175–94.

Fisher, F.H. and Simmons, V.P. (1977) Sound absorption in sea water. *J. acoust. Soc. Am.*, **62**, 558–64.

Fletcher, H. (1940) Auditory patterns. *Rev. Mod. Phys.*, **12**, 47–56.

Foldy, L.L. and Primakoff, H. (1945) General theory of passive linear electroacoustic transducers and the electroacoustic reciprocity theorem I. *J. acoust. Soc. Am.*, **17**, 109–20.

Foldy, L.L. and Primakoff, H. (1947) General theory of passive linear electroacoustic transducers and the electroacoustic reciprocity theorem II. *J. acoust. Soc. Am.*, **19**, 50–8.

Foote, K.G. (1978a) Analysis of empirical observations on the scattering of sound by encaged aggregations of fish. *FiskDir. Skr. Ser. Havunders.*, **16**, 423–56.

Foote, K.G. (1978b) Theoretical estimation of the mean echo intensity - fish number density relation for encaged saithe in the dorsal aspect. *FiskDir. Skr. Ser. Havunders.*, **16**, 457–64.

Foote, K.G. (1979a) On representing the length dependence of acoustic target strengths of fish. *J. Fish. Res. Bd Can.*, **36**, 1490–6.

Foote, K.G. (1979b) Fish target strength-to-length regressions for application in fisheries research, in Proceedings of Ultrasonic International 79, Graz, Austria, 15–17 May 1979, pp. 327–33.

Foote, K.G. (1980a) Effect of fish behaviour on echo energy: the need for measurements of orientation distributions. *J. Cons. perm. int. Explor. Mer*, **39**, 193–201.

Foote, K.G. (1980b) Importance of the swimbladder in acoustic scattering by fish: a comparison of gadoid and mackerel target strengths. *J. acoust. Soc. Am.*, **67**, 2084–9.

Foote, K.G. (1981a) Energy in acoustic echoes from fish aggregations. *Fish. Res.*, **1**, 129–40.

Foote, K.G. (1981b) Absorption term in time-varied-gain functions. *FiskDir. Skr. Ser. Havunders.*, **17**, 191–213.

Foote, K.G. (1981c) Evidence for the influence of fish behaviour on echo energy, in *Meeting on Hydroacoustical Methods for the Estimation of Marine Fish Populations*, 25–29 June 1979, Vol. 2 (ed. J.B. Suomala), published by The Charles Stark Draper Laboratory, Cambridge, Mass., USA, pp. 201–28.

Foote, K.G. (1982) Optimizing copper spheres for precision calibration of hydroacoustic equipment. *J. acoust. Soc. Am.*, **71**, 742–7.

Foote, K.G. (1983) Linearity of fisheries acoustics, with addition theorems. *J. acoust. Soc. Am.*, **73**, 1932–40.

Foote, K.G. (1985) Rather-high-frequency sound scattering by swimbladdered fish. *J. acoust. Soc. Am.*, **78**, 688–700.

Foote, K.G. (1986) A critique of Goddard and Welsby's paper 'The acoustic target strength of live fish'. *J. Cons. perm. int. Explor. Mer*, **42**, 212–20.

Foote, K.G. (1987) Fish target strengths for use in echo integrator surveys. *J. acoust. Soc. Am.*, **82**, 981–7.

Foote, K.G. (1990a) Correcting acoustic measurements of scatterer density for extinction. *J. acoust. Soc. Am.*, **88**, 1543–6.

Foote, K.G. (1990b) Spheres for calibrating an 11-frequency acoustic measurement system. *J. Cons. perm. int. Explor. Mer*, **46**, 284–6.

Foote, K.G. and MacLennan, D.N. (1984) Comparison of copper and tungsten carbide spheres. *J. acoust. Soc. Am.*, **75**, 612–16.

Foote, K.G. and Nakken, O. (1978) Dorsal-aspect target-strength functions of six fishes at two ultrasonic frequencies. *Fisker. og Havet. Ser. B*, 1978 (3), 95 pp.

Foote, K.G. and Traynor, J.J. (1988) Comparison of walleye pollock target strength estimates determined from in situ measurements and calculations based on swimbladder form. *J. acoust. Soc. Am.*, **83**, 9–17.

Foote, K.G., Aglen, A. and Nakken, O. (1986) Measurement of fish target strength with a split-beam echo sounder. *J. acoust. Soc. Am.*, **80**, 612–21.

Foote, K.G., Everson, I., Watkins, J.L. and Bone, D.G. (1990) Target strengths of Antarctic krill (*Euphausia superba*) at 38 and 120 kHz. *J. acoust. Soc. Am.*, **87**, 16–24.

Foote, K.G., Knudsen, H.P. and Vestnes, G. (1983) Standard calibration of echo sounders and integrators with optimal copper spheres. *FiskDir. Skr. Ser. Havunders.*, **17**, 335–46.

Foote, K.G., Knudsen, H.P., Vestnes, G., MacLennan, D.N. and Simmonds, E.J. (1987) Calibration of acoustic instruments for fish density estimation: a practical guide. Int. Coun. Explor. Sea Coop. Res. Rep. no. **144**, 57 pp.

Forbes, S.T. (1985) Progress in dual-beam target-strength measurement on herring and blue whiting. Int. Coun. Explor. Sea CM 1985/B:22, 5 pp. (mimeo).

Forbes, S.T. and Nakken, O. (eds) (1972) Manual of methods for fisheries resource survey and appraisal, Part 2: The use of acoustic instruments for fish detection and abundance estimation. *FAO Man. Fish. Sci.*, no. **5** (138 pp).

Francis, R.I.C.C. (1985) Two acoustic surveys of pelagic fish in Hawke Bay, New Zealand, 1980. *N.Z. J. mar. freshwat. Res.*, **19**, 375–89.

Francois, R.E. and Garrison, G.R. (1982) Sound absorption based on ocean measurements. Part II: Boric acid contribution and equation for total absorption. *J. acoust. Soc. Am.*, **72**, 1879–90.

Freon, P. and Gerlotto, F. (1988) Methodological approach to study the biases induced by the fish behaviour during hydroacoustic surveys. Int. Coun. Explor. Sea CM 1988/B:52, 16 pp. (mimeo).

Furusawa, M. (1988) Prolate spheroidal models for predicting general trends of fish target strength. *J. acoust. Soc. Japan* (E), **9**, 13–24.

Furusawa. M., Ishii, K., Miyanohana, Y. and Maniwa, Y. (1984) Experimental investigation of an acoustic method to estimate fish abundance using culture nets. *Jpn J. appl. Phys.*, **23** (Supp. 23–1), 101–3.

Gaudet, D.M. (1990) Enumeration of migrating salmon populations using fixed-location sonar counters. *Rapp. P.-v. Réun. Cons. perm. int. Explor. Mer*, **189**, 197–209.

Gerlotto, F. and Freon, P. (1988) Influence of the structure and behaviour of fish schools on acoustic assessment. Int. Coun. Explor. Sea CM 1988/B:53, 27 pp. (mimeo).

Goddard, G.C. and Welsby, V.G. (1986) The acoustic target strength of live fish. *J. Cons. perm. int. Explor. Mer*, **42**, 197–211.

Goertner, J.F. (1978) Dynamical model for explosion injury to fish. Naval Surface Weapons Center, White Oak Lab. Tech. Rep. NSWC/W02/TR76–155, 136 pp.

Greene, C.H., Wiebe, P.H. and Burczynski, J. (1989) Analysing zooplankton size distributions using high-frequency sound. *Limnol. Oceanogr.*, **34**, 129–39.

Greene, G.D., Engelhardt, F.R. and Paterson, R.J. (eds) (1985) Proceedings of the workshop on effects of explosive use in the marine environment, 29–31 January 1985, Halifax. Tech. Rep. no. 5, Canada Oil and Gas Lands Administration, Ottawa, Ontario, 383 pp.

Greenlaw, C.F. (1977) Backscattering spectra of preserved zooplankton. *J. acoust. Soc. Am.*, **62**, 44–52.

Greenlaw, C.F. (1979) Acoustical estimation of zooplankton populations. *Limnol. Oceanogr.*, **24**, 226–42.

Guillard, J., Gerdeaux, D. and Chautru, J.M. (1990) The use of geostatistics for abundance estimation by echo integration in lakes: the example of Lake Annecy. *Rapp. P.-v. Réun. Cons. perm. int. Explor. Mer*, **189**, 410–14.

Gulland, J.A. (1983) *Fish Stock Assessment, a Manual of Basic Methods (FAO/Wiley Series on Food and Agriculture*, Vol. 1), Wiley, Chichester, 223 pp.

Hall, J.D. and Johnson, C.S. (1972) Auditory thresholds of a killer whale, *Orcinus orca*. *J. acoust. Soc. Am.*, **51**, 515–17.

Hall, M.V. (1989) A comprehensive model of wind-generated bubbles in the ocean and predictions of the effects on sound propagation at frequencies up to 40 kHz. *J. acoust. Soc. Am.*, **86**, 1103–17.

Halldorsson, O. (1983) On the behaviour of the Icelandic summer spawning herring (*C. harengus* L.) during echo surveying and depth dependence of acoustic target strength in situ. Int. Coun. Explor. Sea CM 1983/H:36, 35 pp. (mimeo).

Halldorsson, O. and Reynisson, P. (1983) Target strength measurement of herring and capelin in-situ at Iceland. *FAO Fish. Rep.*, **300**, 78–84.

Hamilton, D., Lozow, J., Suomala, J., jun. and Werner, R. (1977) A hydroacoustic measurement program to examine target quantification methods. *Rapp. P.-v. Réun. Cons. perm. int. Explor. Mer*, **170**, 105–21.

Harden Jones, F.R. (1968) *Fish Migration*, Edward Arnold, London, 325 pp.

Harden Jones, F.R. and Scholes, P. (1981) The swimbladder vertical movements and the target strength of fish, in *Meeting on Hydroacoustical Methods for the Estimation of Marine Fish Populations*, 25–29 June 1979, Vol. 2 (ed. J.B. Suomala), published by The Charles Stark Draper Laboratory, Cambridge, Mass., USA, pp. 157–82.

Harden Jones, F.R., Margetts, A.R., Greer Walker, M. and Arnold, G.P. (1977) The efficiency of the Granton otter trawl determined by sector scanning sonar and acoustic transponding tags. *Rapp. P.-v. Réun. Cons. perm. int. Explor. Mer*, **170**, 45–51.

Harris, G.G. (1964) Considerations on the physics of sound production by fishes, in *Marine Bio-acoustics* (ed. W.N. Tavolga), Pergamon Press, Oxford, pp. 233–47.

Haslett, R.W.G. (1970) Acoustic echoes from targets underwater, in *Underwater Acoustics* (ed. R.W.B. Stephens), Wiley, Chichester, UK, pp. 129–97.

Hawkins, A.D. (1977) Fish sizing by means of swimbladder resonance. *Rapp. P.-v. Réun. Cons. perm. int. Explor. Mer*, **170**, 122–9.

Hawkins, A.D. (1981) Some biological sources of error in the hydroacoustical methods for the estimation of marine fish populations, in *Meeting on Hydroacoustical Methods for the Estimation of Marine Fish Populations*, 25–29 June 1979, Vol. 2 (ed. J.B. Suomala), published by The Charles Stark Draper Laboratory, Cambridge, Mass., USA, pp. 1983–200.

Hawkins, A.D. (1986) Underwater sound and fish behaviour, in *The Behaviour of Teleost Fishes* (ed. T.J. Pitcher), Croom Helm, London, pp. 114–51.

Hawkins, A.D. and Chapman, C.J. (1975) Masked auditory thresholds in the cod, *Gadus morhua*. *J. comp. Physiol.*, **103A**, 209–26.

Hawkins, A.D. and Johnstone, A.D.F. (1978) The hearing of the Atlantic salmon, *Salmo salar*. *J. Fish Biol.*, **13**, 655–73.

Hawkins, A.D. and MacLennan, D.N. (1976) An acoustic tank for hearing studies on fish, in *Sound Reception in Fish* (eds A. Schuijf and A.D. Hawkins), Elsevier, Amsterdam, pp. 149–70.

Hawkins, A.D. and Myrberg, A.A., jun. (1983) Hearing and sound communication under water, in *Bioacoustics, a Comparative Approach* (ed. B. Lewis), Academic Press, London, pp. 347–405.

Hawkins, A.D. and Sand, O. (1977) Directional hearing in the median vertical plane by the cod. *J. comp. Physiol.*, **122A**, 1–8.

Hawkins, A.D., MacLennan, D.N., Urquhart, G.G. and Robb, C. (1974) Tracking cod *Gadus morhua* L. in a Scottish sea loch. *J. Fish Biol.*, **6**, 225–36.

He, P. and Wardle, C.S. (1988) Endurance at intermediate swimming speeds of Atlantic mackerel (*Scomber scombrus*), herring (*Clupea harengus*) and saithe (*Pollachius virens*). *J. Fish Biol.*, **33**, 255–66.

Hersey, J.B., Backus, R.H. and Hellwig, J. (1962) Sound-scattering spectra of deep scattering layers in the western North Atlantic Ocean. *Deep Sea Res.*, **8**, 196–210.

Hewitt, R.P., Smith, P.E. and Brown, J.C. (1976) Development and use of sonar mapping for pelagic stock assessment in the California Current area. *Fish. Bull.*, **74**, 281–300, US Dept. of Commerce, Seattle, Washington.

Hickling, R. (1962) Analysis of echoes from a solid elastic sphere in water. *J. acoust. Soc. Am.*, **34**, 1582–92.

Holliday, D.V. (1977a) The use of swimbladder resonance in the sizing of schooled pelagic fish. *Rapp. P.-v. Réun. Cons. perm. int. Explor. Mer*, **170**, 130–5.

Holliday, D.V. (1977b) Two applications of the Doppler effect in the study of fish schools. *Rapp. P.-v. Réun. Cons. perm. int. Explor. Mer*, **170**, 21–30.

Holliday, D.V. (1980) Use of acoustic frequency diversity for marine biological measurements, *Advanced Concepts for Marine Biology* (ed. F.P. Dierner), University of South Carolina Press, Columbia, SC, pp. 423–60.

Holliday, D.V. and Peiper, R.E. (1980) Volume scattering strengths and zooplankton distributions at acoustic frequencies between 0.5 and 3 MHz. *J. acoust. Soc. Am.*, **67**, 135–46.

Holliday, D.V., Peiper, R.E. and Kleppel, G.S. (1989) Determination of zooplankton size and distribution with multifrequency acoustic technology. *J. Cons. perm. int. Explor. Mer*, **46**, 52–61.

Huang, K. and Clay, C.S. (1979) Backscattering cross-sections of live fish: PDF and aspect. *J. acoust. Soc. Am.*, **67**, 795–802.

Hubbs, C.L. and Rechnitzer, A.B. (1952) Report on experiments designed to determine effects of underwater explosions on fish life. *Calif. Fish Game*, **38**, 333–66.

ICNAF, (1963) *The Selectivity of Fishing Gear. Spec. Publs int. Comm. NW Atlant. Fish.*, no. **5**, 1–225.

Jacobs, D.W. and Hall, J.D. (1972) Auditory thresholds of a freshwater dolphin, *Inia geoffrensis*. *J. acoust. Soc. Am.*, **51**, 530–3.

Jacobson, P.T., Clay, C.S. and Magnuson, J.J. (1990) Size, distribution, and abundance by deconvolution of single-beam acoustic data. *Rapp. P.-v. Réun. Cons. perm. int. Explor. Mer*, **189**, 304–11.

Jakobsson, J. (1983) Echo surveying of the Icelandic summer spawning herring 1973–1982. *FAO Fish. Rep.*, **300**, 240–8.

Jakupsstovu, H.i. (Rapporteur) (1984) Survey results and applications. Summary of papers presented at the Fisheries Acoustics Symposium, Bergen, 21–24 June 1982. *Rapp. P.-v. Réun. Cons. perm. int. Explor. Mer*, **184**, 113–18.

Jefferts, K., Burczynski, J.J. and Pearcy, W.G. (1987) Acoustical assessment of squid *Loligo opalescens* off the Central Oregon coast. *Can. J. Fish. Aquat. Sci.*, **44**, 1261–7.

Johanesson, K.A. and Losse, G.F. (1977) Methodology of acoustic estimating of fish abundance in some UNDP/FAO resource survey projects. *Rapp. P.-v. Réun. Cons. perm. int. Explor. Mer*, **170**, 296–318.

Johannesson, K.A. and Mitson, R.B. (1983) Fisheries acoustics: a practical manual for biomass estimation. *FAO Fish. tech. Pap.*, no. **240** (249 pp.).

Johnson, B.D. and Cooke, R.C. (1979) Bubble population and spectra in coastal waters: a photographic approach. *J. geophys. Res.*, **84**, 3761–6.

Johnson, C.S. (1968) Masked tonal thresholds in the bottlenose dolphin. *J. acoust. Soc. Am.*, **44**, 965–7.

Johnson, C.S., McManus, M.W. and Skaar, D. (1989) Masked tonal hearing thresholds in the beluga whale. *J. acoust. Soc. Am.*, **85**, 2651–4.

Johnson, R.K. (1977) Sound scattering from a fluid sphere revisited. *J. acoust. Soc. Am.*, **61**, 375–7.

Johnston, S.V. and Hopelain, J.S. (1990) The application of dual-beam target tracking and Doppler-shifted echo processing to assess upstream salmonid migration in the Klamath River, California. *Rapp. P.-v. Réun. Cons. perm. int. Explor. Mer*, **189**, 210–22.

Johnstone, C.G. (1967) Sound detection thresholds in marine mammals, in *Marine Bioacoustics*, Vol. 2 (ed. W.N. Tavolga), Pergamon Press, New York, pp. 247–60.

Jolly, G.M. and Hampton, I. (1990) Some problems in the statistical design and analysis of acoustic surveys to assess fish biomass. *Rapp. P.-v. Réun. Cons. perm. int. Explor. Mer*, **189**, 415–20.

Kendall, M.G. and Buckland, W.R. (1971) *A Dictionary of Statistical Terms*, 3rd edn, Longman, London, 166 pp.

Kieser, R. and Ehrenberg, J.E. (1990) An unbiased, stochastic echo counting model. *Rapp. P.-v. Réun. Cons. perm. int. Explor. Mer*, **189**, 65–72.

Kieser, R. and Mulligan, T.J. (1984) Analysis of echo counting data: a model. *Can. J. Fish. Aquat. Sci.*, **41**, 451–8.

Kimura, K. (1929) On the detection of fish-groups by an acoustic method. *J. imp. Fish. Inst., Tokyo*, **24**, 41–5.

Kimura, K.D. and Lemberg, N.A. (1981) Variability of line intercept density estimates (a simulation study of the variance of hydro-acoustic biomass estimates). *Can. J. Fish. Aquat. Sci.*, **38**, 1141–52.

Kinsler, L.E. and Frey, A.R. (1950) *Fundamentals of Acoustics*, Wiley, New York, 480 pp.

Kish, L. (1965) *Survey Sampling*, Wiley, New York, 643 pp.

Kostyuchenko, L.P. (1971) Effects of elastic waves generated in marine seismic prospecting on fish eggs in the Black Sea. *Hydrobiol. J.*, **9**, 45–8.

Kristensen, A. and Dalen, J. (1986) Acoustic estimation of the distribution and abundance of zooplankton. *J. acoust. Soc. Am.*, **80**, 601–11.

Lawrie, R.G. (1984) Equipment for acoustic surveys. *Scott. Fish. Bull.*, **48**, 26–30.

Leroy, C.C. (1969) Development of simple equations for accurate and more realistic calculation of the speed of sound in sea water. *J. acoust. Soc. Am.*, **46**, 216–26.

Lévénez, J.J., Gerlotto, F. and Petit, D. (1990) Reaction of tropical coastal pelagic species to artificial lighting and implications for the assessment of abundance by echo integration. *Rapp. P.-v. Réun. Cons. perm. int. Explor. Mer*, **189**, 128–34.

Lindem, T. (1981) The application of hydroacoustical methods in monitoring the spawning migration of whitefish *Coregonus lavaretus* in Lake Randsfjorden, Norway, in *Meeting on Hydroacoustical Methods for the Estimation of Marine Fish Populations*, 25–29 June 1979, Vol. 2 (ed. J.B. Suomala), published by The Charles Stark Draper Laboratory, Cambridge, Mass., USA, pp. 925–40.

Lockwood, S.J. (1988) *The Mackerel its Biology, Assessment and the Management of a Fishery*, Fishing News Books, London, 181 pp.

Love, R.H. (1971) Dorsal-aspect target strength of an individual fish. *J. acoust. Soc. Am.*, **49**, 816–23.

Love, R.H. (1977) Target strength of an individual fish at any aspect. *J. acoust. Soc. Am.*, **62**, 1397–1403.

Love, R.H. (1981) A model for estimating distributions of fish school target strengths. *Deep Sea Res.*, **28A**, 705–25.

Løvik, A. and Hovem, J.M. (1979) An experimental investigation of swimbladder resonance in fishes. *J. acoust. Soc. Am.*, **66**, 850–4.

Lytle, D.W. and Maxwell, D.R. (1983) Hydroacoustic assessment in high density fish schools. *FAO Fish. Rep.*, **300**, 157–71.

McCartney, B.S. and Stubbs, A.R. (1971) Measurements of the acoustic target strength of fish in dorsal aspect, including swimbladder resonance. *J. Sound Vib.*, **15**, 397–420.

Mackenzie, K.V. (1981) Nine-term equation for sound speed in the oceans. *J. acoust. Soc. Am.*, **70**, 807-12.

MacLennan, D.N. (1981a) The target strength of cod at 38 kHz as a function of fish length. Int. Coun. Explor. Sea CM 1981/B:27, 11 pp. (mimeo).

MacLennan, D.N. (1981b) The theory of solid spheres as sonar calibration targets. *Scott. fish. res. Rep.*, no. **22**, 17 pp.

MacLennan, D.N. (1982) Target strength measurements on metal spheres. *Scott. fish. res. Rep.*, no. **25**, 20 pp.

MacLennan, D.N. (1987) Time-varied-gain functions for pulsed sonars. *J. Sound Vib.*, **110**, 511–22.

MacLennan, D.N. (1990) Acoustical measurement of fish abundance. *J. acoust. Soc. Am.*, **87**, 1–15.

MacLennan, D.N. and Armstrong, F. (1984) Tungsten carbide calibration spheres. *Proc. Inst. Acoust.*, **6**, 68–75.

MacLennan, D.N. and Dunn, J. (1984) Estimation of sound velocities from resonance measurements on tungsten carbide calibration spheres. *J. Sound Vib.*, **97**, 321–31.

MacLennan, D.N. and Forbes, S.T. (1984) Fisheries acoustics: a review of general principles. *Rapp. P.-v. Réun. Cons. perm. int. Explor. Mer*, **184**, 7–18.

MacLennan, D.N. and Forbes, S.T. (1987) Acoustic methods of fish stock estimation, in *Developments in Fisheries Research in Scotland* (eds R.S. Bailey and B.B. Parrish), Fishing News Books, London, pp. 40–50.

MacLennan, D.N. and MacKenzie, I.G. (1988) Precision of acoustic fish stock estimates. *Can. J. Fish. Aquat. Sci.*, **45**, 605–16.

MacLennan, D.N. and Pope, J.A. (1983) Analysis procedure for the inter ship calibration of echo integrators. Int. Coun. Explor. Sea CM 1983/B:22, 7 pp. (mimeo).

MacLennan, D.N. and Svellingen, I. (1989) Simple calibration technique for the split-beam echosounder. *Fisk.Dir. Skr. Ser. Havunders.*, **18**, 365–79.

MacLennan, D.N., Armstrong, F. and Simmonds, E.J. (1990a) Further observations on the attenuation of sound by aggregations of fish. *Proc. Inst. Acoust.*, **12**, 99–106.

MacLennan, D.N., Hollingworth, C.E. and Armstrong, F. (1989) Target strength and the tilt angle distribution of caged fish. *Proc. Inst. Acoust.*, **11**(3), 11–20, Institute of Acoustics, St Albans, England.

MacLennan, D.N., Magurran, A.E., Pitcher, T.J. and Hollingworth, C.E. (1990b) Behavioural determinants of fish target strength. *Rapp. P.-v. Réun. Cons. perm. int. Explor. Mer*, **189**, 245–53.

Mais, K.F. (1977) Acoustic surveys of northern anchovies in the California Current System 1966–1972. *Rapp. P.-v. Réun. Cons. perm. int. Explor. Mer*, **170**, 287–95.

Marler, P. and Terrace, H.S. (1984) *The Biology of Learning* (Dahlem Konferenzen), Springer-Verlag, Berlin, 740 pp.

Mason, W.P. (1964) *Physical Acoustics, Principles and Methods*, Vol. 1, Pt. A. Academic Press, London, 515 pp.

Mesiar, D.C., Eggers, D.M. and Gaudet, D.M. (1990) Development of techniques for the application of hydroacoustics to counting migratory fish in large rivers. *Rapp. P.-v. Réun. Cons. perm. int. Explor. Mer*, **189**, 223–32.

Midttun, L. (1984) Fish and other organisms as acoustic targets. *Rapp. P.-v. Réun. Cons. perm. int. Explor. Mer*, **184**, 25–33.

Midttun, L. and Hoff, I. (1962) Measurements of the reflection of sound by fish. *FiskDir. Skr. Ser. Havunders.*, **13**, 1–18.

Midttun, L. and Nakken, O. (1977) Some results of abundance estimation studies with echo integrators. *Rapp. P.-v. Réun. Cons. perm. int. Explor. Mer*, **170**, 253–8.

Midttun, L. and Saetersdal, G. (1957) On the use of echo sounder observations for estimating fish abundance. Paper 29 presented at the Joint Scientific Meeting of ICNAF, ICES and FAO, Lisbon 1957. *Spec. Publ. Int. Commn. NW Atlant. Fish.*, no. **2**, 4 pp. (mimeo).

Misund, O.A. and Beltestad, A.K. (1989) School sizing of small pelagic species off Mozambique by density estimation and acoustic dimensioning. *Proc. Inst. Acoust.*, **11**, 260–6.

Misund, O.A. and Øvredal, J.T. (1988) Acoustic measurements of schooling herring; estimation of school biomass and target strength. Int. Coun. Explor. Sea CM 1988/B:26, 16 pp. (mimeo).

Mitson, R.B. (1984a) *Fisheries Sonar*, Fishing News Books, London, 287 pp.

Mitson, R.B. (1984b) Review of high-speed sector-scanning sonar and its application to fisheries research. *Proc. IEE*, **131** Part F, 257–69.

Miyanohana, Y., Ishii, K. and Furusawa, M. (1986) Effect of beam pattern and fish behaviour on averaged target strength. *Bull. Natl. Res. Inst. Fish. Eng.*, (Japan)/Suikoken Hokoku, **7**, 87–96. (In Japanese with English summary and figure legends)

Møhl, B. and Andersen, S. (1973) Echolocation: high frequency component in the click of the harbour porpoise (*Phocoena ph.* L.). *J. acoust. Soc. Am.*, **54**, 1368–72.

Moose, P.H. and Ehrenberg, J.E. (1971) An expression for the variance of abundance estimates using a fish echo integrator. *J. Fish. Res. Bd Can.*, **28**, 1293–1301.

Morse, P.M. (1948) *Vibration and Sound*, 2nd edn, McGraw-Hill, New York, 468 pp.

Mozgovoy, V.A. (1986) Determining the parameters of fish in sound scattering layer and their behaviour during migration based on scattered acoustic signal spectra. *Oceanology*, **26**, 567–74.

Mulligan, T.J. and Kieser, R. (1986) Comparison of acoustic population estimates of salmon in a lake with a weir count. *Can. J. Fish. Aquat. Sci.*, **43**, 1373–85.

Myrberg, A.A. jun., Ha, S.J., Walewski, S. and Banbury, J.C. (1972) Effectiveness of acoustic signals in attracting epipelagic sharks to an underwater sound source. *Bull. mar. Sci.*, **22**, 926–49, University of Miami Press, Coral Gables, Florida, USA.

Nachtigall, P.W. (1980) Odontocete echolocation performance on object size, shape and material, in *Animal Sonar Systems* (eds R.G. Busnel and J.F. Fish), Plenum, New York, pp. 71–95.

Nakken, O. and Dommasnes, A. (1975) The application of an echo integration system in investigations of the stock strength of the Barents Sea capelin 1971–1974. Int. Coun. Explor. Sea CM 1975/B:25, 20 pp. (mimeo).

Nakken, O. and Olsen, K. (1977) Target strength measurements of fish. *Rapp. P.-v. Réun. Cons. perm. int. Explor. Mer*, **170**, 52–69.

Nakken, O. and Ulltang, O.A. (1983) Comparison of the reliability of acoustic estimates of fish stock abundance and estimates obtained by other assessment methods in the northeast Atlantic. *FAO Fish. Rep.*, **300**, 249–61.

Neubauer, W.G., Vogt, R.H. and Dragonette, L.R. (1974) Acoustic reflection from elastic spheres. *J. acoust. Soc. Am.*, **55**, 1123–9.

Nickerson, T.B. and Dowd, R.G. (1977) Design and operation of survey patterns for demersal fishes using the computerised echo counting system. *Rapp. P.-v. Réun. Cons. perm. int. Explor. Mer*, **170**, 232–6.

Novarini, J.C. and Bruno, D.R. (1982) Effects of the sub-surface bubble layer on sound propagation. *J. acoust. Soc. Am.*, **72**, 501–14.

Nunnallee, E.P. (1983) Scaling of an echo integrator using echo counts, and a comparison of acoustic and weir count estimates of a juvenile sockeye salmon population. *FAO Fish. Rep.*, **300**, 262–8.

Nunnallee, E.P. (1990) An alternative to thresholding during echo-integration data collection. *Rapp. P.-v. Réun. Cons. perm. int. Explor. Mer*, **189**, 92–4.

Officer, C.B. (1958) *Introduction to the Theory of Sound Transmission*, McGraw-Hill, New York, 284 pp.

Offutt, C.G. (1970) Acoustic stimulus perception by the American lobster, *Homarus. Experientia*, **26**, 1276–8.

O'Keefe, D. (1985) A computer model for predicting the effects of underwater explosions on swimbladder fish and marine mammals, pp. 324–53 in Greene *et al.* (1985), q.v.

Olsen, K. (1969) Directional response in herring for sound and noise stimuli. Int. Coun. Explor. Sea CM 1969/B:20, 8 pp (mimeo).

Olsen, K. (1990) Fish behaviour and acoustic sampling. *Rapp. P.-v. Réun. Cons. perm. int. Explor. Mer*, **189**, 147–58.

Olsen, K., Angell, J. and Løvik, A. (1983a) Quantitative estimations of the influence of fish behaviour on acoustic abundance. *FAO Fish. Rep.*, **300**, 139–49.

Olsen, K., Angell, J., Pettersen, F. and Løvik, A. (1983b) Observed fish reactions to a surveying vessel with special reference to herring, cod, capelin and polar cod. *FAO Fish. Rep.*, **300**, 131–8.

Olsen, S. (1969) A note on estimating school size from echo traces. *FAO Fish. Rep.*, **78**, 37–48.

Ona, E. (1984) In situ observation of swimbladder compression in herring. Int. Coun. Explor. Sea CM 1984/B:18, 15 pp. (mimeo).

Ona, E. (1990) Physiological factors causing natural variations in acoustic target strength of fish. *J. mar. biol. Ass. UK*, **70**, 107–27.

Ona, E. and Eger, K. (1987) Sonar observations of trawl performance. Paper no. 99, Int. Symp. Fisheries Acoustics, Seattle, WA, 22–26 June 1987, 10 pp. (mimeo).

Ona, E. and Hansen, K. (1986) In-situ target strength observations on haddock. Int. Coun. Explor. Sea CM 1986/B:39, 13 pp. (mimeo).

Otis, L.S., Cerf, J.A. and Thomas, G.J. (1957) Conditioned inhibition of respiration and heart rate in the goldfish. *Science, N.Y.*, **126**, 263–4.

Payne, R.S. and McVay, S. (1971) Songs of the humpback whales. Science, **173**, 587–97.

Pennington, M. (1983) Efficient estimators for fish and plankton surveys. *Biometrics*, **39**, 281–6.

Penrose, J.D. and Kaye, G.T. (1979) Acoustic target strengths of marine organisms. *J. acoust. Soc. Am.*, **65**, 374–80.

Pieper, R.E. (1979) Euphausiid distribution and biomass determined acoustically at 102 kHz. *Deep Sea Res.*, **26**, 687–702.

Pieper, R.E. and Holliday, D.V. (1984) Acoustic measurements of zooplankton distributions in the sea. *J. Cons. perm. int. Explor. Mer.* **41**, 226–38.

Pieper, R.E., Holliday, D.V. and Kleppel, G.S. (1990) Quantitative zooplankton distributions from multifrequency acoustics. *J. Plankton Res.*, **12**, 433–41.

Pilleri, G., Zbinden, K., Gihr, M. and Kraus, C. (1976) Sonar clicks, directionality of the emission field and echolocating behaviour of the Indus River dolphin (*Platinista indi* Blyth, 1859), in *Investigations on Cetacea*, Vol. VII (ed. G. Pilleri), Brain Anatomy Institute, Berne, Switzerland, pp. 13–43.

Pilleri, G., Zbinden, K. and Kraus, C. (1979) The sonar field of *Inia geoffrensis*, in *Investigations on Cetacea*, Vol. X (ed. G. Pilleri), Brain Anatomy Institute, Berne, Switzerland, pp. 157–78.

Pitcher, T.J. and Partridge, B.L. (1979) Fish school density and volume. *Mar. Biol.*, **54**, 383–94.

Popper, A.N. (1972) Pure tone auditory thresholds for the carp *Cyprinus carpio*. *J. acoust. Soc. Am.*, **52**, 1714–17.

Popper, A.N., Rogers, P.H., Saidel, W.M. and Cox, M. (1988) Role of the fish ear in sound processing, in *Sensory Biology of Aquatic Animals* (eds J. Atema, R.R. Fay, A.N. Popper and W.N. Tavolga), Springer-Verlag, New York, pp. 687–710.

Powell, M.J.D. (1974) Piecewise quadratic surface fitting for contour plotting, in *Software for Numerical Mathematics* (ed. D.J. Evans), Academic Press, London, 451 pp.

Purves, P.E. and Pilleri, G.E. (1983) *Echolocation in Whales and Dolphins*, Academic Press, London, 261 pp.

Quinn, T.J. (1985) Line transect estimators for schooling populations. *Fish. Res.*, **3**, 183–99.

Raitt, R.W. (1948) Sound scatterers in the sea. *J. mar. Res.*, **7**, 393–409.

Rayleigh, Lord J.W.S. (1945) *The Theory of Sound.* Dover, New York.

Reynisson, P.A. (1990) A geometric method for measuring the directivity of hull-mounted transducers. *Rapp. P.-v. Réun. Cons. perm. int. Explor. Mer*, **189**, 176–81.

Richter, K.E. (1985) Acoustic scattering at 1.2 MHz from individual zooplankters and copepod populations. *Deep Sea Res.*, **32**, 149–61.

Ricker, W.E. (1973) Linear regressions in fishery research. *J. Fish. Res. Bd Can.*, **30**, 409–34.

Robinson, B.J. (1982) An in-situ technique to determine fish target strength with results for blue whiting *Micromesistius poutassou* (Risso). *J. Cons. perm. int. Explor. Mer*, **40**, 153–60.

Robinson, B.J. (1983) In-situ measurements on the target strength of pelagic fishes. *FAO Fish. Rep.*, **300**, 99–111.

Robinson, B.J. (1984) Calibration of equipment. *Rapp. P.-v. Réun. Cons. perm. int. Explor. Mer*, **184**, 62–7.

Robinson, B.J. and Hood, C. (1983) A procedure for calibrating acoustic survey systems with estimates of obtainable precision and accuracy. *FAO Fish. Rep.*, **300**, 59–62.

Robotham, V.H. and Castillo, J. (1990) The bootstrap method: an alternative for estimating confidence intervals of resources surveyed by means of hydroacoustic techniques. *Rapp. P.-v. Réun. Cons. perm. int. Explor. Mer*, **189**, 421–4.

Rogers, P.H. and van Buren, A.L. (1978) A new approach to constant beam-width transducers. *J. acoust. Soc. Am.*, **64**, 38–43.

Rose, G.A. and Leggett, W.C. (1988) Hydroacoustic signal classification of fish schools by species. *Can. J. Fish. Aquat. Sci.*, **45**, 597–604.

Rottingen, I. (1976) On the relation between echo intensity and fish density. *FiskDir. Skr. Ser. Havunders.*, **16**, 301–14.

Rottingen, I. (1978) Field intercalibrations of echo integrator systems. Int. Coun. Explor. Sea CM 1978/B:25, 23 pp. (mimeo).

Rschevkin, S.N. (1963) *A Course of Lectures on the Theory of Sounds*, Pergamon Press, London, 464 pp.

Rudstam, L.G., Clay, C.S. and Magnuson, J.J. (1987) Density and size estimates of Cisco (*Coregonus artedi*) using analysis of echo peak PDF from a single-transducer sonar. *Can. J. Fish. Aquat. Sci.*, **44**, 811–21.

Rudstam, L.G., Lindem, T. and Hansson, S. (1988) Density and in-situ target strength of herring and sprat: a comparison between two methods of analysing single-beam sonar data. *Fish. Res.*, **6**, 305–15.

Rusby, J.S.M. (1977) Long range survey of a herring fishery by side-scan sonar. *Rapp. P.-v. Réun. Cons. perm. int. Explor. Mer*, **170**, 7–14.

Sakaguchi, S., Fukuhara, O., Umezawa, S., Fuhiya, M. and Ogawa, T. (1976) The influence of underwater explosions on fishes. Bull. Nansei Reg. Fish. Res. Lab., **9**, 33–56, Hiroshima, Japan. (In Japanese, summary and figure legends in English)

Sameoto, D. (1980) Quantitative measurements of euphausids using a 120 kHz sounder and their in-situ orientation. *Can. J. Fish. Aquat. Sci.*, **37**, 693–702.

Sameoto, D. (1982) Zooplankton and micronekton abundance in acoustic scattering layers on the Nova Scotian slope. *Can. J. Fish. Aquat. Sci.*, **39**, 760–77.

Sameoto, D., Cochrane, N. and Herman, A. (1990) Use of multiple frequency acoustics and other methods in estimating copepod and euphausiid abundances. Int. Coun. Explor. Sea CM 1990/L:81, 20 pp. (mimeo).

Sand, O. and Hawkins, A.D. (1973) Acoustic properties of the cod swimbladder. *J. exp. Biol.*, **58**, 797–820.

Schevill, W.E., Watkins, W.A. and Backus, R.H. (1954) The 20-cycle signals and *Balaenoptera* (fin whales), in *Marine Bio-acoustics* (ed. W.N. Tavolga), Pergamon Press, New York, pp. 147–52.

Schuijf, A. and Buwalda, R.J.A. (1980) Underwater localisation – a major problem in fish acoustics, in *Comparative Studies of Hearing in Vertebrates* (eds A.N. Popper and R.R. Fay), Springer-Verlag, New York, pp. 43–77.

Shibata, K. (1971) Experimental measurement of target strength of fish, in *Modern Fishing Gear of the World*, Vol. 2, (ed. H. Kristjonsson), Fishing News (Books), London, pp. 104–8.

Shotton, R. (1981) Acoustic survey design, in *Meeting on Hydroacoustical Methods for the Estimation of Marine Fish Populations*, 25–29 June 1979, Vol. 2 (ed. J.B. Suomala), published by The Charles Stark Draper Laboratory, Cambridge, Mass., USA, pp. 629–88

Shotton, R. and Bazigos, G.P. (1984) Techniques and considerations in the design of acoustic surveys. *Rapp. P.-v. Réun. Cons. perm. int. Explor. Mer*, **180**, 34–57.

Shulkin, M. and Marsh, H.W. (1963) Absorption of sound in sea water. *J. Br. Inst. Radio Eng.*, **25**, 493–9.

Siler, W. (1969) Near and far fields in a marine environment. *J. acoust. Soc. Am.*, **46**, 483–4.

Simmonds, E.J. (1984a) A comparison between measured and theoretical equivalent beam angles for seven similar transducers. *J. Sound Vib.*, **97**, 117–28.

Simmonds, E.J. (1984b) The effect of mountings on the equivalent beam angle of acoustic survey transducers. Int. Coun. Explor. Sea CM 1984/B:32, 5 pp. (mimeo).

Simmonds, E.J. (1990) Very accurate calibration of a vertical echo sounder: a five-year assessment of performance and accuracy. *Rapp. P.-v. Réun. Cons. perm. int. Explor. Mer*, **189**, 183–91.

Simmonds, E.J. and Armstrong, F. (1990) A wideband echo sounder: measurements on cod, saithe, herring, and mackerel from 27 to 54 kHz. *Rapp. P.-v. Réun. Cons. perm. int. Explor. Mer*, **189**, 381–7.

Simmonds, E.J. and Copland, P.J. (1986) A wide band constant beam width echosounder for fish abundance estimation. *Proc. Inst. Acoust.*, **8**, 173–80.

Simmonds, E.J. and Copland, P.J. (1989) Species recognition: results from a wide band echo-sounder 27–54 kHz. *Proc. Inst. Acoust.*, **11**, 54–60.

Simmonds, E.J. and MacLennan, D.N. (1988) Survey and data analysis procedures. Int. Coun. Explor. Sea CM 1988/B:58, 15 pp. (mimeo).

Simmonds, E.J. and MacLennan, D.N. (1990) Seabed following: efficient data collection for echo-integration. Int. Coun. Explor. Sea CM 1990/B:33, 8 pp. (mimeo).

Simmonds, E.J., Petrie, I.B., Armstrong, F. and Copland, P.J. (1984) High precision calibration of a vertical sounder system for use in fish stock estimation. *Proc. Inst. Acoust.*, **6**, 129–38.

Skudrzyk, E.J. (1971) *The Foundations of Acoustics*, Springer-Verlag, New York, 790 pp.

Smith, P.E. (1970) The horizontal dimensions and abundance of fish schools in the upper mixed layer as measured by sonar, in *Proceedings of the International Symposium on Biological Sound Scattering in the Ocean* (ed. G. Brooke Farquhar), Maury Center for Ocean Science, Dep. Navy, Washington, DC, pp. 563–91.

Smith, P.E. (1977) The effect of internal waves on fish school mapping with sonar in the California Current area. *Rapp. P.-v. Réun. Cons. perm. int. Explor. Mer*, **170**, 223–31.

Solomon, D.J. and Hawkins, A.D. (1981) Fish capture and transport, in *Aquarium Systems* (ed. A.D. Hawkins), Academic Press, London, pp. 197–221.

Sorokin, M.A., Perkin, S.I. and Lebedeva, A.N. (1988) On the distance of directional sound discrimination by fishes. *Vop. Ikhtiol.*, **2**, 341–3. (In Russian)

Spindel, R.C. and McElroy, P.T. (1973) Level and zero crossings in volume reverberation signals. *J. acoust. Soc. Am.*, **53**, 1417–26.

Squire, J.L. jun. (1978) Northern anchovy school shapes as related to problems in school size estimation. *Fish. Bull.*, **76**, 443–4, US Dept. of Commerce, Seattle, USA.

Staal, P.R. (1985) Acoustic effects of underwater explosive discharges. pp. 89–110 in Greene *et al.* (1985), q.v.

Stanton, T.K. (1982) Effects of transducer motion on echo-integration techniques. *J. acoust. Soc. Am.*, **72**, 947–9.

Stanton, T.K. (1983) Multiple scattering with applications to fish-echo processing. *J. acoust. Soc. Am.*, **73**, 1164–9.

Stanton, T.K. (1984) Effects of second-order scattering on high resolution sonars. *J. acoust. Soc. Am.*, **76**, 861–6.

Stanton, T.K. (1985a) Volume scattering: echo peak PDF. *J. acoust. Soc. Am.*, **77**, 1358–66.

Stanton, T.K. (1985b) Density estimate of biological sound scatterers using sonar echo peak PDFs. *J. acoust. Soc. Am.*, **78**, 1868–73.

Stanton, T.K. (1988) Sound scattering by cylinders of finite length. I. Fluid cylinders. *J. acoust. Soc. Am.*, **83**, 55–63.

Stanton, T.K. (1990) Sound scattering by zooplankton. *Rapp. P.-v. Réun. Cons. perm. int. Explor. Mer*, **189**, 353–62.

Stanton, T.K. and Clay, C.S. (1986) Sonar echo statistics as a remote sensing tool: volume and seafloor. *IEEE J. Oceanic Eng.*, **11**, 79–99.

Stolyrenko, D.A. (1988) Local integral measurement: an advanced spline approximation method for trawl, echosounding and television surveys. Int. Coun. Explor. Sea CM 1988/D:1, 8 pp. (mimeo).

Strange, E.S. (1981) An introduction to commercial fishing gear and methods used in Scotland. *Scottish Fisheries Information Pamphlet*, no. 1, 35 pp, SOAFD, Marine Laboratory, PO Box 101, Victoria Road, Aberdeen, Scotland.

Stromme, T. and Saetersdal, G. (1987) Consistency of acoustic biomass estimates tested by repeated survey coverages. Paper no. 126, Int. Symp. Fisheries Acoustics, Seattle, WA, 22–26 June 1987, 10 pp. (mimeo).

Sund, O. (1935) Echo sounding in fishery research. *Nature*, **135**, 953.

Swingler, N. and Hampton, I. (1981) Investigation and comparison of current theories for the echo-integration technique of investigating fish abundance and of their verification by experiment, in *Meeting on Hydroacoustical Methods for the Estimation of Marine Fish Populations*, 25–29 June 1979, Vol. 2 (ed.

J.B. Suomala), published by The Charles Stark Draper Laboratory, Cambridge, Mass., USA, pp. 97–156.

Tavolga, W.N. (1962) Mechanisms of sound production in the ariid catfishes *Galeichthys* and *Bagre*. *Bull. Am. Mus. nat. Hist.*, **124**, 1–30.

Tavolga, W.N. (1974) Signal/noise ratio and the critical band in fishes. *J. acoust. Soc. Am.*, **55**, 1323–33.

Tavolga, W.N. (1976) Acoustic obstacle detection in the sea catfish *Arius felis*, in *Sound Reception in Fish* (eds A. Schuijf and A.D. Hawkins), Elsevier, Amsterdam, pp. 185–204.

Tavolga, W.N., Popper, A.N. and Fay, R.R. (eds) (1981) *Hearing and Sound Communication in Fishes*, Springer-Verlag, New York, 608 pp.

Teleki, G.C. and Chamberlain, A.J. (1978) Acute effects of underwater construction blasting on fishes in Long Point Bay, Lake Erie. *J. Fish. Res. Bd Can.*, **35**, 1191–8.

Terhune, J.M. and Ronald, K. (1975) Underwater hearing sensitivity of two ringed seals. *Can. J. Zool.*, **53**, 227–31.

Thorne, R.E. (1971) Investigations into the relation between integrated echo voltage and fish density. *J. Fish. Res. Bd Can.*, **28**, 1269–73.

Thorne, R.E. (1979) Hydroacoustic estimates of adult sockeye salmon (*Oncorhynchus nerka*) in Lake Washington 1972–1975. *J. Fish. Res. Bd Can.*, **36**, 1145–9.

Thorne, R.E. (1983a) Application of hydroacoustic assessment techniques to three lakes with contrasting fish distributions. *FAO Fish. Rep.*, **300**, 269–77.

Thorne, R.E. (1983b) Assessment of population abundance by hydroacoustics. *Biol. Oceanogr.*, **2**, 253–62.

Thorne, R.E. (1988) An empirical evaluation of the duration-in-beam technique for hydroacoustic estimation. *Can. J. Fish. Aquat. Sci.*, **45**, 1244–8.

Thorne, R.E. and Thomas, G.L. (1984) Recent applications of hydroacoustics to assessment of limnetic fish abundance and behaviour, in Proc. NALMS 1983 International Symposium on Lake and Reservoir Management, USEPA 400/5/84–001, pp. 305–9.

Thorne, R.E., Hedgepeth, J. and Campos, J. (1990) The use of stationary hydroacoustic transducers to study diel and tidal influences on fish behaviour. *Rapp. P.-v. Réun. Cons. perm. int. Explor. Mer*, **189**, 167–75.

Thorne, R.E., Trumble, R.J., Lemberg, N.A. and Blankenbeckler, D. (1983) Hydroacoustic assessment and management of herring fisheries in Washington and southeastern Alaska. *FAO Fish. Rep.*, **300**, 217–22.

Traynor, J.J. and Williamson, N.J. (1983) Target strength measurements of walleye pollock (*Theraga chalcogramma*) and a simulation study of the dual beam method. *FAO Fish. Rep.*, **300**, 112–24.

Trout, G.C., Lee, A.J., Richardson, I.D. and Harden Jones, F.R. (1952) Recent echosounder studies. *Nature*, **170** (4315), 71–2.

Tucker, D.G. and Gazey, B.K. (1966) *Applied Underwater Acoustics*, Pergamon Press, London, 244 pp.

Turl, C.W. and Penner, R.H. (1989) Differences in echolocation click patterns of the beluga (*Delphinapterus leucas*) and the bottlenose dolphin (*Tursiops truncatus*). *J. acoust. Soc. Am.*, **86**, 497–502.

Turl, C.W., Penner, R.H. and Au, W.W.L. (1987) Comparison of the target detection capability of the beluga (*Delphinapterus leucas*) and Atlantic bottlenose dolphin (*Tursiops truncatus*) in masking noise. *J. acoust. Soc. Am.*, **82**, 1487–91.

Urick, R.J. (1983) *Principles of Underwater Sound for Engineers*, 3rd edn, McGraw-Hill, New York, 384 pp.

Venema, S.C. (1985) A selected bibliography of acoustics in fisheries research and related fields. *FAO Fish. Circ.*, no. **748**, Revision 1, FIRM/C748.

von Brandt, A. (1984) *Fish Catching Methods of the World*, 3rd edn, Fishing News Books, London, 418 pp.

Walker, R.A. (1963) Some intense low frequency underwater sounds of wide geographic distribution, apparently of biological origin. *J. acoust. Soc. Am.*, **35**, 1816.

Wardle, C.S. (1983) Fish reactions to towed gears, in *Marine Biology at Sea* (eds A. MacDonald and I.G. Priede), Academic Press, London, pp. 167–95.

Weimer, R.T. and Ehrenberg, J.E. (1975) Analysis of threshold-induced bias inherent in acoustic scattering cross-section estimates of individual fish. *J. Fish. Res. Bd Can.*, **32**, 2547–51.

Welsby, V.G. and Hudson, J.E. (1972) Standard small targets for calibrating underwater sonars. *J. Sound Vib.*, **20**, 399–406.

Wenz, G.M. (1972) Review of underwater acoustic research: noise. *J. acoust. Soc. Am.*, **51**, 1010–24.

Weston, D.E. (1989) On the losses due to storm bubbles in oceanic sound transmission. *J. acoust. Soc. Am.*, **86**, 1546–53.

Weston, D.E. and Andrews, H.W. (1990a) Seasonal sonar observations of the diurnal shoaling times of fish. *J. acoust. Soc. Am.*, **87**, 673–80.

Weston, D.E. and Andrews, H.W. (1990b) Monthly estimates of fish numbers using a long-range sonar. *J. Cons. perm. int. Explor. Mer*, **47**, 104–11.

Weston, D.E. and Ching, P.A. (1989) Wind effects in shallow water acoustic transmission. *J. acoust. Soc. Am.*, **86**, 1530–45.

Whitehead, P.J.P. and Blaxter, J.H.S. (1989) Swimbladder form in the clupeoid fishes. *Zool. J. Linn. Soc.*, **97**, 299–372.

Widener, M.W. (1980) The measurement of transducer efficiency using self reciprocity techniques. *J. acoust. Soc. Am.*, **67**, 1058–60.

Wiley, M.L., Gaspin, J.B. and Goertner, J.F. (1981) Effects of underwater explosions on fish with a dynamical model to predict fish kill. *Ocean Sci. Eng.*, **6**, 223–84.

Williamson, N.J. (1982) Cluster sampling estimation of the variance of abundance estimates derived from quantitative echo sounder surveys. *Can. J. Fish. Aquat. Sci.*, **39**, 229–31.

Williamson, N.J. and Traynor, J.J. (1984) In-situ target-strength estimation of Pacific whiting (*Merluccius productus*) using a dual-beam transducer. *J. Cons. perm. int. Explor. Mer*, **41**, 285–92.

Yelverton, J.., Richmond, D.R., Hicks, W., Saunders, K. and Fletcher, E.R. (1975) The relationship between fish size and their response to underwater blast. Rep. DNA 3677 T, Defense Nuclear Agency, Washington, DC, 39 pp.

Zakharia, M.E. (1990) A prototype wideband sonar for fisheries in lakes and rivers. *Rapp. P.-v. Réun. Cons. perm. int. Explor. Mer*, **189**, 394–7.

Species index

Author index

Subject index